Rethinking Human Evolution

Vienna Series in Theoretical Biology

Gerd B. Müller, editor-in-chief

Thomas Pradeu, Katrin Schäfer, associate editors

The Evolution of Cognition, edited by Cecilia Heyes and Ludwig Huber, 2000

Origination of Organismal Form, edited by Gerd B. Müller and Stuart A. Newman, 2003

Environment, Development, and Evolution, edited by Brian K. Hall, Roy D. Pearson, and Gerd B. Müller, 2004

Evolution of Communication Systems, edited by D. Kimbrough Oller and Ulrike Griebel, 2004

Modularity: Understanding the Development and Evolution of Natural Complex Systems, edited by Werner Callebaut and Diego Rasskin-Gutman, 2005

Compositional Evolution: The Impact of Sex, Symbiosis, and Modularity on the Gradualist Framework of Evolution, by Richard A. Watson, 2006

Biological Emergences: Evolution by Natural Experiment, by Robert G. B. Reid, 2007

Modeling Biology: Structure, Behaviors, Evolution, edited by Manfred D. Laubichler and Gerd B. Müller, 2007

Evolution of Communicative Flexibility, edited by Kimbrough D. Oller and Ulrike Griebel, 2008

Functions in Biological and Artificial Worlds, edited by Ulrich Krohs and Peter Kroes, 2009

Cognitive Biology, edited by Luca Tommasi, Mary A. Peterson and Lynn Nadel, 2009

Innovation in Cultural Systems, edited by Michael J. O'Brien and Stephen J. Shennan, 2010

The Major Transitions in Evolution Revisited, edited by Brett Calcott and Kim Sterelny, 2011

Transformations of Lamarckism, edited by Snait B. Gissis and Eva Jablonka, 2011

Convergent Evolution: Limited Forms Most Beautiful, by George McGhee, 2011

From Groups to Individuals, edited by Frédéric Bouchard and Philippe Huneman, 2013

Developing Scaffolds in Evolution, Culture, and Cognition, edited by Linnda R. Caporael, James Griesemer, and William C. Wimsatt, 2014

Multicellularity: Origins and Evolution, edited by Karl J. Niklas and Stuart A. Newman, 2015

Vivarium: Experimental, Quantitative, and Theoretical Biology at Vienna's Biologische Versuchsanstalt, edited by Gerd B. Müller, 2017

Landscapes of Collectivity in the Life Sciences, edited by Snait B. Gissis, Ehud Lamm, and Ayelet Shavit, 2017

Rethinking Human Evolution, edited by Jeffrey H. Schwartz, 2017

Rethinking Human Evolution

edited by Jeffrey H. Schwartz

The MIT Press
Cambridge, Massachusetts
London, England

This book was set in Times New Roman by Westchester Publishing Services.

Library of Congress Cataloging-in-Publication Data

Names: Schwartz, Jeffrey H., editor.
Title: Rethinking human evolution / edited by Jeffrey H. Schwartz.
Description: Cambridge, MA : The MIT Press, [2018] | Series: Vienna series in theoretical biology |
    Includes bibliographical references and index.
Identifiers: LCCN 2017027095 | ISBN 9780262037327 (hardcover : alk. paper), 9780262546744 (pb)
Subjects: LCSH: Human evolution.
Classification: LCC GN281 .R417 2018 | DDC 599.93/8—dc23
LC record available at https://lccn.loc.gov/2017027095

In memory of Werner Callebaut

Contents

Series Foreword

Biology is a leading science in this century. As in all other sciences, progress in biology depends on the interrelations between empirical research, theory building, modeling, and societal context. But whereas molecular and experimental biology have evolved dramatically in recent years, generating a flood of highly detailed data, the integration of these results into useful theoretical frameworks has lagged behind. Driven largely by pragmatic and technical considerations, research in biology continues to be less guided by theory than seems indicated. By promoting the formulation and discussion of new theoretical concepts in the biosciences, this series intends to help fill important gaps in our understanding of some of the major open questions of biology, such as the origin and organization of organismal form, the relationship between development and evolution, and the biological bases of cognition and mind. Theoretical biology has important roots in the experimental tradition of early-twentieth-century Vienna. Paul Weiss and Ludwig von Bertalanffy were among the first to use the term *theoretical biology* in its modern sense. In their understanding the subject was not limited to mathematical formalization, as is often the case today, but extended to the conceptual foundations of biology. It is this commitment to a comprehensive and cross-disciplinary integration of theoretical concepts that the Vienna Series intends to emphasize. Today, theoretical biology has genetic, developmental, and evolutionary components, the central connective themes in modern biology, but it also includes relevant aspects of computational or systems biology and extends to the naturalistic philosophy of sciences. The Vienna Series grew out of theory-oriented workshops organized by the KLI, an international institute for the advanced study of natural complex systems. The KLI fosters research projects, workshops, book projects, and the journal *Biological Theory*, all devoted to aspects of theoretical biology, with an emphasis on—but not restriction to—integrating the developmental, evolutionary, and cognitive sciences. The series editors welcome suggestions for book projects in these domains.

Gerd B. Müller, Thomas Pradeu, Katrin Schäfer

Preface

In September 2015, a workshop was held at the Konrad Lorenz Institute for Evolution and Cognition Studies in which participants were presented with the challenge of considering for their own fields of study the questions: *Is paleoanthropology an evolutionary science? Or, Are analyses of human evolution biological?* The invitation to this gathering began:

Given the pronouncements about human evolution that dominate anthropology textbooks and frequent the pages of newspapers and science e-news websites, it would seem that the major questions in paleoanthropology have been answered. Indeed, it is commonplace to read that a new fossil or molecular analysis supports fully or tweaks only a tiny bit scenarios of who's related to whom and how, when, where, and why one species of human relative (hominid) transformed seamlessly into another. If this is so, one may ask: Why bother trying to find more fossils, identify another molecule, or seek evidence pertaining to the life-history and persistence of any hominid, or to pursue these inquiries with the latest technology, if you already know the story?

In light of the impact pronouncements on human evolution have on the public and non-biologically savvy academics, it seems appropriate to convene a workshop that focuses on the disconnect between human evolutionary studies and the theoretical and methodological standards and practice that inform the rest of evolutionary biology. The result of this workshop will hopefully be a broad-based publication that will bring to light assumptions and misconceptions as well as positive and biologically viable aspects of human evolutionary studies. In turn, such wide-ranging collaboration should at the very least make apparent to scholars who assume that the study of human evolution is both biologically and theoretically sound that this is not necessarily or universally correct. More optimistically, such an endeavor—indeed, challenge—may provide a spark of intellectual curiosity among paleoanthropologists and their academic kin that could have long-lasting, positive effects on their disciplines. Although ambitious, I would also hope that some of the insights and recommendations that will emerge from this workshop will become known to the media and disseminated to the public that, after all, accepts pronouncements on human evolution as biologically sound fact.

The intention of this workshop was not to dwell on negative aspects of one's area of study, or of human evolutionary studies at large, but for each participant to reflect on her own research and the discipline in which it is situated, to delineate what "works," and to consider how she and that field of inquiry might move forward in different, if not also new, ways. The questions *Is paleoanthropology an evolutionary science? Or, Are analyses of human evolution biological?* were thus meant to provoke personal as well as disciplinary introspection. Consequently, the title of the resulting collective effort—*Rethinking Human Evolution*—reflects a participant's "rethinking" of the "state of the appropriate art" as she perceives it.

In organizing this meeting of minds, I sought representation of a diversity of intellectual pursuits, not only in the more common realms of human evolutionary exploration, but in the presence of "outsiders" who, through the eyes of a historian (Richard Delisle) and philosopher of science (Siobhan Mc Manus), can offer perspectives and raise questions about aspects of the field that are often unappreciated by practitioners, who often proceed along a certain intellectual path. Specifically, Delisle reviews the history of procedural idiosyncrasies that continue to characterize much of paleoanthropology, and suggests that one way in which the field could become more scientific is by articulating specific "rules of engagement" that would be open to hypothesis testing and falsification. Mc Manus melds philosophy of science with philosophy of biology and cultural studies of society in taking a philosophy-of-mind approach to understanding human evolution: If the complexities of mind and society are part of (some component of) human evolution, at what point should we embrace these human innovations in epistemological considerations of our evolutionary past? In their chapters, Delisle and Mc Manus set the stage for the contributions that follow.

With regard to inquiries involving the physicality of human evolution, mammalian paleontologists John de Vos and Jelle Reumer remind us that, whatever special qualities one attributes to living humans and perhaps also to their extinct relatives, humans and their relatives are first and foremost mammals, and therefore should be approached phylogenetically, not as special cases as in paleoanthropology, but as a paleontologist would deal with other large mammals. As such, they argue that the evolutionary processes that shaped artiodactyls, perrisodactyls, and proboscideans also obtain to hominids. They offer examples of island dwarfing and adaptations to increasing grasslands during the Pleistocene, and also make the case for Late Pleistocene hominids, like other large mammals, being endowed with a woolly pelage.

Following suit in treating humans as another mammal, I review the history of and the assumptions underlying how paleoanthropologists came and continue to interpret and allocate human fossils to the few taxa that are "allowed" to encompass the entire human fossil record: Why paleoanthropologists deny to humans the possibility of taxic diversity that characterizes the evolutionary past of other mammals. I also discuss the lack of regard many paleoanthropologists have for the *International Code of Zoological Nomenclature* that every vertebrate paleontologist follows, including the scientific validity of type specimens as the bases for testing the naming of species. Lastly, I review the assumptions that have informed the interpretation of molecular data vis-à-vis human evolution.

Rob DeSalle and colleagues pursue the claims derived from genomic analyses on the geographic distribution and evolutionary history of present-day human populations, including their conception as discrete biological units. In doing so, they identify three major problems underlying these claims: the disparity between the large genomic data sets from living humans and the relatively small sampling of (relatively recent) fossil hominids; assumptions regarding the origin of human populational diversity; and the assumptions that inform the algorithms from which the clustering of human populations derive.

In arguing for integrating genetics with paleontology, archaeology, and paleoclimatology in attempting to understand human evolution, Alan Templeton emphasizes the importance of hypothesis testing, which, unfortunately, is often overlooked or just ignored in paleoanthropological considerations. From a geneticist's perspective, he uses the "out-of-Africa" (OoA) replacement model to illustrate how the reliance on what he identifies as noninformative data and pseudohypotheses has sustained an untestable hypothesis. In a similar vein, Michael Petraglia and Huw Groucutt discuss other problems with OoA models, especially their basis in historicity and focus on "who, what, when, and where" questions. Using the disparate histories of populations in the Arabian Peninsula and India as examples, they discuss the problems inherent in drawing direct connections between the past and present, and argue for an "evolutionary" approach that incorporates and integrates environmental and demographic perspectives.

The discussions by DeSalle and colleagues and Templeton also remind us not only to be aware of, but more importantly to understand, the differences between computational approaches to phylogenetic reconstruction. In this regard the chapters by Peter Waddell, Fred Bookstein, and Markus Bastir demonstrate how one's choice of algorithm—which itself subsumes evolutionary assumptions—impacts one's analysis both of molecular and of morphological data. Waddell tackles the OoA versus Multiregional (MR) models for the emergence of modern human populational diversity using primarily data on complete genomic sequences and aspects of skull shape. Interpreting similarity and difference in terms of possible introgression versus divergence, he calculates that *Homo sapiens* became distinct in south or southeastern Africa between 150 and 200 kya, and dispersed in a stepwise (not MR) process throughout Africa before migrating out 60 to 100 kya. Echoing the sentiment of many contributors, Waddell calls for a more complete integration of biological information in addressing a still-unsatisfactory conception and definition of species *sapiens*.

Bookstein comes at the problem from the perspective of arithmetic and a philosophy of scientific—specifically paleoanthropological—method and rhetoric. He disagrees not only with Waddell's approach, but in general with the use of correlation, regression, principal components, and other metaphors of "distance." He provokes rethinking of common practice by describing paleoanthropology. "In an effort to move paleoanthropology forward in the realm of the arithmetic, Bookstein emphasizes that, whatever the method, a reliable quantitative science must also include an "organized skepticism."

Bastir approaches a more integrative approach to doing paleoanthropology through not only geomorphometrics but also from a philosophical and theoretical perspective. In reviewing the pros and cons of "traditional" descriptive morphology versus a morphometrics, he argues for a melding (i.e., integration, in keeping with the general theme of the volume) of these seemingly disparate and irreconcilable ways of dealing with topology; that is, of analyzing three-dimensional morphology, which is what specimens, whether fossil or recent, present to the paleoanthropologist. Paralleling my point that by maintaining a "molecules versus morphology" mentality one is falsely dichotomizing a biological continuum, Bastir argues that a step forward in phylogenetic analysis would be the melding of the strength of geomorphometrics in deciphering morphological size-shape relationships with the morphologist's focus on the morphological detail necessary to distinguish between primitiveness and derivedness.

Robin Dennell recounts the history of paleoanthropology in China to illustrate how sociopolitical history can impact scenarios of human evolution. Specifically, conceptions changed from a broad biological context, in which the important events of vertebrate (including humans) diversification were thought to have occurred in East Asia, to Weidenreich's multiregional, *in situ* evolutionary model, Movius' claim that East Asia was a human evolutionary backwater, and, with the rise of ethno-nationalism after the founding of the People's Republic of China, to embracing the Zhoukoudian hominid as the direct, ancient ancestor of all Chinese.

Going deeper in time, Gabrielle Macho tackles the problem of interpreting hominid origins by highlighting that humans are the sole survivors of a speciose group. She questions the validity of choosing a specific ape as a model from which to reconstruct the last common ancestor of all hominids (which also makes clear that our closest living relative is not necessarily our closest relative). Toward a more comprehensive endeavor, Macho argues that paleoanthropologists should focus on the plasticity of closely related and ecologically similar species, and integrate the study of morphology, development, geography, paleoclimatology, and genetics.

Archaeology, of course, also redounds on the understanding of early human behavior, which Thomas Plummer and Emma Finestone consider in terms of an adaptive shift in human evolution as represented by the appearance of the Olduwan stone tool technology at East African sites dated approximately 2.6 to 1.7 mya, and the implications for hominid manipulation of a variety of other resources. Beyond acknowledging that early hominids transported and used high-quality raw material for producing stone flakes, they highlight how little is known about the function of these tools. They call for research that integrates the interpretation of human fossils with archaeologically derived insights about hominid adaptation and behavior.

Dietrich Stout discusses the question: Should human evolutionary studies stress the uniqueness of human nature, and thus a unique human evolutionary history (also addressed by Mc Manus and Cohen), or try to be predictive and based in general "laws" (see also Delisle)? Using Robin Dunbar's "social brain" hypothesis and derivative scenarios of human brain evolution (brain size constrains group size; thus increase in group size provokes increase in brain size) as a platform from which to consider this question, he argues not for one side or the other, but both sides of the coin. Stout concludes that a better understanding of human brain evolution must come from an integration of archaeology and paleontology with comparative and evolutionary biology.

In his consideration of increase in brain size, Ian Tattersall points out that while it is evident that this event occurred at least three times in separate hominid lineages, it remains unclear why or how this occurred and whether the pattern was the same in each instance. While acknowledging that there is a correlation between complex behavior and increased brain volume, he argues that modern symbolic reasoning is not only unique to *Homo sapiens*, but that its emergence was epiphenomenal, that is, after the appearance of morphologically identifiable humans. Specifically, Tattersall points to the invention of language as the likely *agent provocateur* of symbolic reasoning and its more "economical" use of brain matter.

To conclude this volume, Claudine Cohen approaches human evolution with an eye toward the influences of reproductive physiology and behavior. In doing so, she emphasizes the importance of female anatomy and physiology—such as the menstrual cycle, the absence of estrus (concealed ovulation), and life after the reproductive phase—that, perhaps even from the very time of emergence of hominids, have contributed to the pattern of human evolution. Cohen argues further that understanding reproductive behavior provides a window on the importance of interaction between cultural as well as biological factors in human evolution.

We all hope these thoughts and introspections will contribute both to explicating as well as to broadening the intellectual diversity that characterizes the study of human evolutionary.

Jeffrey H. Schwartz
24 December 2016

1 The Deceptive Search for "Missing Links" in Human Evolution, 1860–2010: Do Paleoanthropologists Always Work in the Best Interests of Their Discipline?

Richard G. Delisle

Introduction

A key question underlying this paper is: *Do those paleoanthropologists who discover and pronounce upon fossils follow good and sound scientific practices that are in the best interests of their field?*[1] To quote White (2000, 289): "Modern paleoanthropology . . . is like an inverted ecological pyramid. Armchair commentators abound. . . . Actual producers of fossil data are increasingly rare." Because, more broadly, this is often the case in science, historians of science tend to focus on the "discoverers," since they are initially responsible for enlarging knowledge through the addition of novel empirical evidence.

However, as Thomas Kuhn convincingly presented in *The Structure of Scientific Revolutions* (1970), historians of science have become increasingly aware of the sociological fact that new discoveries are not introduced into an intellectual vacuum. Rather, discoverers present their discoveries to an opinionated scientific community that then evaluates them in the context of received wisdom. In order for it to be recognized as legitimate, a new empirical fact (or conceptual breakthrough) must ultimately be accepted by a sufficient number of scholars within a particular area.

Although not always immediately appreciated by its presumed practitioners, the production of scientific knowledge is organized around a complex intellectual interplay that relies on scholars who attempt to introduce new components, and those who evaluate, analyze, synthesize, and interpret new data on their own as well as in the context of past knowledge. On this view, as much as we may be grateful to fossil-discovering paleoanthropologists for raising money, organizing expeditions, and perhaps finding new fossils and contingent information (e.g., environment and its exploitation), history reveals an alarming adherence to debatable scientific practice, such as claiming discovery of a "missing link," or, more specifically, the ancestor of a known species (its disputable identification notwithstanding) or of a specific hominid clade.

Here, as a historian of science, I assume the role not only of commentator of paleoanthropological practice, but as *agent provocateur* of those who also comment on this discipline. This paper is directed at both its producers and commentators.

But before continuing, it is necessary to make clear that: (1) this contribution does not espouse or reject any particular perspective on human evolution; (2) mention of practicing paleoanthropologists (past or present) is not intended to be a statement on scientific "misconduct" (i.e., mention of individual perspectives serves only to illustrate what appears to be *collective practice* in the field); and (3) the review presented here is offered as an invitation to practicing paleoanthropologists to reflect on their procedures. While it is legitimate for a historian to extend such an invitation, he or she must stop at the threshold of the practice itself, and thus maintain distance between the historian of the discipline and its scientists.

Missing Links: A Historical Overview Beginning in the Late Nineteenth Century

The following discoveries, which were originally claimed to be missing links, will drive home the point made previously.

1891–1892 (Java): *Pithecanthropus erectus*

Eugène Dubois conceived that living hominoids, including humans, emerged from a hypothetical Tertiary gibbonlike form called *Prothylobates*. The human lineage had gone through a *Pithecanthropus* stage that gave rise to *Homo sapiens* (Dubois 1894, 1896a, 1896b). Since the fossil material underlying this claim was limited to a skullcap, femur, and molar with questionable association, Dubois' claim was subject to heated debate (see Theunissen 1989, 53–108).

1880s–1900s (Argentina): *Tetraprothomo argentinus*, *Diprothomo platensis*, *Homo pampaeus*, and *Homo pliocenicus*

According to Florentino Ameghino (1906, 1907, 1909), all of these Miocene and Pliocene forms led either directly or tangentially to the emergence of living humans. Nevertheless, few scholars embraced Ameghino's claims, especially after demonstration of the fossils' recency (see Inés Baffi and Torres 1997).

1908–1912 (England): *Eoanthropus dawsoni* (Piltdown)

Following the original description of the discovery by Charles Dawson and A. Smith Woodward (Dawson and Smith Woodward 1912; Smith Woodward 1913), Grafton Elliot Smith (1912, 1913, 1924) became associated with the discovery through his reconstruction of the skull. In keeping with his belief that human evolution was tied to evolution of the brain, Smith supplanted *Homo neanderthalensis* and *H. heidelbergensis* with the Piltdown hominid as the basal species in the lineage of genus *Homo*. Beginning with the application of fluorine analysis to the Piltdown bones, this "hominid" had been shown without doubt to be a fake: a composite of cranial bones and a lower canine of a recent human with the mandible of a recent orangutan (see Spencer 1990).

1924 (South Africa): *Australopithecus africanus*

Raymond Dart (1925, 1926, 1929) was steadfast in presenting the Taung child skull as a taxon intermediate between apes and extant humans. Nevertheless, the juvenility of this specimen has led to debate concerning its taxonomic, and also phylogenetic, status (see Delisle 2007, 222–231).

1932 (Kenya): *Homo kanamensis*

Louis Leakey (1934, 207) considered a mandibular fragment proof that the roots of modern *Homo sapiens* lay in the very early Pleistocene. Unfortunately, Leakey lost geographical and geological data concerning this specimen (see Morell 1995, chapter 5).

1958 (Italy): *Oreopithecus bambolii*

Although first known only from dentognathic specimens, the recovery of a nearly complete skeleton provoked Johannes Hürzeler (1958, 1960) to argue that this Miocene hominoid, with its relatively small canines, orthognathic face, and apparent bipedality, was more central to human evolution than the more apelike East African *Proconsul* and European dryopithecines.

1959 (Tanzania): *Zinjanthropus boisei*

Found by Mary Leakey in Upper Bed I, Olduvai Gorge, Louis Leakey (1959, 1960, 1961a) argued that this very robust Lower Pleistocene australopith, and not other australopiths, was basal to the human lineage, which was a view difficult to defend in light of his own claim that the divergence of human and australopith lineages had been quite early (see Lewin 1987, 137–141).

1960–1963 (Tanzania): *Homo habilis*

The discovery of gracile specimens of early Pleistocene hominid at Olduvai Gorge that Leakey and colleagues (1964) allocated to the new species *Homo habilis* served to demonstrate that *Zinjanthropus* was not the basal ancestor of *Homo*, and that human and australopith lineages had long been separated. Although Leakey and colleagues thought their *H. habilis* the ancestor of all *Homo*, they may not have seen it as a direct ancestor of *H. erectus* (see Reader 2011, 317–323).

1961–1967 (Kenya): *Kenyapithecus africanus* **and** *K. wickeri*

Restless in his search for hominid "missing links," Louis Leakey (1961b, 1967, 1970) turned to specimens of *K. africanus* and *K. wickeri* from the Miocene site Fort Ternan, Kenya, which he argued were closely related, if not ancestral to the human lineage. In addition to bolstering this assertion on the basis of non-great-apelike features (e.g. reduced upper canines, not very prognathic face), Leakey (1968) claimed that *Kenyapithecus* had made stone tools.

1973–1975 (Ethiopia and Tanzania): *Australopithecus afarensis*

Donald Johanson and colleagues (1979) conceived their Pliocene species as being sufficiently primitive and apelike to stand as the ancestor from which separate australopith and *Homo* lineages diverged, thereby displacing *A. africanus* as the otherwise preferred common ancestor.

1989 (Greece): *Ouranopithecus* **(***Graecopithecus***)** *macedoniensis*

First known only from a partial mandible with a worn molar, the 1980s witnessed discovery of numerous upper and lower jaws, as well as a partial lower face of this markedly sexually dimorphic hominoid. Based on the larger sample, Louis de Bonis and colleagues (1990) argued that this late Miocene hominoid was a better potential ancestor of hominids than other Miocene hominoids, for example *Proconsul*, *Lufengpithecus*, and *Sivapithecus*.

1992–1993 (Ethiopia): *Ardipithecus* **(***Australopithecus***)** *ramidus*

Tim White and associates (1994, 1995) maintain that this mid-Pliocene form may well be the long-sought link between the living humans and their African ape ancestors.

1994–1995 (Spain): *Homo antecessor*

José Bermudéz de Castro and colleagues (1997) argued that this Lower Pleistocene hominid was the likely most recent common ancestor of Neanderthals and modern humans.

1996 (Ethiopia): *Australopithecus garhi*

Berhane Asfaw and colleagues (1999) suggested that this Plio-Pleistocene hominid is a descendant of *A. afarensis* and the ancestor of the *Homo* lineage.

1999 (Kenya): *Kenyanthropus platyops*

Meave Leakey et al. (2001) argued that this middle Pliocene hominid, and not *A. afarensis*, was ancestral to *Homo*.

2000 (Kenya): *Orrorin tugenensis*

According to Brigitte Senut and colleagues (2001) this late Miocene hominid was the ancestor of all others, thus sidelining *Ardipithecus* and all australopiths.

2001 (Chad): *Sahelanthropus tchadensis*

Michel Brunet et al. (2002) claimed that their Upper Miocene specimen was both a hominid and the ancestor of all hominids.

2008 (South Africa): *Australopithecus sediba*

Lee Berger and colleagues (2010) presented this early Pleistocene species as a descendant of *A. africanus* and the ancestor of *Homo*.

What Makes Paleoanthropologists Tick?

While the above list of discoveries is not exhaustive, it is sufficient to permit identification of a common paleoanthropological practice: namely, the twofold strategy of claiming that one's discovery is likely a direct evolutionary link to living humans, and of displacing other specimens from this position (if necessary). There appear to be several closely related motivations for this practice.

Scientific fame (prestige)

Without doubt, the discovery of a claimed "missing link" attracts more attention than discovering a specimen that is deemed an "evolutionary dead end." Indeed, the pursuit of recognition within and beyond the boundaries of one's discipline is a common feature of scientific endeavors, paleoanthropology being one. As Jacek Tomczyk (2004, 234) summarized the situation: "Paleoanthropologists are in no way different from other people: they want to be popular, they are desirous of fame and they compete against each other. The interpretation of fossil material provides ample opportunities for such contests."

Media attention

The media—for example, radio, television, documentaries, popular science magazines, semipopular books, and even high-impact scholarly magazines and journals—are likely to cover an event announcing the discovery of a new "missing link," especially if it impacts views of human evolution (Larsen 2000, 2; Lewin 1987, 13–18). This is so even at the risk of distorting the scientific message in order to attract public attention (White 2000, 288–289; 2009, 127–128). The advent of the Internet, and its uncontrolled exploitation by individuals wanting to promote their perspective at the expense of other and often more informed presentations has only exacerbated the problem of fame first, science second (Cartmill 2000).

Funding imperatives

Funding agencies are usually more generous when significant discoveries, such as those dealing with missing links, are involved. Of course, the notion that finding "missing links" is more significant than finding fossils deemed "dead ends" is misguided. After all, if the goal of science is to reveal the complete story, the discovery of "evolutionary dead ends" is as crucial to understanding this picture as discovering presumed "missing links." Unfortunately, given increasingly limited financial resources, funding agencies are forced to weigh the potential impact of the research projects they subsidize. Consequently, the search for potential missing links is intrinsically more appealing than adding another specimen to a known fossil record, especially if this merely corroborates the identity of evolutionary dead ends.

Just being lucky

Whether consciously or unconsciously, those who discover fossils find themselves in the position of taking liberties with scientific practice that result in stretching interpretation to include "missing links." An excellent example of this intellectual contortion is provided by Louis Leakey's behavior in the late 1950s and early 1960s. As then (but not previously[2]) a proponent of a "moderate" pre-*sapiens* hypothesis, Leakey consciously sought to demonstrate that the lineage leading to *Homo sapiens* had ancient phylogenetic roots, and that its earliest participant were not too apelike or primitive.

Indeed, with the 1959 discovery in Upper Bed I of Olduvai Gorge, Tanzania, of the partial cranium allocated to *Zinjanthropus*, and the recovery of primitive stone tools elsewhere in Upper Bed I, Leakey could not resist the appeal of humanizing an otherwise primitively large-jawed, large-toothed, and small-brained hominid by imbuing it with tool-making behavior. Nevertheless, the challenge of reconciling his pre-sapiens hypothesis with *Zinjanthropus* as tool maker became moot with the discovery in the early 1960s of an array of specimens that Leakey et al. (1964) presented as representing the same hominid, *Homo habilis*, that was perceived overall as being more gracile than *Zinjanthropus* and that may have been the genuine creator of Upper Bed I stone tools. All along, it seemed, Leakey was correct: a pre-sapiens hominid (*Homo habilis*), and not something *Zinjanthropus*-like, was the progenitor of the human lineage. Perhaps there is a lesson to be learned here.

Although most discoverers will err in their phylogenetic interpretations (and specification of missing links), some will be correct. Proposing the discovery of a missing link has, therefore, a lot to do with being lucky. In other words, one may eventually be right, but not always for the right reasons.

For all the reasons listed above, paleoanthropologists have a strong incentive to find missing links. Unless the paleoanthropological community eventually agrees on rules of engagement that bind all scholars with respect to fossil discoveries, there seems little hope that things will change. To sum up: (1) scientists in human evolution are often driven by extra-scientific considerations, including fame, media attention, funding, and being lucky (along with a few other reasons); and (2), much of this is due more to the sociology of the sciences than to scientific or epistemic rigor. One need not be alarmed that science has a sociological dimension,[3] but one should be worried when this dimension predominates. That discoverers repeatedly claim to find missing links, even though most of them will be wrong— as *they themselves probably suspect*—is troubling, and it reveals paleoanthropology's lack of rigor and scientific maturity (a responsibility also shared with nondiscoverers, as will be shown). Apparently, there is still room for improving paleoanthropology's procedures.

The "Cunning of Reason" (Hegel): How Paleoanthropology Achieved Its Goals Thus Far Despite Self-Interested Scholars

Leaving aside the normative enterprise that revising paleoanthropology's procedures would require, I would like to examine the practice of paleoanthropology as revealed by its history.

It is my contention that preparing the ground for a more rigorous paleoanthropology can only be done in light of its past. In a nutshell, progress on conceiving human evolution was significant between the 1860s and the 1970s, but has been less so since. Why such a difference in the rate of progress? Contrary to what one might expect, the number of fossil discoveries is not the only important factor. Scholars in the early phases of development of the field were committed to amazingly divergent views of human evolution, views that went well beyond what can be imagined today. To look back at this early phase is not unlike visiting the Twilight Zone. At this juncture, another aspect of this paper's thesis should be mentioned: *Commentators (like producers) in the area of human evolution are not always acting in the best interest of their field by sometimes adopting views based on wild guesses and preconceptions.*

These guesses are best seen in the disparity of positions once taken by scholars on four key debates: (1) Early phylogenetic perspectives permitted searching for humankind's closest living relatives among the entire primate spectrum (hominoids, Old and New World monkeys, and even prosimians). This diversity of views was gradually narrowed to apply to the great apes, then to African apes, and now only to the chimpanzee. (2) It was once debated whether humans and their closest living relatives were descended from a common ancestor that was more humanlike or more apelike in conformation. The debate was eventually resolved in favor of the latter view. (3) Previously, the time frame suggested for the emergence of humankind from its nearest living relative varied considerably, encompassing the entire

Tertiary Period (Pleistocene, Pliocene, Miocene, Oligocene, Eocene, and Paleocene) and sometimes even beyond (Cretaceous). This time range has been substantially compressed to the very last portion of the Tertiary (mid-Miocene and after). (4) The geographical range proposed as humankind's cradle used to include nearly the entire surface of the planet (North America, South America, Europe, Asia, and Africa). This is now restricted to the tropical zones of the Old World.

With this original disparity of hypotheses, progress eventually came because a number of fossils (or their persistent absence in some regions and geological horizons) tipped the balance in each of these four debates. Since about 1980, paleoanthropology has rested on what could be called a "near consensus." Indeed, the scientific framework used to think about human evolution ever since has been confined to the following: A human lineage descended from something apelike that inhabited the tropical regions of the Old World post-mid-Miocene. Although this still leaves room for disagreement, the progress made to get there must be appreciated at its full value. In the current age of near consensus, it must also be realized that progress will be harder to achieve in the future. For, while it is relative easy to choose between two widely different hypotheses—as they often presented themselves in the earlier history of the field—it is much more difficult to choose between hypotheses sharing many similarities, as is now the case. The practice of paleoanthropology in the context of a near consensus requires more powerful tools of resolution then in the past.

Before proceeding to review the history of the field, let us explore further the various impediments on the road of paleoanthropology. Let us insist on two main points:

The first point: Until today debates in paleoanthropology have been conducted in a rather loose and undisciplined fashion. Scholars have been free to express their personal preferences and preconceptions in a number of ways:

Paleoanthropologists identifying so-called missing links (as already discussed above)

Subscribing to a family tradition

Perhaps the most obvious case of this pitfall is represented by the Leakeys, who consistently opposed placing australopiths at the base of the *Homo* lineage (L. Leakey 1953, 1966, 1971; R. Leakey 1976, 1981, 1989; R. Leakey and Lewin 1992; M. Leakey and Walker 1997, 2003; M. Leakey et al. 2001).[4]

Following an institutional tradition

Those closely associated with the same institution sometimes adopt what seems to be an "institutional line," holding onto the same view for as long as possible. For example, I would mention the ancestral positioning of *Australopithecus afarensis* as a form eventually leading to the genus *Homo*, which has been central to the Institute of Human Origins (IHO) since the 1970s (Johanson and White 1979; White, Johanson, and Kimbel 1981; Kimbel, White, and Johanson 1984; Rak 1985; Johanson 1989, 1996; Asfaw et al. 1999). The logo of the IHO is quite revealing on that count. Another case was the dogged insistence on the single-species hypothesis during the 1960s and the early 1970s by members of the Department of Anthropology at the University of Michigan (Brace 1964, 1967; Wolpoff 1968, 1971, 1973).

Promoting a national tradition

Countrymen are sometimes inclined to share a similar view, being united as heirs to a national tradition. In South Africa, for instance, and in the face of a changing fossil record, is a defense of *Australopithecus africanus* as a form leading to the genus *Homo* that has become a kind of tradition (Dart 1925, 1926, 1929; Tobias 1966, 1968, 1980; Berger et al. 2010). Of course, the South African tradition is equally an institutional one when the scholars in question are all closely associated with the University of the Witwatersrand, Johannesburg. In France, a national tradition has long been in existence through a version of a "pre-*sapiens*" hypothesis that states that the human lineage avoided an evolutionary phase too committed to a specialized or primitive-looking apelike state, which was an evolutionary phase also believed to be geologically early (Arambourg 1943, 1957; Piveteau 1957, 1962; Genet-Varcin 1963, 1969, 1979; Coppens 1983; Senut 1996, 2001; Senut et al. 2001).

Embracing a nationalistic view

The quest for the cradle of human origins has a long history, sometimes enshrined in nationalistic pride[5] and imbued with ideological overtones and propaganda (see Dennell, this volume).[6] Claims of missing links on national territory, whether in China (*Lufengpithecus* [Wu, Xu, and Lu 1986; Wu 1987]), Spain (*Homo antecessor* [Bermúdez de Castro et al. 1997]), South Africa (*Australopithecus africanus*), and so on, are not always bereft of nationalistic flavor. In fact, nationalism may well be an underlying factor whenever significant evolutionary discoveries, and thus a phase of human evolution, can be attributed to a particular country (e.g., *Homo georgicus*, Republic of Georgia [Gabunia et al. 2001]).

The consistencies of views observed within these family, institutional, national, and nationalistic traditions are undoubtedly scientifically questionable. Taken individually and in isolation, such practices are fairly benign, since the rest of the scientific community is not bound by them. However, collectively and at the current scale, they introduce a level of distortion that pulls the field in multiple directions. Indeed, these practices are so entrenched in paleoanthropology that they are apparently believed to be normal, at least judging from the fact that they are implicitly tolerated. While some of these views may eventually be proven right and may not necessarily be driven by personal preferences and preconceptions, when taken collectively, they reveal a significant number of personal preferences and preconceptions that remain at play in paleoanthropology.

The second point: While many paleoanthropologists have been guilty of repeatedly (mis)-identifying missing links, others (let us call them "interpreters") have felt anything but bound by these original assessments. On the contrary, interpreters have gone on with the business of identifying their own missing links in the fossil record, sometimes even postulating the existence of entirely hypothetical forms that they predict will eventually be uncovered. Here, presented in the context of the limited fossil record that prevailed until the 1930s, are some examples.

Haeckel (1868) created the genus and species *Pithecanthropus alalus* to represent a speechless form intermediate between an ape and a human that had a humanlike torso but an apelike brain, long skull, long and strong upper arms, short and thin lower limbs, thick wooly hair, and dark skin; "alalus" referred to its inability to speak. *Anthropopithecus* was a Tertiary creature with an endocranial capacity intermediate between ape and human, a low forehead, and a salient supraorbital torus; it was capable of manufacturing crude stone and bone tools (De Mortillet 1873; Hovelacque 1877). In addition there was *Prothylobates* (a Tertiary gibbonlike form at the base of all hominoids [humans and all apes; Dubois 1896]); *Proanthropus* (a somewhat humanlike ancestor of all hominoids, that was arboreal and characterized by a semi-erect posture, legs and arms of equal length, prehensile extremities [hands and feet], and a relatively large and rounded skull [Klaatsch 1902]); and *Homosimius precursor* (a humanlike ancestor of all hominoids based on ontogenetic commonalities, which may not have been too far removed from the something like Piltdown "man," with a rounded and well-developed braincase and moderate simian features in the jaw and teeth [Hill-Tout 1921]). All of this only adds to the rather idiosyncratic development of the field in its particular proclivity for seeking missing links, with "missing" being literally interpreted. Although one might think that this mindset was eventually put to rest, one may ask: Is this still the case, but in a more concealed form, in phylogenetic analysis?

As one reviews the ways knowledge has been gathered since the inception of paleoanthropology, it would appear that the guiding epistemology could be provocatively summarized as "anything goes." This was certainly the case in the early phase of the discipline when scholars were largely unconstrained in their interpretations. Examples of this early disparity between views will be reviewed shortly. Yet, as much as a lack of guiding rules constitutes an impediment to sound scientific practices, the fact remains that this undisciplined development yielded decent results, especially during the period from 1860 to 1980. Indeed, while scholars have always been driven to suggest a plethora of widely divergent hypotheses, the slow yet inexorable accumulation of empirical facts—the combined product of an improving fossil record and more sophisticated studies of comparative anatomy and genetics—gradually forced paleoanthropologists to abandon hypotheses that were no longer compatible with scientific reality. Thus, the space necessary to express phylogenetic views was considerably reduced as the twentieth century progressed. As an analogy, it is tempting to refer to the notion of the "cunning of reason" that Hegel presented in his *Lectures on the Philosophy of World History* (1837): egoistic historical actors acting only in their own interest, not realizing that the clash of their contradictory, self-interested actions are helping to realize a secret plan of History that coincides with the rise of freedom (see Dray 1964, chapter 6). Likewise, selfish paleoanthropologists, driven by preconceptions, idiosyncratic ideas, ideological agendas, pursuit of fame, and response to traditions, were unconsciously pushing the field forward by proposing all sorts of alternatives. In so doing, new fossils and interpretations eventually led to proving that many of these evolutionary hypotheses were not scientifically acceptable. A review of the historical development of paleoanthropology since 1860 serves to illustrate.

Paleoanthropology, 1860–1980: A Sketch

What follows is a sketch for which more details can be found elsewhere (Delisle 2007, 2012a, 2012b, 2012c). Hopefully, this endeavor will be sufficiently suggestive to permit the rediscovery of both the original pluralism of perspective that then prevailed and the way it was gradually constrained.

The period from 1860 to 1890 saw a surprising range of suggestions regarding the relationship between living humans and nonhuman primates. Four main positions existed. First, humankind was believed to have originated either from a primitive form unrelated to the anthropoid (apes and monkeys) stem, or from a nonprimate ancestor (Owen 1857, 1861, 1863; Mivart 1873; de Quatrefages 1870, 1877, 1894). Second, the common ancestor of all primates was presumed to be a primitive anthropoid that was perhaps related to monkeys (Schmidt 1884, 1887; Topinard 1888, 1891, 1892). Third, the rise of the human lineage was envisioned as being either within or basal to the larger radiation of hominoids (Huxley 1863; Haeckel 1868; Darwin 1871; Wallace 1889). (It should not necessarily be assumed that these views implied a fairly recent time of divergence, for Darwin even speculated that the human line might possibly have emerged as early as the Eocene.) The scholars thus far reviewed were, by virtue of their defending the notion of "monophyletism," at least in agreement on the question of the unity of humankind. In contrast were scholars who espoused "polyphyletism," in which human evolution was literally intertwined with the hominoid apes, with specific human populations being closer, phylogenetically, to great apes than to other human populations (Vogt 1864; Schaaffhausen 1868; Hovelacque and Hervé 1887).

While scholars were engaged in debates over humankind's place in the primate family tree, they were simultaneously involved in interpreting the human fossil record in the context of the prehistoric peopling of Europe. Since lack of space prevents going into every detail, what must be pointed out is that, during the period from 1860 to 1890, the human fossil record consisted primarily of fairly modern-humanlike specimens that were geographically restricted to Europe. Given these circumstances, the most a scholar could do was propose a partial phylogeny. Indeed, a common approach to interpreting human fossils at the time is noted in the remarkable synthesis of unprecedented breadth presented by de Quatrefages and Hamy (1873, 1874a, 1874b, 1875), which they eventually unified in the first part of their *Crania Ethnica* (de Quatrefages and Hamy 1882). There, de Quatrefages and Hamy identified no fewer than three morphological types of human that existed in Europe during the Quaternary. The most ancient type—the *Canstadt race*—which included skulls and jaws from Canstadt, Eguisheim, Brüx, Feldhofer Grotto, Gibraltar, Denise, Staengenaes, Olmo, La Naulette, Arcy, Clichy, and Goyet, was distributed primarily throughout the substratum of Europe. De Quatrefages and Hamy described this highly variable race as dolichoplatycephalic and prognathic, with a more or less developed supraorbital ridge and a low, receding forehead. The occasional presence of this physical type also in living populations throughout a large part of the Old World, especially Australia and India, was attributed to atavism. The second prehistoric type was the *Cro-Magnon race*, which de Quatrefages and Hamy thought was contemporaneous with the Canstadt race. This race was restricted to western Europe, and included specimens from Cro-Magnon, Laugerie-Basse, Bruniquel, Menton, Isola del Liri, Grenelle, Solutré, Engis, La Madeleine, Engihoul, and Smeermaas. In addition to being dolichocephalic (but not platycephalic), the forehead and facial region of the Cro-Magon race differed from the Canstadt race in being, respectively, more vertical and less protrusive. The appearance of the Cro-Magnon type in populations from northwest Africa, and sporadically in living Europeans, was also attributed to atavism. The third prehistoric type was actually a mixture of races whose cranial shape ranged from brachycephalic to mesaticephalic, and included specimens from Furfooz, La Truchère, Grenelle, Moulin-Quignon, and Nagy-Sap. This type was believed to have lived in contemporaneity with both the Canstadt and Cro-Magnon types, having left numerous traces among living populations.

Although scholars of the period from 1860 to 1890 could not reconcile the debate on the peopling of Europe with that of humankind's place among the primates, this gap began to close between 1890 and 1935, although pluralistic perspectives did not. That is, as scholars produced increasing diverse hypotheses about humankind's place among the primates, this endeavor encompassed debates on human evolution itself that went beyond the Eurocentric focus of the pre-1890 period. Further, as the new empirical context came to include regions beyond Europe, the number of scenarios increased and, so, too, did level of disagreement.

In further contrast from the pre-1890 period, when nonhuman fossil primates were largely seen as not providing insight into humankind's place among the primates (but see, e.g., Lydekker [1879], Cope [1888], and Schlosser [1888]), the new perspective embraced them (e.g., *Anaptomorphus, Dryopithecus, Pliopithecus, Propliopithecus, Palaeopithecus, Palaeosimia, Sivapithecus, Gryphopithecus, Neopithecus, Pliohylobates, Hesperopithecus, Pitheculites,* and *Homunculus*). In addition, the period from 1890 through 1935 witnessed an increasing acceptance of the New World as a participant in human origins. Among the array of speculation, four phylogenetic hypotheses dominated that period.

First is the notion that human evolution was something apart from the evolution of primates in general (Adloff 1908; Sergi 1914). In the second, humans branched off near the base of the primate family tree, without a hominoid stage of evolution (Boule 1921; Wood Jones 1929). The third acknowledged a relationship between humans and living apes, although the issues of which ape or apes (gibbon, orangutan, gorilla, or chimpanzee) were involved, and whether the common human/ape ancestor was morphologically more human- or apelike, remained unresolved (Hubrecht 1897; Klaastch 1902; Ameghino 1906; Keith 1915; Osborn 1915; Pilgrim 1915; Hill-Tout 1921; Gregory 1922; Elliot Smith 1924; Schultz 1930; Hooton 1931; Weinert 1932; Le Gros Clark 1934). Lastly was the popular polyphyletic hypothesis, which subsumed two distinct hypotheses: living humans were linked evolutionary either exclusively with apes (Sergi 1908; Melchers 1910; Gray 1911; Buttel-Reepen 1913; Klaastch 1923; Kurz 1924; Frasetto 1927; Crookshank 1931), or, more broadly, apes, Old and even New World monkeys, and sometimes even prosimians (Sergi 1909–1910; Horst 1913; Arldt 1915; Sera 1918; figures 1.1–1.4).

With regard to debates on the details of human evolution, the period from 1890 to 1935 witnessed not only the continued discovery of specimens that were similar to living humans (e.g., from Cro-Magnon), as well as of specimens that expanded the Neanderthal realm, but also of un-*sapiens*-like hominids (real or fictitious), for example, *Pithecanthropus*, *Sinanthropus*, *Eoanthropus*, *Homo heidelbergensis*, and *H. rhodesiensis*. These discoveries led to three kinds of pluralistic hypotheses.

One was conceiving human evolution as being organized around a series of more or less independent and parallel evolutionary lines, each rooted in specific specimens and each giving rise to a distinct line ("race"; Dixon 1923; Pycraft 1925; Taylor 1927). Another envisioned human evolution as a process in which a single lineage gradually morphed into living *Homo sapiens* (Dubois 1896; Schwalbe 1906; Verneau 1906; Keith 1911; Mahoudeau 1912; Hrdlička, 1927). Lastly was a multilinear hypothesis in which only a few specimens were embraced as being directly in the line leading to living humans; the rest were put on side branches that eventually went extinct (Keith 1915; Boule 1921; Hooton 1931; Leakey 1934; figure 1.5).

With the eventual rejection of polyphyletism in the 1930s, paleoanthropology began to take on its current configuration of research. Further, the improved fossil record had the effect of disentangling hominids from any known ape or monkey, and positing a more ancient origin. Specifically relevant to details of human evolution was the discovery in South African limestone cave sites of australopiths, beginning with Raymond Dart's (1925) *Australopithecus africanus* (based on the partial skull and mandible of the child from Taung) and then Robert Broom's (1936, 1946) work at Kromdraai, Sterkfontein, and Swartkrans that led to his naming *A. prometheus*, *Paranthropus robustus*, and *P. crassidens*. By the early 1960s,

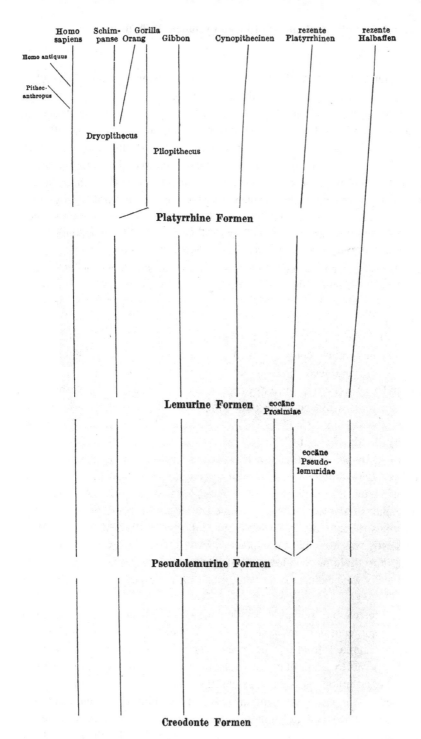

Figure 1.1
Paul Adloff's (1908) parallel tree, illustrating the unique, nonprimate origin of the human lineage.

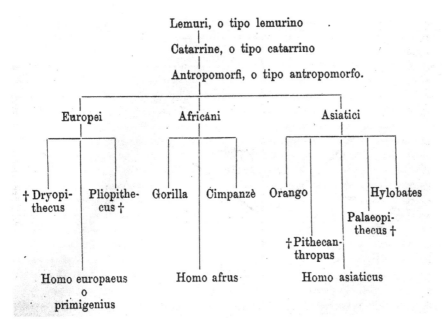

Figure 1.2
Giuseppe Sergi's (1908) polyphyletic tree and the intertwining of apes and various human populations.

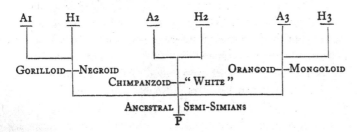

Figure 1.3
Francis Crookshank's (1931) polyphyletic tree and the intertwining of the great apes and various human populations.

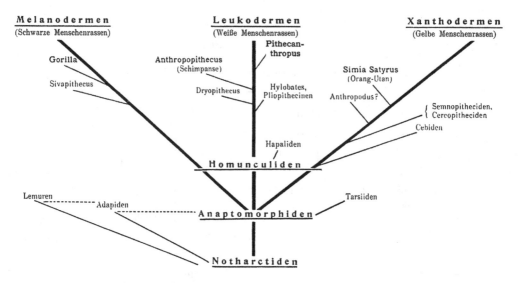

Figure 1.4
Theodor Arldt's (1915) polyphyletic tree and the intertwining of various human populations, apes, and monkeys.

John Robinson (1954, 1956, 1962), who succeeded Broom, had convinced previously skeptical paleoanthropologists of the existence of two groups of australopiths that were distinguished in perceived notions of "gracility" versus "robustness." The former were allocated to *Australopithecus* (with two species) and the latter to *Paranthropus* (also with two species). Further, Robinson's fact-based arguments for australopiths-as-early-hominids constrained and focused hypotheses about human evolution. Indeed, whereas the pre-1935 period witnessed an increase in the number of divergent hypotheses and levels of pluralism, this trend was reversed in the post-1935, with rejection not only of the New World as a possible player, but also eventually of two prominent pre-1935 notions of human origins (from either an early, generalized primate [Schepers 1946; Wood Jones 1948; Straus 1949; Osman Hill 1954; Piveteau 1957; Genet-Varcin 1963], or a distant "generalized" ape [Arambourg 1943; Schultz 1950; Leakey 1953; Hürzeler 1954; Le Gros Clark 1955; Vallois 1955; Heberer 1959]). This left the hypothesis that humans evolved fairly recently from a specific ape or ape group (Hooton 1946; Weidenreich 1946; Washburn 1950; Gregory 1951; Patterson 1954; Simons 1961; Koenigswald 1962), which most paleoanthropologists accepted in large part as a result of continuing discoveries of hominoids from late Eocene–early Oligocene deposits in the Fayum Depression (Egypt), and from Miocene and even Pliocene and Pleistocene deposits in Europe (France, Germany, Italy, and now also Spain and Greece), the Siwalik Mountains of Indo-Pakistan, southern China, and East Africa (Tanzania, Kenya, Uganda).

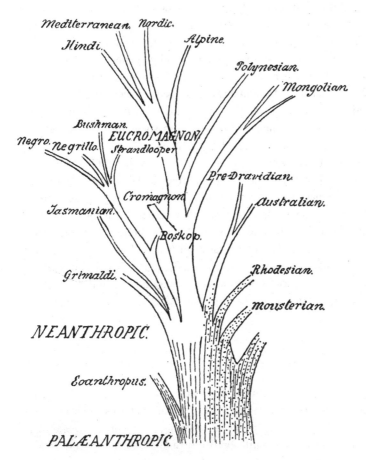

Figure 1.5
William Pycraft's (1925) parallel tree showing both modern-looking and primitive-like lineages giving rise to living human populations.

The period from 1965 to 1980 witnessed solidification of the notion that hominids diverged from a Miocene hominoid, as well as the identification of the earliest hominid, most popularly in the Siwalik *Ramapithecus* (Simons and Pilbeam 1965; Aguirre 1972; Eckhardt 1972; Simons 1977; Xu and Lu 1979; Greenfield 1980; Schwartz 1984; Wu 1984; Pilbeam 1985). Empirically speaking, while the fossils embedded in the variants of this scenario remained the same, the interpretation of some changed (e.g., *Ramapithecus*, *Sivapithecus*, *Dryopithecus*, *Gigantopithecus*, and *Oreopithecus*). But hypotheses of human-ape relationships and dating the emergence of hominids began to change dramatically in the 1960s with the advent of molecular anthropology.

Among the pioneers of this new discipline was Morris Goodman (1962a, 1962b) who studied various blood serum proteins (via electrophoresis and serum-antiserum reactivity). Based on a small sampling of primates and the assumption that, because molecules were supposed to continually change, overall similarity reflected phylogenetic propinquity, he rejected the dominant theory of humans being related to a great-ape group (orangutans and a *Pan-Gorilla* sister group), and asserted that humans were closely related to the African apes only. A plethora of genetic studies followed, all of which excluded the orangutan from a close evolutionary relationship with humans, with some disassociating *Gorilla* and *Pan* and claiming that the latter ape alone was closely related to humans (Goodman 1963; Sarich and Wilson 1966, 1967a, 1967b; Benveniste and Todaro 1976; Miller 1977; Sibley and Ahlquist 1987; Ruvolo 1997; Takahata and Satta 1997). Further, and guided by the assumption that molecules are always changing, many molecular anthropologists proposed dates at which the various hominoids, and humans and chimpanzees in particular, diverged from their common ancestors. Although Goodman tried to accommodate paleontological derived dates, which, via *Ramapithecus*, put the split between humans and *Pan* well into the Miocene, by suggesting that the rate of molecular change varied, others—most notably Sarich and Wilson—insisted that it was constant. As far as Sarich (1970) was concerned, humans and chimpanzees diverged no earlier than 5 mya, and that nothing earlier than that could be related to humans, no matter what it looked like.

Breaking the Deadlocks After 1980?

Interestingly, until about 1980, the undisciplined approaches in paleoanthropology were oddly productive. It would, however, be inaccurate to identify paleoanthropological practice in general as an "epistemology," inasmuch as it was anything but a concerted and well-articulated effort by a theoretically and philosophically unified scientific community. Indeed, the essence of paleoanthropology then is reducible to this: The formulation of every hypotheses imaginable, at any time in the history of the field. Nevertheless, as empirical paleontological evidence mounted, so, too, did the field of scientifically acceptable hypotheses.

One may contrast this history with Karl Popper's (1963) method of conjectures and refutations, which consists of proposing a hypothesis, thoroughly testing it, and discarding it if and when it ceases to be corroborated (i.e., is falsified) and fails to fit the conceptual-empirical context. Although there are limitations to a Popperian epistemology (see Chalmers 1999), until its value for paleoanthropologists is fully investigated, an intellectual process of testing and eliminating hypotheses will continue to lag far behind the paleoanthropological proclivity to propose as many different hypotheses as imaginable, and the unspoken expectation that the last hypothesis standing will be the correct one. In a sense, then, paleoanthropologists have long been spared the necessity of adopting precise rules concerning their procedures. This may, however, have been an easy victory. For it has not been difficult, and with only a few fossils to lend support to choose from among the array of widely divergent hypotheses those one feels should be discarded, in large part because they stray far afield from the accepted norm. However, in the current age of near consensus, it is much more difficult to choose from among evolutionary hypotheses which are similar. Indeed, the purported uncrossable chasm between current phylogenetic hypotheses is pure illusion when compared, for instance, to pre-1950 views. In truth, since the 1980s, the similarities between key hypotheses and debates have been sufficiently similar that they devolve into intellectual impasses with no obvious resolution or significant progress in sight. Let us briefly consider three debates.

Debates about early hominids

As noted above, approximately between 1960 and 1980, the discovery of dryopithecines and australopiths (widely construed), in addition to the advent of molecular anthropology, served to compress the taxonomic and temporal frameworks in which debates about human origins and evolutionary relationships could maneuver. Since then, and beyond *Homo habilis*, *H. rudolfensis*, *Australopithecus africanus*, and *A. afarensis* an impressive number of so-called missing links have been proposed: *A. sediba*, *A. garhi*, *Ardipithecus ramidus*, *Kenyanthropus platyops*, *Orrorin tugenensis*, *Sahelanthropus tchadensis*, *Ouranopithecus macedoniensis*, and *Lufengpithecus lufengensis*. The consequence is that this rash of proposed ultimate ancestors and missing links obscures common reference points; it thus makes it impossible to make sense of them both specifically and in the greater scheme of human origins and evolution. In the previous period, when hypotheses were more diverse and overlapped less, this would not have been the case.

Debates about later hominids

The pre-1980 discovery of specimens of australopith and "primitive" *Homo* (*H. habilis*, East African *H. erectus*), and reconsideration and rejection of the relevance of various specimens to later human evolution (e.g., specimens from Piltdown, Swancombe, Galley Hill, Fontéchevade, Kanam, and Kanjera; see Oakley 1964), led to the notion that living humans evolved from some primitive hominid of the early Pleistocene. In the context of this limitation on the scope of debate, some paleoanthropologists attempted to break the deadlock by shuffling and reshuffling fossils within two different models: Out of Africa (i.e., after a single ancestor left Africa, humans diversified into present-day populations) versus Multiregional (i.e., *in situ* emergence of diverse modern human populations from geographically widespread and morphologically different Pleistocene hominids [Thorne and Wolpoff 1981; Wolpoff, Wu, and Thorne 1984; Stringer and Andrews 1988; Stringer 1990; see also Delisle 2001]). Yet, to view the debate between the Multiregional and Out of Africa models as between two distinct theories or paradigms, as is sometimes done, is to assume, incorrectly, that the two are cognitively and logically distinct (Willermet and Clark 1995). Indeed, this assumption breaks down in the face of intermediate phylogenetic hypotheses that combine various elements from both models (Bräuer 1992; F. Smith 1992; S. Smith and Harrold 1997). In this context, the Out of Africa and Multiregional models are actually opposite poles of a continuum of defensible hypotheses. In my view, this debate will not be resolved by appealing to a specific theory/model because there will always be competing alternatives that will be consistent with the current empirical context. If paleoanthropology's present-day procedures are used, it could take a long time before the empirical reality improves sufficiently so as to permit the rejection of a number of theories/models that are no longer consistent with it.

Molecular anthropology and living humans

In contrast to the relative consistency over the past three decades with which molecular analyses have claimed the phylogenetic relationships of *Pan, Gorilla, Pongo*, and extant *Homo sapiens*, consensus on the origin of living humans has been less than unanimous. The apparent reason is that, when one is dealing with *taxonomically distinct* entities (humans, chimpanzees, gorillas, orangutans), one is dealing with genetic separation on the scale of millions of years. Since, however, living human populations derive from a recent common ancestor, they have not accrued taxonomically significant genetic differences, and are thus potentially capable of exchanging genes. In this *instraspecific* context, wherein genetic entity is not clear cut, there are myriad opportunities for "noise" to be a factor in molecular analyses: that is, variables such as population size, mutation rate, amount of gene flow, migration, hybridization, and selection (Excoffier and Roessli 1990; Langaney, van Blyenburgh, and Nadot 1990; Stoneking 1993; Relethford 1995, 1999; Crubézy and Braga 2003). Consequently, these molecular studies essentially mimic the level of disagreement that characterizes morphological attempts to decipher the emergence and identity of, and inter-relationships between, living human populations.

Conclusion

If we can learn anything about paleoanthropology by understanding its intellectual history, it is that, while its approaches to evolutionary questions in general, and the phylogeny of human and non-human primates in particular, have been and still are undisciplined, until about 1980 the field at large produced surprisingly good results. Since then, however, key debates on human phylogeny have essentially come to a standstill. This appears to be because the driving forces of the past, which were often due to personal preferences and biases, are no longer sufficient to move paleoanthropology forward. In an age when exponents of various extreme postmodernist approaches are only too happy to claim that science is a "social construct," paleoanthropologists may want to avoid exposing their field to facile and superficial criticism. Without prejudging what the future will look like, a reasonably modest suggestion would be that paleoanthropology should reduce reliance on personal preferences and preconceptions, and move toward more rigorous operational and analytical practice and procedure, elaboration of common rules of engagement, and more open and accessible discourse. Perhaps then paleoanthropology may enter its second epistemological phase of development.

Acknowledgments

I thank Jeffrey Schwartz for his kind invitation to join both the 32nd Altenberg Workshop in Theoretical Biology and the Vienna Series in Theoretical Biology.

Special thanks are also extended to the KLI's Eva Lackner, Isabella Sarto-Jackson, and Gerd Müller. I thank F. Mc Manus for critical comments leading to revision and clarification of this manuscript. Lastly, I thank James Tierney (Yale University) and Jeffrey Schwartz for assistance in improving the English version of this paper.

Notes

1. The notion of "fossil discoverer" is understood here in its broadest sense, that is, the fossil discoverers themselves; the first to study finds and publish the results; or scholars very closely involved in the discovery, analysis, and description of new finds.

2. Leakey's slow and complex shift between the early 1950s and the early 1970s from a strong to a weaker pre-sapiens view is summarized in Delisle (2007: 279–281).

3. For a balanced overview of the achievements in the sociology of science, see S. Shapin (1995).

4. With the exception of Louis Leakey of the short-lived episode of *Zinjanthropus* in 1959–1960. On the views of Mary D. Leakey, see Morell (1995: 441–442, 479–481).

5. I was always struck by the pride Africans have for what are so far the oldest known hominids. A similar pride was likely present during the very nationalistic (and racial) context of Europe in the 19th and 20th centuries. One can imagine the pride of the French at having a form so modern-looking as Cro-Magnon, or the British for having in Piltdown "man," with its apelike mandible and teeth but modern humanlike large and vaulted braincase, a rival to Dubois' *Pithecanthropus erectus*. The same can also be said of the Argentinian Florentino Ameghino and his claims of having found on his country's soil many potential human ancestors (i.e., *Tetraprothomo argentinus*, *Diprothomo platensis*, *Homo pampaeus*, and *Homo pliocenicus*).

6. Regarding China, for instance, see Robin W. Dennell (this volume). As Dennell rightly concludes: "Non-Chinese researchers in paleoanthropology have to understand that their Chinese counterparts operate within a very different social and political context from their own. At the same time, non-Chinese researchers need also to examine their own belief frameworks and academic traditions as critically as they examine those of their Chinese colleagues."

References

Adloff, P. 1908. *Das Gebiss des Menschen und der Anthropomorphen*. Berlin: Julius Springer.

Aguirre, E. 1972. "Les rapports phylétiques de *Ramapithecus* et de *Kenyapithecus* et l'origine de l'homme." *L'Anthropologie* 76: 501–523.

Ameghino, F. 1906. "Les formations sédimentaires du Crétacé supérieur et du Tertiaire de Patagonie." *Anales del Museo Nacional de Buenos Aires*, 3rd series, 8: 1–568.

Ameghino, F. 1907. "Notas preliminares sobre el *Tetraprothomo argentinus*: Un precuror del hombre del Mioceno superior de Monte Hermoso." *Anales del Museo Nacional de Buenos Aires*, 3rd series, 9: 107–242.

Ameghino, F. 1909. "Le *Diprothomo platensis*: Un précurseur de l'homme du Pliocène inférieur de Buenos Aires." *Anales del Museo Nacional de Buenos Aires*, 3rd series, 12: 107–209.

Arambourg, C. 1943. *La genèse de l'humanité*. Paris: Presses Universitaires de France.

Arambourg, C. 1957. *La genèse de l'humanité*. 5th ed. Paris: Presses Universitaires de France.

Arldt, T. 1915. "Die Stammesgeschichte der Primaten und die Entwicklung der Menschenrassen." *Fortschritte der Rassenkunde* 1: 1–52.

Asfaw, B., T. White, O. Lovejoy, B. Latimer, S. Simpson, and G. Suwa. 1999. "*Australopithecus garhi*: A New Species of Early Hominid From Ethiopia." *Science* 284: 629–635.

Benveniste, R. E. and G. J. Todaro. 1976. "Evolution of the Type C Viral Genes: Evidence for an Asian Origin of Man." *Nature* 261: 101–108.

Berger, L. R., D. J. de Ruiter, S. E. Churchill, P. Schmid, K. J. Carlson, P. H. Dirks, and J. M. Kibii. 2010. "*Australopithecus sediba*: A New Species of *Homo*-Like Australopith From South Africa." *Science* 328: 195–204.

Bermudez de Castro. J., J. L. Arsuaga, E. Carbonell, A. Rosas, I. Martinez, and M. Mosquera. 1997. "A Hominid From the Lower Pleistocene of Atapuerca, Spain: Possible Ancestor to Neanderthals and Modern Humans." *Science* 276:1392–1395.

Boule, M. 1921. *Les hommes fossiles*. Paris: Masson.

Brace, C. L. 1964. "The Fate of the 'Classic' Neanderthals: A Consideration of Hominid Catastrophism." *Current Anthropology* 5: 3–43.

Brace, C. L. 1967. *The Stages of Human Evolution*. New Jersey: Prentice Hall.

Bräuer, G. 1992. "Africa's Place in the Evolution of *Homo sapiens*." In *Continuity or Replacement: Controversies in* Homo sapiens *Evolution*, edited by G. Bräuer and F. H. Smith, 83–98. Rotterdam: A. A. Balkema.

Broom, R. 1936. "A New Fossil Anthroid Skull from South Africa." *Nature* 138: 486–488

Broom, R. 1946. *The South African Fossil Ape-Men: The Australopithecinae, Part I*. Pretoria: Transvaal Museum Memoir, No. 2.

Brunet, M., F. Guy, D. Pilbeam, H. T. Mackaye, A. Likius, D. Ahounta, A. Beauvilain, C. Blondel, H. Bocherens, J.-R. Boisserie, L. de Bonis, Y. Coppens, J. Dejax, C. Denys, P. Duringer, V. Eisenmann, G. Fanone, P. Fronty, D. Geraads, T. Lehmann, F. Lihoreau, A. Louchart, A. Mahamat, G. Merceron, G. Mouchelin, O. Otero, P. P. Campomanes, M. P. De Leon, J.-C. Rage, M. Sapanet, M. Schuster, J. Sudre, P. Tassy, X. Valentin, P. Vignaud, L. Viriot, A. Zazzo, and C. Zollikofer. 2002. "A New Hominid From the Upper Miocene of Chad, Central Africa." *Nature* 418: 145–151.

Buttel-Reepen, von H. 1913. *Man and His Forerunners*. London: Longmans, Green and Co.

Cartmill, M. 2000. "A View on the Science: Physical Anthropology at the Millennium." *American Journal of Physical Anthropology* 113: 145–149.

Chalmers, A. F. 1999. *What Is This Thing Called Science?* 3rd ed. Indianapolis: Hackett.

Cope, E. D. 1888. "Archaeology and Anthropology." *American Naturalist* 22: 660–663.

Coppens, Y. 1983. *Le singe, l'Afrique et l'homme*. Paris: Fayard.

Crookshank, F. G. 1931. *The Mongol in Our Midst*. 3rd ed. London: Kegan Paul, Trench, Trubner & Co.

Crubézy, E. and J. Braga. 2003. "Homo sapiens prend de l'âge." *La Recherche* 368: 30–35.

Dawson, C. and A. Smith Woodward. 1912. "On the Discovery of a Palaeolithic Human Skull and Mandible in a Flint-Bearing Gravel Overlying the Wealden (Hastings Beds) at Piltdown, Fletching (Sussex)." *Quarterly Journal of the Geological Society of London* 69: 117–144.

Dart, R. A. 1925. "*Australopithecus Africanus*: The Ape-Man of South Africa." *Nature* 115: 195–199.

Dart, R. A. 1926. "Taungs and Its Significance." *Natural History* 26: 315–326.

Dart, R. A. 1929. "A Note on the Taung Skull." *South African Journal of Science* 26: 648–658.

Darwin, C. 1871. *The Descent of Man*. Vol. 1. London: John Murray.

De Bonis, L., G. Bouvrain, D. Geraads, and G. Koufos. 1990. "New Hominid Skull Material From the Late Miocene of Macedonia in Northern Greece." *Nature* 345: 712–714.

Delisle, R. G. 2001. "Adaptationism Versus Cladism in Human Evolution Studies." In *Studying Human Origins: Disciplinary History and Epistemology*, edited by R. Corbey and W. Roebroeks, 107–121. Amsterdam: Amsterdam University Press.

Delisle, R. G. 2007. *Debating Humankind's Place in Nature, 1860–2000: The Nature of Paleoanthropology*. New Jersey: Pearson/Prentice Hall.

Delisle, R. G. 2012a. "The Disciplinary and Epistemological Structure of Paleoanthropology: One Hundred and Fifty Years of Development." *History and Philosophy of the Life Sciences* 34: 283–330.

Delisle, R. G. 2012b. "Welcome to the Twilight Zone: A Forgotten Early Phase of Human Evolutionary Studies." *Endeavour* 36: 55–64.

Delisle, R. G. 2012c. "Human Evolution, Theories of: Introduction." In *The Oxford Companion to Archaeology*, 2nd ed., edited by N. A. Silberman, 21–26. Oxford: Oxford University Press.

De Mortillet, G. 1873. "Le précurseur de l'homme." *Compte Rendu de l'Association Française pour l'Avancement des Sciences*, 2e session (Lyon), pp. 607–613.

De Quatrefages, A. 1870. *Charles Darwin et ses précurseurs français. Études sur le transformisme*. Paris: Baillière.

De Quatrefages, A. 1877. *L'Espèce humaine*. Paris: Alcan.

De Quatrefages, A. 1894. *Les émules de Darwin*. Vol. 2. Paris: Baillière.

De Quatrefages, A. and E.-T. Hamy. 1873. "Races humaines fossiles—race de Canstadt." *Bulletins de la Société d'Anthropologie de Paris*, 2e série, 8: 518–523.

De Quatrefages, A. and E.-T. Hamy. 1874a. "La race de Cro-Magnon dans l'espace et dans le temps." *Bulletins de la Société d'Anthropologie de Paris*, 2e série, 9: 260–266.

De Quatrefages, A. and E.-T. Hamy. 1874b. "Races humaines fossiles mésaticéphales et brachycéphales." *Bulletins de la Société d'Anthropologie de Paris*, 2e série, 9: 819–826.

De Quatrefages, A. and E.-T. Hamy. 1875. "Les Crania Ethnica." *Bulletins de la Société d'Anthropologie de Paris*, 2e série, 10: 612–619.

De Quatrefages, A. and E.-T. Hamy. 1882. *Crania Ethnica: Les crânes des races humaines*. Paris: Baillière.

Dixon, R. B. 1923. *The Racial History of Man*. New York: Charles Scribner's Sons.

Dray, W. H. 1964. *Philosophy of History*. New Jersey: Prentice Hall.

Dubois, E. 1894. *Pithecanthropus erectus, eine menschenähnliche Uebergangsform aus Java*. Batavia.

Dubois, E. 1896a. "On *Pithecanthropus Erectus*: A Transitional Form Between Man and the Apes." *Scientific Transactions of the Royal Dublin Society* 6: 1–18.

Dubois, E. 1896b. "The Place of *Pithecanthropus* in the Genealogical Tree." *Nature* 53: 245.

Eckhardt, R. B. 1972. "Population Genetics and Human Origins." *Scientific American* 226: 94–103.

Elliot Smith, G. 1912. "President's Address, Anthropology Section." *Report of the British Association for the Advancement of Science*, 1912: 575–598.

Elliot Smith, G. 1913. "The Piltdown Skull and Brain Cast." *Nature* 92: 267–268, 318–319.

Elliot Smith, G. 1924. *Essays on the Evolution of Man*. London, Oxford University Press.

Excoffier, L. and D. Roessli. 1990. "Origine et évolution de l'ADN mitochondrial humain: la paradigme perdu." *Bulletins et Mémoires de la Société d'Anthropologie de Paris*, new series, 2: 25–42.

Frasetto, F. 1927. "New Views on the Dawn Man of Piltdown (Sussex)." *Man* 27: 121–124.

Gabunia, L., S. C. Anton, D. Lordkipanidze, A. Vekua, A. Justus, and C. C. Swisher III. 2001). "Dmanisi and Dispersal." *Evolutionary Anthropology* 10: 158–170.

Genet-Varcin, E. 1963. *Les singes actuels et fossiles*. Paris: N. Boubée et Cie.

Genet-Varcin, E. 1969. *À la recherche du primate ancêtre de l'homme*. Paris: N. Boubée et Cie.

Genet-Varcin, E. 1978. "Réflexion sur l'origine des hominidés." In *Les origines humaines et les époques de l'intelligence*, 13–36. Paris: Masson.

Genet-Varcin, E. 1979. *Éléments de primatologie: Les hommes fossiles*. Paris: N. Boubée et Cie.

Goodman, M. 1962a. "Evolution of the Immunologic Species Specificity of Human Serum Proteins." *Human Biology* 34: 104–150.

Goodman, M. 1962b. "Immunochemistry of the Primates and Primate Evolution." *Annals of the New York Academy of Sciences* 102: 219–234.

Goodman, M. 1963. "Man's Place in the Phyloneny of the Primates as Reflected in Serum Proteins." In *Classification and Human Evolution*, edited by S. L. Washburn, 204–234. Chicago: Aldine.

Gray, J. 1911. "The Differences and Affinities of Palaeolithic Man and the Anthropoid Apes." *Man* 11: 117–120.

Greenfield, L. O. 1980. "A Late Divergence Hypothesis." *American Journal of Physical Anthropology* 52: 351–365.

Gregory, W. K. 1922. *The Origin and Evolution of the Human Dentition*. Baltimore: Williams & Wilkins.

Gregory, W. K. 1951. *Evolution Emerging*. Vol. 1. New York: Macmillan.

Haeckel, E. 1868. *Natürliche Schöpfungsgeschichte*. Berlin: Georg Reimer.

Heberer, G. 1959. "The Descent of Man and the Present Fossil Record." *Cold Spring Harbor Symposia on Quantitative Biology* 24: 235–244.

Hill-Tout, C. 1921. "The Phylogeny of Man from a New Angle." *Transactions of the Royal Society of Canada* 15: 47–82.

Hooton, E. A. 1931. *Up from the Ape*. New York: Macmillan.

Hooton, E. A. 1946. *Up from the Ape*. 2nd ed. New York: Macmillan.

Horst, M. *Die natürlichen Grundstämme der Menschheit*. Hildburghausen: Thüringische Verlags-Anstalt.

Hovelacque, A. 1877. "Notre ancêtre: recherches d'anatomie et d'ethnologie sur le précurseur de l'homme." *Revue d'Anthropologie* 6: 62–99.

Hovelacque, A. and G. Hervé. 1887. *Précis d'Anthropologie*. Paris: Delahaye et Lecrosnier.

Hrdlička, A. 1927. "The Neanderthal Phase of Man." *Journal of the Royal Anthropological Institute* 57: 249–274.

Hubrecht, A. A. W. 1897. *The Descent of the Primates*. New York: Charles Scribner's Sons.

Hürzeler, J. 1954. "Contribution à l'odontologie et à la phylogénèse du Genre *Pliopithecus* Gervais." *Annales de Paléontologie* 40: 5–63.

Hürzeler, J. 1958. "*Oreopithecus bambolii* Gervais: A Preliminary Report." *Verhandlungen der Naturforschenden Gesellschaft in Basel* 69: 1–58.

Hürzeler, J. 1960. "Signification de l'Oréopithèque dans la phylogénie humaine." *Triangle* 4: 164–174.

Hürzeler, J. 1978. "Les racines paléontologiques de l'humanité." In *Les origines humaines et les époques de l'intelligence*, 5–12. Paris: Masson.

Huxley, T. H. 1863. *Evidence as to Man's Place in Nature*. London: William & Norgate.

Inés Baffi, E. and M. F. Torres. 1997. "Ameghino, Florentino (1854–1911)." In *History of Physical Anthropology: An Encyclopedia*, Vol. 1, edited by F. Spencer, 54–56. New York: Garland.

Johanson, D. C. 1989. "The Current Status of *Australopithecus*." In *Hominidae*, edited by G. Giacobini, 77–96. Milan: Jaca Book.

Johanson, D. C. 1996. "Face-to-Face With Lucy's Family." *National Geographic* 189(3): 96–117.

Johanson, D. and T. White. 1979. "A Systematic Assessment of Early African Hominids." *Science* 203: 321–330.

Keith, A. 1911. *Ancient Types of Man*. London: Harper.

Keith, A. 1915. *The Antiquity of Man*. London: William and Norgate.

Kimbel, W. H., T. D. White, and D. C. Johanson. 1984. "Cranial Morphology of *Australopithecus afarensis*: A Comparative Study Based on a Composite Reconstruction of the Adult Skull." *American Journal of Physical Anthropology* 64: 337–388.

Klaastch, H. 1902. "Entstehung und Entwickelung des Menschengeschlechtes." In *Weltall und Menschheit*, Vol. 2, edited by H. Kraemer, 1–338. Berlin: Bong.

Klaastch, H. 1923. *The Evolution and Progress of Mankind*. London: T. Fisher Unwin.

Koenigswald, G. H. R. von. 1962. *The Evolution of Man*. Ann Arbor: University of Michigan Press.

Kuhn, T. S. 1970. *The Structure of Scientific Revolutions*. 2nd ed. Chicago: University of Chicago Press.

Kurtén, B. 1972. *Not From the Apes*. New York: Pantheon Books.

Kurz, G. E. 1924. "Das Gehirn des Gelben und die mehrstämmige Abkunft der Menschenarten." *Anatomischer Anzeiger* 58: 107–117.

Langaney, A., N. M. van Blyenburgh, and R. Nadot. 1990. "L'histoire génétique des mille derniers siècles et ses mécanismes: une revue." *Bulletins et Mémoires de la Société d'Anthropologie de Paris*, new series, 2: 43–56.

Larsen, C. P. 2000. "A View on the Science: Physical Anthropology at the Millennium." *American Journal of Physical Anthropology*, 111: 1–4.

Leakey, L. S. B. 1934. *Adam's Ancestors: An Up-To-Date Outline of What is Known About the Origin of Man.* 2nd ed. London: Methuen.

Leakey, L. S. B. 1953. *Adam's Ancestors.* 4th ed. London: Methuen.

Leakey, L. S. B. 1959. "A New Fossil Skull From Olduvai." *Nature* 184: 491–493.

Leakey, L. S. B. 1960. "The Origin of the Genus *Homo*." In *Evolution After Darwin*, Vol. 2, edited by S. Tax, 17–32. Chicago: University of Chicago Press.

Leakey, L. S. B. 1961a. "Africa's Contribution to the Evolution of Man." *South African Archaeological Bulletin* 16: 3–7.

Leakey, L. S. B. 1961b. "A New Lower Pliocene Fossil Primate from Kenya." *Annals and Magazine of Natural History* 4: 689–696.

Leakey, L. S. B. 1966. "*Homo habilis, Homo erectus* and the Australopithecines." *Nature* 209: 1279–1281.

Leakey, L. S. B. 1967. "An Early Miocene Member of the Hominidae." *Nature* 213: 155–163.

Leakey, L. S. B. 1968. "Bone Smashing by Late Miocene Hominidae." *Nature* 218: 528–530.

Leakey, L. S. B. 1970. "The Relationship of African Apes, Man, and Old World Monkeys." *Proceedings of the National Academy of Sciences* 67: 746–748.

Leakey, L. S. B. 1971. "*Homo Sapiens* in the Middle Pleistocene and the Evidence of *Homo Sapiens*' Evolution." In *The Origin of Homo sapiens*, edited by F. Bordes, 25–29. Paris: Unesco.

Leakey, L. S. B., P. V. Tobias, and J. R. Napier. 1964. "A New Species of the Genus *Homo* from Olduvai Gorge." *Nature* 20: 7–9.

Leakey, M. G. and A. Walker. 1997. "Early Hominid Fossils from Africa." *Scientific American* 276(6): 74–79.

Leakey, M. G. and A. Walker. 2003. "Early Hominid Fossils from Africa." *Scientific American*, special edition, 13(2): 14–19.

Leakey, M. G., F. Spoor, F. H. Brown, P. N. Gathogo, C. Kiari, L. N. Leakey, and I. McDougall. 2001. "New Hominin Genus From Eastern Africa Shows Diverse Middle Pliocene Lineages." *Nature* 410: 433–440.

Leakey, R. E. F. 1976. "Hominids in Africa." *American Scientists* 64: 174–179.

Leakey, R. E. F. 1981. *The Making of Mankind.* New York: E.P. Dutton.

Leakey, R. E. F. 1989). "Recent Fossil Finds from East Africa." In *Human Origins*, edited by J. R. Durant, 53–62. Oxford: Clarendon Press.

Leakey, R. E. F. and R. Lewin. 1992. *Origins Reconsidered.* New York: Doubleday.

Le Gros Clark, W. E. 1934. *Early Forerunners of Man.* Baltimore: William Wood.

Le Gros Clark, W. E. 1955. *The Fossil Evidence for Human Evolution.* Chicago: University of Chicago Press.

Lewin, R. 1987. *Bones of Contention: Controversies in the Search for Human Origins.* New York: Simon and Schuster.

Lydekker, R. 1879. "Further Notices of Siwalik Mammalia." *Records of the Geological Survey of India* 11: 33–57.

Melchers, F. 1910. "Der Ursprung der Menschenrassen." *Der Zeitgeist* (Beiblatt zum *Berliner Tageblatt*) 25, 20 Juni: front page.

Mahoudeau, P.-G. 1912. "Le Pithécanthrope de Java." *Revue Anthropologique* 22: 453–472.

Miller, D. A. 1977. "Evolution of Primate Chromosomes." *Science* 198: 1116–1124.

Mivart, St. George. 1873. *Man and Apes*. London: Robert Hardwicke.

Morell, V. 1995. *Ancestral Passions: The Leakey Family and the Quest for Humankind's Beginnings*. New York: Simon and Schuster.

Oakley, K. P. 1964. *Frameworks for Dating Fossil Man*. Chicago: Aldine.

Osborn, H. F. 1915. *Men of the Old Stone Age*. New York: Charles Scribner's Sons.

Osman Hill, W. C. 1954. *Man's Ancestry*. London: W. Heinemann.

Owen, R. 1857. "On the Characters, Principles of Division, and Primate Groups of the Class Mammalia." *Journal of the Proceedings of the Linnean Society of London* (Zoology) 2: 1–37.

Owen, R. 1861. "The Gorilla and the Negro." *The Athenaeum*. 395–396, 467.

Owen R. 1863. "On the Aye-aye." *Transactions of the Zoological Society of London* 5: 33–101.

Patterson, B. 1954. "The Geologic History of Non-Hominid Primates in the Old World." *Human Biology* 26: 191–209.

Pilgrim, G. E. 1915. "New Siwalik Primates and Their Bearing on the Question of the Evolution of Man and the Anthropoidea." *Records of the Geological Survey of India* 45: 1–74.

Pilbeam, D. R. 1985. "Patterns of Hominoid Evolution." In *Ancestors: The Hard Evidence*, edited by E. Delson, 51–59. New York: Alan R. Liss.

Piveteau, J. 1957. *Traité de Paléontologie: Primates et Paléontologie Humaine*, Vol. 7. Paris: Masson.

Piveteau, J. 1962. *L'Origine de l'Homme*. Paris: Hachette.

Piveteau, J. 1983. *Origine et destinée de l'homme*. 2nd ed. Paris: Masson.

Popper, K. 1963. *Conjectures and Refutations*. New York: Routledge & Kegan Paul.

Pycraft, W. P. 1925. "On the Calvaria Found at Boskop, Transvaal, in 1913, and Its Relationship to Cromagnard and Negroid Skulls." *Journal of the Royal Anthropological Institute* 55: 179–198.

Rak, Y. 1985. "Australopithecine Taxonomy and Phylogeny in Light of Facial Morphology." *American Journal of Physical Anthropology* 66: 281–287.

Reader, J. 2011. *Missing Links: In Search of Human Origins*. Oxford: Oxford University Press.

Relethford, J. H. 1995. "Genetics and Modern Human Origins." *Evolutionary Anthropology* 4: 53–63.

Relethford, J. H. 1999. "Models, Predictions, and the Fossil Record." *Evolutionary Anthropology* 8: 7–10.

Robinson, J. T. 1954. "The Genera and Species of the Australopithecinae." *American Journal of Physical Anthropology* 12: 181–200.

Robinson, J. T. 1956. *The Dentition of the Australopithecinae*. Pretoria: Transvaal Museum Memoir, No. 9.

Robinson, J. T. 1962. "The Origin and Adaptive Radiation of the Australopithecines." In *Evolution and Hominisation*, edited by G. Kurth, 120–140. Stuttgart: Gustav Fisher Verlag.

Ruvolo, M. 1997. "Molecular Evolutionary Processes and Conflicting Gene Trees: The Hominoid Case." *American Journal of Physical Anthropology* 94: 89–113.

Sarich, V. M. 1970. "Primate Systematics with Special Reference to Old World Monkeys." In *Old World Monkeys: Evolution, Systematics, and Behavior*, 175–226, edited by J. R. Napier and P. H. Napier. New York: Academic Press.

Sarich, V. M. and A. C. Wilson. 1966. "Quantitative Immunochemistry and the Evolution of Primate Albumins: Micro-Complement Fixation." *Science* 154: 1563–1566.

Sarich, V. M. and A. C. Wilson. 1967a. "Immunological Time Scale for Hominid Evolution." *Science* 158: 1200–1203.

Sarich, V. M. and A. C. Wilson. 1967b. "Rates of Albumin Evolution in Primates," *Proceedings of the National Academy of Sciences of the USA* 58: 142–148.

Schaaffhausen, H. 1868. "On the Primitive Form of the Human Skull." *Anthropological Review* 6: 412–431.

Schepers, G. W. H. 1946. "The South African Fossil Ape-Men: The Australopithecinae, Part II." *Transvaal Memoir* 2: 165–275.

Schlosser, M. 1888. "Die Affen, Lemuren, Chiropteren, Insectivoren, Marsupialier, Creodonten und Carnivoren des Europäischen Tertiärs." *Beiträge zur Paläontologie Österreich-Ungarns und des Orients* 6: 1–224.

Schmidt, O. 1884. *Die Säugethiere in ihrem Verhältnis zur Vorwelt.* Leipzig: Brockhaus.

Schmidt, O. 1887. *The Doctrine of Descent and Darwinism.* 7th ed. London: Kegan Paul, Trench.

Schultz, A. H. 1930. "The Skeleton of the Trunk and Limbs of the Higher Primates." *Human Biology* 2: 303–438.

Schultz, A. H. 1950. "The Specializations of Man and His Place Among the Catarrhine Primates." *Cold Spring Harbor Symposia on Quantitative Biology* 15: 37–53.

Schwalbe, G. 1906. *Studien zur Vorgeschichte des Menschen.* Stuttgart: E. Schweizerbartsche.

Schwartz, J. H. 1984. "The Evolutionary Relationships of Man and Orang-Utans." *Nature* 308: 501–505.

Senut, B. 1996. "Pliocene Hominid Systematics and Phylogeny." *South African Journal of Science* 92: 165–166.

Senut, B. 2001. "L'Émergence de la famille de l'homme." In *Aux origines de l'humanité*, Vol. 1, edited by Y. Coppens and P. Picq, 166–199. Paris: Fayard.

Senut, B., M. Pickford, D. Gommery, P. Mein, K. Cheboi, and Y. Coppens. 2001. "First Hominid From the Miocene (Lukeino Formation, Kenya)." *Comptes Rendus de l'Académie des Sciences de Paris* IIa, no. 332: 137–144.

Sera, G. L. 1918. "I caratteri della faccia e il polifiletismo dei primati." *Giornale per la Morfologia dell'Uomo e dei Primati* 2: 1–296.

Sergi, G. 1908. *Europa.* Milano: Fratelli Bocca.

Sergi, G. 1909–1910. "L'Apologia del mio poligenismo." *Atti della Società Romana di Antropologia* 15: 187–195.

Sergi, G. 1914. *L'Evoluzione Organica e le Origini Umane.* Torino: Fratelli Bocca.

Shapin, S. 1995. "Here and Everywhere: Sociology of Scientific Knowledge." *Annual Review of Sociology* 21: 289–321.

Sibley, C. G. and J. E. Ahlquist. 1987. "DNA Hybridization Evidence of Hominoid Phyloneny: Results From an Expanded Data Set." *Journal of Molecular Evolution* 26: 99–121.

Simons, E. L. 1961. "The Phyletic Position of *Ramapithecus*." *Postilla* 57: 1–9.

Simons, E. L. 1977. *"Ramapithecus."* *Scientific American* 236(5): 28–35.

Simons, E. L. and D. R. Pilbeam. 1965. "Preliminary Revision of the Dryopithecinae (Pongidae, Anthropoidea)." *Folia Primatologia* 3: 81–152.

Smith, F. H. 1992. "The Role of Continuity in Modern Human Origins." In *Continuity or Replacement: Controversies in* Homo Sapiens *Evolution*, edited by G. Bräuer and F. H. Smith, 145–156. Rotterdam: A. A. Balkema.

Smith, S. L. and F. B. Harrold. 1997. "A Paradigm's Worth of Difference? Understanding the Impasse Over Modern Human Origins." *Yearbook of Physical Anthropology* 40: 113–138.

Smith Woodward, A. 1913. "Note on the Piltdown Man (*Eoanthropus Dawsoni*)." *Geological Magazine* 10: 433–434.

Spencer, F. 1990. *Piltdown: A Scientific Forgery*. Oxford: Oxford University Press.

Straus, W. L. 1949. "The Riddle of Man's Ancestry." *Quarterly Review of Biology* 24: 200–223.

Stringer, C. B. 1990. "The Emergence of Modern Humans." *Scientific American* 263: 98–104.

Stringer, C. B. and P. Andrews. 1988. "Genetic and Fossil Evidence for the Origin of Modern Humans." *Science* 239: 1263–1268.

Stoneking, M. 1993. "DNA and Recent Human Evolution." *Evolutionary Anthropology* 2: 60–73.

Takahata, N. and Y. Satta. 1997. "Evolution of the Primate Lineage Leading to Modern Humans: Phylogenetic and Demographic Inferences From DNA Sequences." *Proceedings of the National Academy of Sciences* (USA) 94: 4811–4815.

Taylor, G. 1927. *Environment and Race: A Study of the Evolution, Migration, Settlement, and Status of the Races of Man*. Oxford: Oxford University Press.

Theunissen, B. 1989. *Eugène Dubois and the Ape-Man from Java: The History of the First "Missing Link" and its Discoverer*. Dordrecht: Kluwer.

Thorne, A. G. and M. H. Wolpoff. 1981. "Regional Continuity in Australasian Hominid Evolution." *American Journal of Physical Anthropology* 55: 337–349.

Tobias, P. V. 1966. "The Distinctiveness of *Homo Habilis*." *Nature* 209: 953–957.

Tobias, P. V. 1968. "The Taxonomy and Phylogeny of the Australopithecines." In *Taxonomy and Phylogeny of Old World Primates With References to the Origin of Man*, edited by B. Chiarelli, 277–315. Torino: Rosenberg & Sellier.

Tobias, P. V. 1980. "'*Australopithecus afarensis*' and *A. africanus*: Critique and an Alternative Hypothesis." *Palaeontologia Africana* 23: 1–17.

Tomczyk, J. 2004. "Hominids' Taxonomy: Three Levels of Discussion." *Human Evolution* 19: 227–238.

Topinard, P. 1888. "Les dernières étapes de la généalogie de l'homme." *Revue d'Anthropologie* 3: 298–332.

Topinard, P. 1891. *L'Homme dans la nature*. Paris: Félix Alcan.

Topinard, P. 1892. "De l'évolution des molaires et prémolaires chez les primates et en particulier chez l'homme." *L'Anthropologie* 3: 641–710.

Vallois, H. V. 1955. "L'ordre des primates." In *Traité de zoologie*, Tome XVII, 2nd part, edited by P-P. Grassé, 1854–2206. Paris: Masson.

Verneau, R. 1906. "La race de Spy ou de Néanderthal." *Revue de l'École d'Anthropologie de Paris* 16: 388–400.

Vogt, C. 1864. *Lectures on Man*. London: Longman, Green, Longman, and Roberts.

Wallace, A. 1889. *Darwinism*. London: Macmillan.

Washburn, S. L. 1950. "The Analysis of Primate Evolution with Particular Reference to the Origin of Man." *Cold Spring Harbor Symposia on Quantitative Biology* 15: 67–78.

Weidenreich, F. 1946. *Apes, Giants and Man*. Chicago: University of Chicago Press.

Weinert, H. 1932. *Ursprung der Menschheit*. Stuttgart: F. Enke.

White, T. D. 2000. "A View on the Science: Physical Anthropology at the Millennium." *American Journal of Physical Anthropology* 113: 287–292.

White, T. D. 2009. "Ladders, Bushes, Punctuations, and Clades: Hominid Paleobiology in the Late Twentieth Century." In *The Paleobiological Revolution*, edited by D. Sepkoski and M. Ruse, 122–148. Chicago: University of Chicago Press.

White, T. D., D. C. Johanson, and W. H. Kimbel. 1981. "*Australopithecus africanus*: Its Phyletic Position Reconsidered." *South African Journal of Science* 77: 445–470.

White, T. D., G. Suwa, and B. Asfaw. 1994–1995. "*Australopithecus [Ardipithecus] ramidus*, a New Species of Early Hominid From Aramis, Ethiopia." *Nature* 371: 306–312; *Nature* 375: 88.

Willermet, C. M. and G. A. Clark. 1995. "Paradigm Crisis in Modern Human Origins Research." *Journal of Human Evolution* 29: 487–490.

Wood Jones, F. 1929. *Man's Place among the Mammals*. London: Edward Arnold.

Wood Jones, F. 1948. *Hallmarks of Mankind*. London: Baillière, Tindall and Cox.

Wolpoff, M. H. 1968. "*Telanthropus* and the Single Species Hypothesis." *American Anthropologist* 70: 477–493.

Wolpoff, M. H. 1971. "Competitive Exclusion among Lower Pleistocene Hominids." *Man* 6: 601–617.

Wolpoff, M. H. 1973. "The Evidence for Two Australopithecine Lineages in South Africa." *Yearbook of Physical Anthropology* 17: 113–139.

Wolpoff, M. H., X. Wu, and A. G. Thorne. 1984. "Modern *Homo sapiens* Origins: A General Theory of Hominid Evolution Involving the Fossil Evidence From East Asia." In *The Origin of Modern Humans*, edited by F. H. Smith and F. Spencer, 411–483. New York: Alan R. Liss.

Wu, Rukang. 1984. "The Crania of *Ramapithecus* and *Sivapithecus* From Lufeng, China." In *The Early Evolution of Man With Special Emphasis on Southeast Asia and Africa*, edited by P. Andrews and J.L. Franzen, *Courier Forschungsinstitut Senckenberg* 69: 41–48.

Wu, Rukang. 1987. "A Revision of the Classification of the Lufeng Great Apes." *Acta Anthropologica Sinica* 6: 265–271.

Wu, Rukang, Q. Xu, and Q. Lu. 1986. "Relationship Between Lufeng *Sivapithecus* and *Ramapithecus* and Their Phylogenetic Position." *Acta Anthropologica Sinica* 5: 1–30.

Xu, Q. and Q. Lu. 1979. "The Mandibles of *Ramapithecus* and *Sivapithecus* from Lufeng, Yunnan." *Vertebrata Palasiatica* 17: 1–13.

2 Biological Explanations and Their Limits: Paleoanthropology among the Sciences

Siobhan Mc Manus

Introduction

Narratives on human evolution are almost always controversial and contested. This is so because explaining how we, modern humans, became what we are—how we evolved into these ex-apes—confronts us with several fundamental but politically loaded questions. First, it seems to demand the selection or construction of an idea or image of who is this "we" to be explained. This is not a trivial task because we run the risk of conflating our culturally situated "we" with the entire species. For example, the "we" produced by Western science is most likely different than the "we" produced by those who belong to the First Nations of the Americas (e.g., Viveiros de Castro [2004] describes how the indigenous perspectivism in Latin America is different from any Western view on nature and humanity).

Second, societies and cultures are obviously not homogenous and coherent wholes. There is variation and difference at their core. Class, "race," gender, sexual orientation, nationality, ethnicity, religion, and political affiliation must certainly affect the kinds of images used to interpret who we are. Of course, there is no reason to believe that all of these images should be granted the same epistemic status, but this fact does not solve the problem of how situated are the accounts of ourselves that we take to be universal.

Third, any description of this "we" is going to be informed by the specific language and theoretical resources of concrete fields and disciplines and, thus, is going to be prone to another set of biases, this time in terms of disciplinary cultures. Again, selecting among the myriad of options is not trivial because, by constructing this "we" as an *explanandum* in light of specific fields and disciplines, we implicitly privilege some accounts over others as more suitable for the task of explaining us, as more suitable for providing a successful *explanans*. Historically, this has led to many unpleasant chapters in the so-called Science Wars, in which the social sciences and humanities have engaged in deep criticism of the natural sciences for de-historicizing and naturalizing some aspects of specific groups that are taken to be a consequence of human nature, or even human nature itself (see, for example, Mc Manus 2013).

Fourth and related, even if we embrace a radical interdisciplinary approach, we face the challenge of articulating an architectonics of how different disciplines and fields, and the explanations rooted in them, should be connected. Here, we should bear in mind that the construction of an ontologically enriched *theory* is not the same as constructing an inter-disciplinary *dialogue* among fields in which no theory claims to be all encompassing, and no discipline pretends to be the architect of knowledge.

Indeed, it is this architectonic problem that usually leads to reductive or eliminativistic perspectives of how we should connect the social and the natural sciences. Again, this is not a minor inconvenience for scientists because, to offer a relevant example, even if we embrace an enriched ontology produced by an expanded evolutionary synthesis in which models of cultural evolution incorporate many important insights from social sciences, we still run the risk of assaulting the autonomy of social sciences and humanities or their accounts on the phenomena to be explained. This would, indeed, be counterproductive, because a dialogue might be more fecund than the view from a single, although complex, theory. Therefore, paying attention to the architectonics of knowledge is not a menial task because it might be the only way to avoid falling prey to what I call the *Epistemologies of Replacement*.

Fifth and finally, any choice we take is going to have effects beyond science. Whether we like it or not, sciences produce cosmologies that have become hegemonic in most parts of the world. To examine the usefulness and validity—as well as the limits and shortcomings—of biological explanations of human evolution demands recognizing both the situatedness of the images thus produced and their capacity to circulate beyond their context of origin and beyond their intended scope.

Accordingly, examining a particular discipline, such as paleoanthropology, demands paying attention to the five points mentioned above. More specifically, to inquire whether paleoanthropology is an evolutionary science and/or if analyses of human evolution are biological,[1] necessarily invokes, on one hand, addressing the architectonics of knowledge previously mentioned and, on the other, how we understand ourselves and how this affects our understanding of the role of paleoanthropology within this architecture. Furthermore, it requires attending to the specifics of the image of the human that may lie at the core of paleoanthropology and how this image can be complemented by other disciplines. We may find ourselves surprised if we allow ourselves to engage in such a project; we may find paleoanthropology's aims radically expanded and transformed.

This paper serves as an invitation to these reflections. It is divided into the following sections. First, a discussion rooted in classical topics from philosophy of science, that is, unity of science vs. pluralism. My aim in this section is to engage in a reflection regarding the place of paleoanthropology among the sciences in light of the debates between the unity of science and pluralism. In section two, I turn to philosophy of biology in order to discuss whether humans are a special kind of *explanandum* that necessitates particular explanations informed by our own particular biology, which, for some, has given place to a new ontology. Third, I elaborate on some of the bio-political consequences of the previous sections and critically examine how they affect paleoanthropology as a discipline. Finally, I conclude with general remarks on how to answer the questions at hand: Is paleoanthropology an evolutionary science and/or are analyses of human evolution biological?

Paleoanthropology and Philosophy of Science: Unity or Pluralism?

The two questions raised in the introduction—whether paleoanthropology is an evolutionary science and/or if analyses of human evolution are biological—are deeply connected with a more general question regarding the place that paleoanthropology should occupy among the sciences. Indeed, it is clear that this more general concern has much to do with an old, but still lively, topic within philosophy of science: Do important differences exist between social and natural sciences? Further, if these differences exist, are they purely methodological, or do they rest in deeper ontological differences regarding the types of *explananda* found in these disciplines? That is, if such a divide does exist, where should we locate paleoanthropology? Is it a natural science or does it lie somewhere in the middle?

For example, for philosophers such as Dilthey (1989) or von Wright (2004), natural sciences deal with causal phenomena, and thus offer *causal explanations* in terms of *efficient causes*. In opposition, social sciences deal with meaningful phenomena, and thus seek to *understand* them in terms of goals, intentions, norms, or values held by intentional agents. Hence, the distinction is not only methodological (causal explanation vs. hermeneutical interpretation), but it is also ontological (causally structured phenomena vs. actions or meanings held by intentional agents). With specific regard to paleoanthropology, it might seem at first glance that it fits squarely into the realm of natural sciences. But, then, what do we do with cultural evolution, tool fabrication and use, funerary rituals, and so on? Those phenomena seem more similar to those studied by social sciences and traditionally accounted for in terms of hermeneutical interpretation.

On the other hand, it is far from clear that paleoanthropology is internally homogeneous,[2] or that the types of explanations it offers are all of the same kind. Delisle (2012), for example, claims that, if we come to embrace paleoanthropology as the science of human evolution, we should distinguish between explanations centered on the *pattern* of human evolution and those centered on the evolutionary *process of anthropogenesis*.[3] The former type would then be clearly biological—and, hence, natural in the sense just mentioned above—and very much in line with phylogenetics as practiced in different domains of biology, such as zoology, botany, or microbiology.

Nevertheless, it would be more difficult to judge the degree of similarity or dissimilarity in the latter case because those explanations in which humans—or, more precisely, the processes of anthropogenesis—figure as the *explanandum* are going to be more deeply affected by the five points raised in the introduction. In other words, if humans are the *explanandum*, and we are interested in offering an explanation on how the process of anthropogenesis occurred, we then need to offer an image of what humans are in biological terms. This, as one might imagine, is anything but trivial because we have a number of traits (e.g., language, self-awareness, collective intentionality, cumulative cultural evolution, the use and fabrication of tools, complex social structures, and so on) that historically have been taken as being exclusively human attributes that set the boundaries between *Nature and Culture*. These attributes bring us closer to the phenomena studied by social sciences. Although recent studies from evolutionary biology and philosophy of biology dispute this claim, the controversy is anything but over.

Indeed, since the nineteenth century, philosophers such as Frederich Engels (1934 [1876]) have argued that labor, for example, sets humans apart from other animals because the former, and not the latter, are quite literally—and materially—World-constructers and Self-constructers. For Engels, labor was part and parcel of anthropogenesis. These ideas are similar to those espoused by philosophers such as Dilthey or Heidegger, who, in giving priority to language—instead of labor, as Engels did—considered that humans differ significantly from other animals.

Be this as it may, as Delisle (2012) points out, nineteenth- and twentieth-century taxonomists were fond of highlighting *human exceptionalism*. In fact, taxonomy as practiced by paleontologists such as G. G. Simpson classified humans apart from the apes precisely because of the presumed deep differences between them.

The point, then, is: Granting Delisle's distinction between pattern explanations and process explanations, it is easier to accept that pattern explanations of human evolution are continuous, or are of the same type as those found in general phylogenetics, than to accept that explanations targeting the process of anthropogenesis are basically of the same type as those found in other species or taxa.[4] This is so because (process) explanations of human evolution must account for traits such as language, collective intentionality, cumulative cultural evolution, material cultures (tools, clothes, installations, and so on), complex social structure, and so forth. However, these explanations also need to incorporate these very same traits, not only as a part of the *explanandum*, but also as relevant elements of the *explanans*. The latter is true because, once these traits appeared, their causal (and symbolic?) effects on human evolution gave rise to a unique process.

Specifically, pattern explanations offer explanations of the same type as those found in other areas of biology because the *explanans* is fundamentally a phylogenetic tree, while the *explanandum* constitutes a set of traits within a matrix of certain taxa. Consequently, not only the *explanandum* and *explanans*, but also the algorithms involved, are essentially of the same kind as those in other areas of biology. This having been said, though, the question is: Do we find a similar situation when we analyze process explanations? Are the *explananda* and *explanantia* specific to this case of the same type as those found in other taxa? The answer depends on how we characterize humans, and this, in turn, depends on our commitment (or lack thereof) to human exceptionalism, namely, both the distinction between explanation and interpretation and the boundary between Nature and Culture.

These points have two major consequences. One is related to the unity vs. pluralism debate that was so central to twentieth-century philosophy of science, while the other is more connected with philosophy of biology and topics such as homology, evolutionary novelty, and evolutionary transitions. The first point will be addressed in the rest of this section, and the second in the following section.

Regarding the first consequence, twentieth-century philosophy of science oscillated between defenders of the idea that science should be unified and those who resisted this claim (Godfrey-Smith 2009). Logical positivists and Popperians belonged to the first group and believed that science could be unified on the basis of the logical structure of explanations organized around lawlike sentences. These sentences would employ the language of physics and, thus, would require reducing any other theory/explanation from other sciences into this language. In opposition, historically minded philosophers such as Kuhn (2012) and Feyerabend (1993) argued that it was both impractical and technically challenging to carry on this reduction.

Although unificationist projects have lost their appeal, their general insights are still valuable. For example, the two questions raised at the beginning of this section can be analyzed in light of the unity of science literature. We may argue that, given the fact that humans are just one among many taxa, the explanations regarding their/our evolution should be epistemologically of the same type that those offered for other groups. That is, if we foreground those ontological aspects that connect us with the rest of the tree of life, it seems natural to offer explanations that are continuous with those coming from other parts of biology. Therefore, under an analysis like this, paleoanthropology would be continuous with the entire field of biology and, thus, part of a unified whole.

Nevertheless, it is anything but trivial to justify a choice like that. Why should we set aside those ontological aspects that seem to be characteristic of humans and demand explanations of another type? Indeed, if we follow the previous discussion based on the pattern vs. process distinction, we might then find that paleoanthropology itself is not homogeneous, and, thus, cannot, as a whole, be subsumed within the domain of evolutionary biology. Pattern explanations can likely be so unified, but maybe it is not desirable to do so with process explanations of anthropogenesis (see previous discussion).

If we embrace this conclusion, the answers to the questions of whether paleoanthropology is an evolutionary science and/or if analyses of human evolution are biological are complex. Paleoanthropology is obviously an evolutionary science, and pattern explanations are clearly of the same type than those of phylogenetics in general. However, it is far from clear that this means, or implies, that paleoanthropology is *merely* a part of a unified whole called biology. Rather, it seems more credible to assert that paleoanthropology embraces a plurality of questions, methods and explanations that connect it with biology and also with social and human sciences.

Perhaps the questions reflect an historical trend within biology and, more recently, philosophy of biology that can be traced to Darwin. Here, it might be noted that Darwin intentionally attempted to forge his explanation of evolution by means of natural selection by following the value of *consilience*, a term coined by William Whewell (Hodge 2000). For Whewell, consilience meant the virtue of conjoining explanations from different disciplines and subdisciplines into a single and unified framework. According to him, we could argue in favor of the truthfulness of a theory when independent and different inductions of multiple series of facts coincide as predicted by the theory.

This value might explain why Darwin insisted that descent with modification and adaptation by natural selection could account for such a variety of disparate phenomena. This reflects a truthfulness of his theory as well as the existence of an underlying common cause: a *vera causa*. In other words, if the previous narrative evokes some sort of epistemic virtue associated with finding common patterns of explanations,[5] this is because consilience would serve as an indicator of common, underlying processes.

Philosophers of science still find Darwin's legacy praiseworthy, as is reflected, for example, in his commitment to consilience. For example, for philosophers such as Philip Kitcher (1989), Darwin exemplifies why we value so much the exercise of finding connections and *unifying patterns of explanations*. Following Michael Friedman (1974), Kitcher believes that what characterizes the scientific enterprise is its capacity not only to provide objective explanations of the phenomena that populate the world, but also its capacity to provide us with an *understanding* of why some things occur or fail to occur.

For Kitcher, "understanding" is a technical term. It denotes the reduction in number of the assumptions that we need to take as given in order to cope with the world. More specifically, we gain understanding when we explain more phenomena by appealing to fewer assumptions. Science does this, according to him, when an explanatory pattern, defined in terms of a minimal theoretical vocabulary and a minimal logical structure, is shown to be applicable beyond its original domain. That is exactly what Darwin did when he expanded the notion of artificial selection beyond the human domain and showed, *mutatis mutandis*, that it was applicable to understanding the fossil record, developmental similarities between organisms, and biogeographical patterns of distribution.

Because the task of unifying leads to understanding by revealing the commonalities between different phenomena, it explains them objectively because these explanatory patterns are not idiosyncratic. They are located in what Kitcher calls a "Store of Explanations" in which they can be accessed by anyone within a scientific domain. Obviously, they are instantiated not only in universities, textbooks, and research facilities, but, more importantly, also in the practice of the scientists who continuously mobilize them. Additionally, the unification of explanations is not tantamount to reductionism because what we unify are structural patterns that share a minimal vocabulary. For example, we do not need to reduce phenotypic evolution into genetic evolution when we claim that they exhibit a similar process and can be accounted for in similar ways; we only need to show how they share these elements.

As a modern descendent of consilience, Kitcher's analysis seems relevant to our discussion. When we ask ourselves if paleoanthropology is an evolutionary science, and if analyses of human evolution are biological, do we have in mind something similar to consilience, something similar to unification? In other words, do we seek to defend the idea that paleoanthropology's patterns of explanation are in the same store as those from the rest of the evolutionary sciences? Do they, in general, share the vocabulary and logical structure of other patterns within evolution?

Further, while paleoanthropological pattern explanations may be part of the same store of explanations as those of general phylogenetics, perhaps paleoanthropological process explanations of anthropogenesis are more *sui generis*. If so, the key lies in our very special ontology, and the way in which it demands different explanations that cannot, and should not, be unified. That is, instead of bringing understanding, they might obscure some of our specificities. Indeed, if we attempt to explain away these specificities, we might very well fall prey to reductionism or to a similar epistemic project in which we replace some types of explanations with others, in the course of which we disregard why, in the first place, we needed different types of explanations.

Perhaps, then, the lesson may lie in the fact that paleoanthropology should not be *merely* an evolutionary science and that analyses of human evolution should not be *merely* biological: Humans are simply not just another organism. Acknowledging this is neither sufficient to avoid falling prey to the risk of enacting a replacing project, nor is it tantamount to embracing a nature vs. nurture dichotomy in which disciplines like paleoanthropology have no place. Clearly, since paleoanthropology deals with cultured natures or nature-cultures, this requires more complex patterns of explanations.

To recognize this should suggest a rejection of any monistic project in which humanity is thought of as just another "beast" within the garden of Nature. This should also suggest the rejection of any project that disregards, eliminates, or reduces the social into the natural. But these *epistemologies of replacement*—to coin a catchy phrase—are not exemplified solely by classical reductionist projects such as the gene-centered view of the Selfish gene, or the equally gene-centered view of sociobiology and evolutionary psychology. There is a very real risk of confronting a new type of epistemology of replacement in which biology's best insights give rise to a theory with a complex and rich ontology that replaces by engulfing and incorporating other disciplines into one single body of knowledge with an architectonics oriented primarily by biologists' concerns.

The risk of these epistemologies of replacement lies in their capacity to combine an attempt to construct a complex ontology with an architectonic project that privileges biology as the organizing science. Here the complex ontology comprises multiple semi-autonomous levels of organization that distance themselves from classical ontological reductionism, and that are explained and understood by different mechanisms that, in turn, lead to the rejection of any epistemological monism. Nevertheless, in these epistemologies of replacement, we find an architectonic project that privileges biology as the organizing science, one in which biology engulfs and consumes the explanations and theories from social sciences, and in which it chooses specific concepts in order to strengthen itself as an integrative science. In doing so, however and paradoxically, this process replaces or even eliminates the very theories, methods, and other voices it has devoured. Paleoanthropology runs the risk of becoming such a monster!

The consequences of this (see part four) could be hazardous. Perhaps these questions may be better managed by a multiplicity of voices and a multitude of positions that permanently question who the "we" being addressed and explained is, and that continually remind us that human nature is legion because we are many. There is, after all, a sociopolitical and epistemic side of this project that is best served by the multitude of this "we," rather than by the single voice of a humanistic and unificationist project. Again, the answers are no and no.

Paleoanthropology and Philosophy of Biology: Are Humans Unique?

So far I have claimed that, ontologically speaking, humans posses a number of traits that set them/us apart from other organisms, thus leading to the necessity of offering a kind of (process) explanation that is different from those applied to other taxa, that is, these traits are not *fully* accountable in light of evolutionary theory. But why should anyone accept this diagnosis and take it as a valid starting point? In this section I defend this diagnosis by recruiting some concepts from theoretical biology and philosophy. My general argument consists of two parts, both of which are centered on the unique character of human traits. In order to explicate why anthropogenesis might indeed give rise to very unique traits, I will briefly revisit the debate on the subjects of evolutionary novelty and major evolutionary transitions.

I also want to deepen this analysis by highlighting some of the consequences of this very particular evolutionary history. This is fundamental because it is not sufficient to show that we have autapomorphic traits. For, after all, every species possesses some. What is required is to show how they produce novel interactions that demand new types of explanations that not only demarcate the boundaries of social sciences, but that are also relevant for paleoanthropologists interested in the study of anthropogenesis. To summarize, the focus of this section lies on: (1) specific human traits, including language, collective intentionality, and complex social structures, (2) the evolutionary processes that gave rise to them, and (3) the consequences of their appearance.

Major Transitions, Evolutionary Innovations and World Construction

It might seem paradoxical to claim that evolutionary theory cannot deal with one of the most important major transitions:[6] The rise of the symbolic and the construction of *Worlds*. I say this sounds paradoxical because evolutionary theory exhibits one particular and very interesting trait: Its scope covers multiple levels of organization, from the molecular to the ecological, and from the instantaneous and ephemeral to the *longue durée* of eons. Indeed, evolutionary theory might be considered extremely special because it remains explanatory, even though it deals with objects showing increasing degrees of complexity. Obviously, this explanatory power is the consequence of substantive revisions in the core of the theory itself. For example, biologists realized long ago that classical gradualist explanations—*à la* Mayr—were unable to account for the appearance of what came to be known as *key evolutionary innovations*.

These were new (phenotypic) traits that led to adaptive breakthroughs usually associated with adaptive radiations and the invasion of multiple new niches. Part of the problem with the explanation of these novelties has to do with the Mayrian dictum that nothing evolves *ex nihilo*. The problem, of course, lies in how to account for events such as the origin of life, DNA, the eukaryotic cell, multicellularity, cooperative behavior, or cognition. These innovations appear to lack concrete and preexistent *homologues*. Indeed, they seem to open new levels of complexity that constitute major transitions.

Human cognition and the rise of the symbolic, for example, provide a very interesting case. We know that the human brain seems to have experienced a major transformation during evolution. But how do we account for this? Did new brain areas just come into being, or are we facing the gradual transformation of a more basic brain morphology? Modern Evo-Devo theory appears to provide an answer. In the words of neurologist Georg Striedter (2005), and *contra* Mayr, it is the development of novelties *ex nihilo* that evolutionary theory prohibits. Witness the fact that developmental primordia can be recruited and transformed into new traits that, even though homologous to previous primordia in terms of the concept of *field homologues*, are not homologous in terms of the more classical *taxic homology* concept. Indeed, radical, evolutionarily significant innovations can emerge through differential co-opting of developmental commonalities.

This demonstrates three important things. First, major transitions can, and have, occurred via key innovations. Second, these innovations can generate new levels of organization, new functions, and new types of relationships by adding, transforming, or refunctionalizing parts or the arrangement of parts: for example, mutations affecting *HOX* gene expression, or the multiplication in mammals on number of cell lineages via duplication of *HOX* gene complexes (e.g., Carroll 2005). Third, evolutionary theory remains explanatory after these transitions because it has enriched its store of explanations with important new patterns of explanations coming from disciplines such as Evo-Devo or Eco-Evo-Devo.

In contrast to other major transitions, however, the rise of the symbolic not only implied a new level of complexity, but also the beginning of a *new metaphysical realm*: The metaphysics of the social.[7] By this, I mean entities such as the State, laws of government, NGOs, and supranational organizations (e.g., the United Nations); in sum, what we call *institutions* (see also Schmitt 2003). Additionally, within this new metaphysical order, we find *socially created entities* and *properties* (e.g., clans, tribes, group and individual identities, nationalities, social roles, etc.; in the last case, roles are usually defined in terms of values that are themselves institutions), as well as ideas and values (e.g., justice, democracy, and objectivity[8]).

In this regard, paleoanthropology is not only interested in properties that we encounter in other mammals. It is interested in a new set of properties that are not defined in the same way, and that are not predicated on objects that possess them irrespective of their (*ex hypothesi* impossible) awareness of having them; it's precisely for this reason the process explanations of anthropogenesis are so unique! On the contrary, these properties, which Heidegger (1962) called *existentiells*,[9] only come into existence because a subject self-reflexively predicates them on him/herself. Modern identities are probably the best examples because, as identities, even if they are grounded in some physical trait, they emerge as identitarian, that is, they acquire epistemic salience for the subject only in light of the social meanings and practices in which the subject is embedded.

If I claim that paleoanthropology is also close to these other sciences, I do it because paleoanthropology aims not only to know the physical and biological environment in which individuals lived. It also aspires to know more than the physical and biological nature of hominins and hominids: that is, their social lives, and their symbolic and material cultures. In other words, paleoanthropology strives to understand what in hermeneutics—a branch of philosophy—is called *Worlds*. By "World" we should understand this interdiscursive and intermaterial plexus in which meanings and matter are intertwined in the form of crafts, artifacts, installations (such as common settlements or houses), clothes, and body modifications (such as tattoos or scarifications), as well as ecological and productive relations that include others (human and nonhuman) and might give rise to institutions, mythologies, or symbolisms. In these Worlds, nature and normativity—epistemic, ethical, political, and even economical—are co-producing each other while storing information and scaffolding our actions[10] as individuals and as collectivities. Indeed, it could be argued that what defines paleoanthropology as a bridge between paleontology and anthropology is the exploration of when and how the environment, understood as a collection of biotic and abiotic elements organized in time and space and interacting in terms of causation, became a World, fully enriched by meanings, intentions, beliefs, desires, norms, institutions, social roles, and more, that constitutes a space in which causation gave rise to actions, to purposeful behaviors motivated by norms, desires, or beliefs.

Obviously, I am neither arguing in favor of an anachronistic and postmodern Paleolithic, nor rejecting the thesis that some mammals, especially primates, might have the rudiments of Worlds, but, as Derrida (2008) reminds us, those animals are *poor-in-world* (*weltarme*, as Heidegger said), even if they are not, as rocks or plants, totally *worldless* (*weltloss*).[11]

I do defend two claims. First, that paleoanthropology's aims of understanding human evolution exceed the realm of properties and enter the realm of *existentiells*. And, second, while understanding how biological evolution gave place to the social is a constitutive goal of paleoanthropology, this demands paying attention to the ontology of the social, which cannot be thought of in terms of the available patterns within the "store of evolution" (*sensu* Kitcher [1989]). Consequently, if this is correct, we are facing a major transition produced by several evolutionary innovations (cognition, language, motor skills, social structures, and so on) that resists any attempt to explain it in purely biological terms.

In a sense, it could be argued that the complex ontology of "the human," both material and organic but also phenomenological and symbolic, serves as a reminder of our unique predicament as an object/subject of inquiry. The fact that "the human" is simultaneously within the realms of the natural sciences, on the one hand, and the social and human sciences, on the other, should be taken as a precautionary advice against any unificationist agenda that emphasizes unity and continuity at the expense of disunity and discontinuity. Paraphrasing Ian Tattersall (2012), there is a gulf between humanity and the rest of nature that has to do with the way in which we live as a fully symbolic organism. We not only exist, we exist for ourselves. We are self-aware and aware of our existence. This is reflected in the way in which our bodies also become the expression of an individual, of an embodied identity that actively transforms the body, shaping it to create and express a Self. This occurs in a social context, full of norms and meanings, full of material scaffoldings, and it represents a major transition in evolution, an evolutionary innovation that goes beyond consciousness because it represents the dawn of World construction and, thus, the birth of the metaphysics of the social. This new metaphysics order is open to myriad possible analyses. Here, I will follow John Searle's (1995) approach based on (human) intentionality and its capacity to articulate the social through the mutual attribution of intentional contents.[12] I have chosen Searle because I think he has shown the relevance of paying attention to intentionality as a necessary, but not sufficient, condition that would permit the rise of the social. Moreover, Searle's account has become deeply influential even within human evolutionary sciences thanks in part to Michael Tomasello's (2014) attempt to accommodate intentionality within an evolutionary narrative.

Parenthetically, it is important to be aware that, even though intentionality[13] seems immaterial, it actually impacts materiality in at least three ways. First, it obviously affects the bodies of intentional agents themselves—humans. Second, through the actions of agents, it affects the environment and shapes a material culture; however, because material cultures scaffold intentionality by guiding actions, storing information in the way of devices or by amplifying human capabilities, this process should be understood as co-productive and not unidirectional. Third, institutions themselves tend to have what Pierre Levy (1995) calls *binding objects*, such as a ball in football games, which help participants to focus their shared attentions and intentionalities in particular material objects. In turn, this focalizes the collective's intentionality and, thus, consolidates and binds the mutual commitment to joint goals and actions.

Be this as it may, intentionality, according to Tomasello, is already present in apes, although not in a propositional manner. Consequently, intentionality would appear to be plesiomorphic for the entire hominoid clade. Nonetheless, during human evolution, intentionality gave rise, first, to what Tomasello calls *joint intentionality* and, later, to *collective intentionality*. Joint intentionality implies the rise of the "we" mode in actions, that is, individuals are capable of having common goals and common strategies and, more importantly, are able to coordinate themselves to achieve these goals through the mutual attributions of desires and beliefs.

A "we" mode is made possible by our capacity not only to empathize with others, but to recognize their situatedness and, therefore, the existence of points of view different from ours. If, for example, someone has a belief about a particular object, we can attribute it to that person, even though we do not share this belief. We also possess the ability to display instrumental rationalities by way of which we come to know that some action is required to achieve some goal. As such, we can understand a situation that we are not directly experiencing and consequently recognize how someone else needs to engage in a particular action to achieve his/her goal, even if that goal is not ours. Finally, we can and usually do suggest to others how to achieve goals that we do not have. These capacities appear to be key evolutionary innovations and the first step to World construction. A World construction that would mature through the second step: collective intentionality. Here, cooperation does not occur between specific and concrete individuals, but between entire collectivities in which particular assumptions are taken for granted when interacting with any of its member. It is, in a way, the first real culture or World (cf. above discussion).

It should be noted that, as suggested by Tomasello and extensively documented by Frans de Waal (2009), apes only act in the "I" mode. Although they seem to exhibit empathy and a basic understanding of the opposition between Ego/Self and Alter/Other, and between the needs of Ego and the needs of Alter, they are not entirely able to comprehend how a specific situation might be experienced in different ways, and give rise to different points of view. More concretely, in competitive scenarios in which apes share a goal with other apes or with researchers, the former do understand that there are different points of view, which illustrates that they can comprehend in terms of instrumental reasons why the other is engaging in certain actions. In noncompetitive scenarios, however, apes fail to behave altruistically through engaging in actions that would lead to the achievement of a goal by other individuals.[14]

In sum, we might read Tomasello's suggestions as the elements of a process explanation of how we became not only a symbolic species, but also World constructers. From this perspective, it is important to acknowledge that Tomasello is providing an explanation that goes beyond the realm of evolutionary theory by incorporating elements drawn from philosophy of mind, philosophy of language, and political philosophy. After all, Searle's account is an actualization of classical social contract theories as developed by Hobbes or Rousseau. Hopefully, this demonstrates why this explanation, although naturalistic and scientific, does not entirely belong to the store of patterns of explanation from evolutionary biology. Finally, I suggest that this model has consequences for paleoanthropology. For example, although endorsing a different model, Kim Sterelny (2012) is also interested in how humans came to be as cooperative as they are. He shows that we cooperate in at least three axes: nutrition, defense, and communication/technique. For him, all three elements depend both on our particular cognitive capacities and on a material culture that leaves traces.[15]

Material culture is reflected, for example, in stone tools used in defense, hunting, gathering, processing of food, and transmitting information about predators, prey, or any other resource. Material culture can also be documented by analyzing the spatial arrangements of settlements, from which a spatial division of labor can be inferred and, consequently, a social structure in which instrumental rationalities were likely at play. Finally, bodies themselves would exhibit the effects of cooperation in terms of a delay in development that led to longer period of childhood and a concomitantly extended learning phase, as well as by evidence of improved mastery of the environment.

To conclude: If the case made in this section is viable, paleoanthropology faces a twofold challenge. It needs to explain in biological terms the emergence of hominins, and, with documentation, explicate the major transition that gave rise to World construction. While the first can be done entirely within the store of evolution, the second task requires conceptual tools coming from social sciences and humanities. In other words, even if we restrict ourselves to process explanations, it seems that anthropogenesis demands distinguishing between the explanations that account for these novel traits and other explanations in which these traits figure, as a part of the *explanandum* and also of the *explanans*.

Indeed, even within a narrow understanding of paleoanthropology as a discipline centered on classifying and explaining the evolution of hominins, both the task of doing systematics and of dealing with issues such as social structure, division of labor, production of tools, patterns of cooperation, cognitive capacities, and so on, would be unavoidable because these endeavors are paramount in reconstructing phylogenies and producing explanations of human evolution. It is, therefore, evident that paleoanthropology requires more than biology in order to operate. This is so because biology as such—including its current most popular theories, like niche construction and the extended synthesis of Eco-Evo-Devo—lacks the elements necessary to tackle this new metaphysics. In a nutshell, niche construction is not yet World construction.

Social Studies of Science: Toward an Ontopolitics of Bodies

So far I have argued in favor of two points. I have strongly advocated against any unificationist project in which paleoanthropology is presented as entirely continuous with the rest of the biological and evolutionary sciences. As I stated, if we come to embrace the distinction between pattern explanations and process explanations, then pattern explanations might indeed be entirely within the realm of phylogenetics. A different situation occurs when we analyze process explanations of anthropogenesis. On the other hand, and as justification of my prior claim, I have maintained that the explanatory patterns of process explanations must include elements from the social and human sciences. This is because our ontology as a species has given rise to a new metaphysical order in which the tools of interpretive/hermeneutical disciplines become relevant if our goal is to understand specific aspects of our evolution, such as cultural evolution, the crafting of tools, our new social and complex structures, and so forth.

In this final section, I will briefly elaborate on some of the risks of failing to appreciate these subtleties. These subtleties should be taken seriously by scholars who study paleoanthropology, bioarcheology, or cultural evolution, because they risk engaging in the sort of architectonics that is erected on the concepts of unity and continuity (see previous discussion). In the best scenario, the explanatory monism of such a project tokenizes or reduces the social sciences and humanities; in the worst, it simply discards and replaces them.[16] Thus, inadvertently, these scenarios provide an image of the world in which the ontologies of biology replace all other ontologies.

A question that arises is: If these are the risks, how should we study ourselves and our close evolutionary relatives? How continuous are we compared not only to the rest of nature and other extant primates (apes included), but also to our extinct relatives? Epistemologically, what kinds of explanations are best suited for these tasks? How can we know *objectively* how similar or dissimilar we all are? Methodologically, which tools from the human, social and/or biological sciences should be recruited if we aim to address these concerns? The core of these questions is the epistemic and political value of *unity*.

In order to address the questions "Is paleoanthropology an evolutionary science, and are analyses of human evolution biological?" one must pay attention to the role of unity and continuity in any answer proffered. Examine, also, as Michel Foucault (1972) would have said, the archaeological and genealogical conditions of possibility for both questions, that is, under which historical situations were these questions asked, why did they become relevant and how were certain explanations chosen as the right answers. For, after all, archaeo-genealogies help us locate both in history, and within different systems of values, the concepts of continuity and unity. This is especially important given the plasticity of these concepts and the ease with which they can be adopted by, and adapted to, a multitude of discourses.

Specifically from an archaeo-genealogical perspective, raising the questions of whether paleoanthropology is an evolutionary science, or if analyses of human evolution are biological, rests on the possibility of the answer being "no," which itself was the result of Charles Darwin and contemporaries (especially Thomas Henry Huxley) arguing that humans and humanity were as much a part of Nature as any other animal and, therefore, explainable in naturalistic terms. This negative response also becomes possible when, through Franz Boas (1940), culture and nurture acquired the capacity to figure as domains of knowledge with their own methods, standards, and ontologies. No less importantly, the possibility of "no" is intimately tied to the notion that "man" constitutes an ontologically demarcated realm of knowledge in which accounts of language and labor are not amenable to conceptions of naturalization (see Foucault 1970). Paradoxically, then, the concerns guiding this reflection have as their conditions of possibility the fact that nineteenth century natural sciences developed an account that posited humans as merely another organism, while, on the other hand, the nascent social and human sciences allowed for the possibility of studying humanity as an independent, although unified, phenomenon in which the commonalities between humans—in contrast to other animals—are given priority over the differences.[17]

Clearly, both the natural and human and social sciences invoke unity and continuity, that is, a division of Ego and Alter that classified humans as either continuous with Nature, or continuous in discontinuity with Nature. In turn, this led to an impossible position because "we"—whoever "we" is—described ourselves simultaneously as continuous and discontinuous with Nature. Consequently, by the twentieth century, the need to explain this tension became unavoidable.

Paleoanthropology needs to be aware of this legacy and challenge because, as a discipline, it works on the fringes of these claims. Obviously, the risks are many: Paleoanthropology might end up being just human paleontology, thus disregarding our World-constructing capacities or tokenizing them as an instance of niche construction. On the other hand, it is not a trivial matter to offer an account of how we should overcome the Nature-Culture dichotomy while, at the same time, paying attention to a complex ontology in which we are interdiscursive and intermaterial bodies inhabiting Worlds that are also, but not merely, habitats.

It would seem that feminist philosophers of biology, such as like Donna Haraway (2007), are aware of all of this. For example, in her claim "We have never been human," she echoes Latour's (1993) dictum "We have never been modern." For Haraway, in keeping with multispecies ethnographies, every human culture constructed itself by forging different relations, materially and symbolically, with their surroundings, including nonhuman animals as well as plants. Consequently, it is impossible to assert that a common human nature ever existed.

Although species co-produce each other, ecologically and *locally*, they do not co-evolve in a biological sense. Our mutual niche construction is not merely *circular causation* (as may be surmised upon reading Laland et al. [2011]). In the sense of Jasanoff (2004), it is also a co-production of materiality at the realm of *material cultures*, for example, technologies, devices, shelters, and diets. Echoing Developmental Systems Theory (DST), our natures, as well as the eco-cultural systems that have long-defined us, are the products of cultures. Understanding these Worlds demands an inextricable pluralistic approach in which explanations coming from the social sciences, humanities and, more importantly, other cultures, are not taken as pure objects to by analyzed, debunked or assimilated into a larger, unified, and coherent whole.

This criticism best exemplifies the productive tension that haunts not only paleoanthropology, but other fringe disciplines, such as bioarchaeology and human ecology. This is so because unity and continuity might be seen politically desirable when they promote environmentalism and antiracist attitudes, or reject nationalism and xenophobia. However, these conceptions might be undesirable when they erase cultural and individual specificities as well as the methodological relevance of the human and social sciences, or serve biological determinism and the reduction of the social into the biological sciences. Further, unity and disunity are not oppositional terms. For, while the unity of humanity dignifies us all, it also engenders a depreciation of Nature and creates an explanatory gap regarding our most salient traits. Nationalism, racism, and xenophobia are also examples of discontinuities that assume unities. Indeed, the very idea of "them" and "us" demands the creation of differences by their erasure.

What, however, should we do if any possible choice seems to have implications beyond science and beyond paleoanthropology? We should either choose unity and continuity at the expense of difference and diversity, or embrace difference and diversity at the expense of a unified framework for those sciences that are interested in human evolution. At the same time, however, we must also free ourselves of any integrated image of the human species.

The problem may lie in the very idea of constructing a sort of super-science of the human. The very questions addressed here (regarding paleoanthropology as an evolutionary science and the nature of the analyses of human evolution) appear to imply that an architectonics of knowledge orchestrated by biology is not only desirable, but also more legitimate and objective. But this is dubious, especially if we give up the values of continuity and unity. For, then, we might see as more desirable an approach that promote a more pluralistic and interdisciplinary agenda in which we do not seek to construct an architectonics of knowledge regarding human evolution under ideological premises. Instead, we would promote a dialogue between disciplines that includes the natural as well as human and social sciences: a dialogue in which the voice of the other is always a reminder of our situatedness, of our limitations and biases and preconceptions. It is a dialogue in which the voice of the other promotes objectivity by forcing us to examine further our assumptions. A single although complex theory cannot do this. Only an embodied science in which the multitude speaks can.[18] Further, we still need to face the challenge posed by disunity and discontinuity at the level of evolutionary theory itself. In other words, if what I have said so far is true, is evolutionary biology able to fully explicate who we are? Indeed, should it even attempt to do this?

Conclusion. The Loyalties of Paleoanthropology: From Niches to Worlds

In the same vein as Jonathan Marks (2011) and others, I have attempted to offer a critical reading of the disciplinary loyalties of paleoanthropology. My primary concern is that a narrow understanding of the aims and explanatory tools of paleoanthropology would lead to a dehistorisizing of modern relationships, roles, and livelihoods, and a reification of them as merely a consequence of a purely biological, although complex and rich, evolutionary process. In addition, I have characterized the challenge to this narrow approach as the consequence of a transition from niches to Worlds, which can also be described as the rise of the hermeneutics of norms or the *transnaturalization* of humans, as conceived by Marxist philosopher Bolívar Echeverría (2000). At base, I have sought to demonstrate how, even after letting the dichotomy between nature vs. nurture wither, human uniqueness still poses a challenge that goes beyond sciences.

I have also advocated a dialogue between the natural and social and human sciences that is in opposition to the construction of an architectonics of knowledge that privileges unification, consilience, unity, and continuity. Paleoanthropology represents one of these voices, but its dialogues are many. I believe the alternative presented here is desirable because:

1. It problematizes the concept of who the "us" that pervades our scientific concerns is. More specifically, it allows us to problematize those parts of evolutionary science, including paleoanthropology and cultural evolution, in which we have inadvertently introduced a liberal and Western conception of the subject[19] that celebrates our underlying unity as one species with one nature at the expense of other world views.

2. It promotes a dialogue within the social sciences and humanities in which their ontologies and epistemologies are neither denied nor subsumed by, or reduced to, biology. In fact, it opens the possibility of introducing the field of social metaphysics into debates on human evolution.

3. By doing this, we will be able to avoid collapsing, as Emily Schultz (2015) has remarked, the distinction between the social and the environmental. Memes, Genes, and Artifacts are different domains, and so, too, is the process of niche construction when compared to the creation of Worlds. We need different disciplines, different tools, and different frameworks to study these phenomena.

This is a critical—and even feminist—text, not because it deals directly with gender, "race," or class, including a gendered and racialized division of labor or sexual orientation. Rather, these topics are, in a sense, a subtext. If we accept the idea of World construction, *existentiells*, and complex bodyscapes, we will then grasp why feminism, critical race studies, Marxism, and Queer Theory have claimed that biologistic approaches to these phenomena are mistargeting their intended *explananda*. Gender, sexual orientation, "race," and class do not have counterparts ("homologues") in the behaviors of other primates, even though they may be hierarchical and vary sociosexually in, for example, same-sex interactions and nonreproductive sex displays. Rather, genders and sexual orientations occur within a cultural practice in which the subject relates to his/her body and his/her self-awareness as an individual through culturally mediated categories. Instead of offering what most texts in critical cultural studies do (such as historical deconstructions of how some scientific facts were produced), I have followed the new materialistic approaches within feminism (Sheldon 2015) and the ecological turn in queer studies (Giffney and Hird 2008), which is to engage in a constructive dialogue of how to nourish biological disciplines in order to promote a more interdisciplinary perspective.

Finally, I suggest that the rise of hermeneutics led to a different understanding of evolvability. That is, how our cognitive plasticity and prosociality, including the plasticity of desire and sexual drives, inaugurated a new way of adapting to heterogeneous and changing environments by making it possible to engage in various social relationships that result in multiple metastable familial and social arrangements whose dynamics are better explained by the social and human sciences. If true, there is, as Haraway claims, no single human nature because, paradoxically, our nature is to co-create ourselves in myriad eco-cultural systems in which different social arrangements are the norm: Human nature is legion because we are many.

Acknowledgments

I would like to thank the KLI and Eva Lackner, Isabella Sarto-Jackson, and Gerd Müller for the wonderful opportunity to talk about human evolution in such a marvelous place. I also want to express my gratitude to Jeff Schwartz for organizing such a splendid workshop and for inviting me to participate. Finally, I want to thank Adam Hochman, Daniel Liu, Jan Baedke, Abigail Nieves, Rubén Madrigal, Alonso Gutiérrez, and Agustín Mercado for their critical and insightful comments on a previous draft; as always, writing is a deeply collective exercise even when one single person is the author.

Notes

1. These two very specific questions were at the center of the workshop organized by Jeff Schwartz at the KLI that led to the present volume. Therefore, I decided to keep them as the organizing questions of my contribution. As a précis, the questions shouldn't be read as expressing any kind of skepticism regarding the fact of human evolution but as an invitation to reflect on the kinds of models, narratives and explanations that paleoanthropology employs in dealing with human evolution.

2. I thank Richard Delisle for pointing out this important matter.

3. The distinction between pattern explanations and process explanations was obviously not coined by Delisle. For example, Elliot Sober (1991) and David Hull (1988) also distinguish these aspects of evolution and consider that explanations of patterns are not explanations of processes and vice-versa. In both cases, these philosophers were following similar ideas developed by taxonomists like Nelson and Platnick (1981), who distinguished among cladograms, phylogenetic trees, and evolutionary scenarios, depending on the degree of information regarding the evolutionary process that these representations included (see, for example, Mc Manus, 2009).

4. Process explanations regarding the evolution of hominins are included within the domain of paleoanthropology precisely because of the reason given in this paragraph.

5. The terminology here can be misleading. *"Patterns of explanations"* refers to the types of explanations we encounter in a given field while the term *"pattern explanations"* refers to a very specific pattern of explanation: explanations of the evolutionary pattern.

6. I use this term in the sense of Maynard-Smith and Szathmáry (2000).

7. However, it would be an error to assume that this metaphysical breach only occurs at the level of language and cognition. In fact, this breach is deeply seated in a new relationship with the body itself and its surrounding environment and, thus, it demands new analytical categories to comprehend these changes. This is so because it implied an entire reshuffling of human ecology through the development of tools and the advent of new social structures and ways of comprehending, communicating, and intervening.

8. There is at least one theory of cultural evolution that acknowledges the heterogeneity of the *explananda* and *explanantia* of human evolution; here I have in mind Wimsatt's (2013) theory in which genes, memes, and artifacts are central factors but also organizations and institutions in a sense very much akin to the one just offered.

9. *Existentiells* do not necessarily encompass all socially mediated entities and properties, but it goes beyond the scope of this paper to address the connections between these two general notions of conceiving the metaphysics of the social.

10. In fact, this intermaterial and interdiscursive arrangement that characterizes these "Worlds" scaffolds our actions and serves as a store of information. It is an extension of our cognitive capacities and, as such, it takes place, at least partially, outside the brain. This explains why these arrangements become entrenched and gave rise to a cumulative cultural evolution famously described in terms of a "ratchet" in which we construct on top of what we inherited—an inheritance that includes the environment but also the material cultures of crafts, artifacts and installations, on the one hand, and the symbolic culture or lore present in myths and rites, in social roles and the division of society, on the other. So, the idea of "Worlds" also exhibits a temporal dimension that is connected with the idea of a culturally embedded and embodied Self/Body that is temporally extended and scaffolded (Bartra 2007).

11. This claim might be contested in light of modern hermeneutics, specifically, in light of modern ontophytology. Authors such as Michael Marder (2013) claim that asserting that only humans have fully developed Worlds and that animals are poor-in-world, is a sort of anthropocentric bias that philosophy should scrutinize. Sadly, given the length, aim, and context of this text I will have to sidestep this issue.

12. I must confess that I find his analysis defective because I think it fails to grasp many important issues like power, hegemony, and class struggle *à la* Foucault, Gramsci, or Marx, and, by doing this, it reifies a liberal conception of the individual. Anyway, my aim in this text is to argue in favor of the rise of the metaphysics of the social as an evolutionary transition. It is not the aim of the text to explore which philosophic tradition better describes it.

13. In philosophy of mind, intentionality does not denote intention or purpose as its primary meaning but the capacity of the human mind to point beyond itself. Intentionality is present in, for example, beliefs or desires; in both cases, we find that something beyond itself is being pointed at. Therefore, these states are also labeled *intentional states*. Traditionally, intentionality has been modeled in terms of propositions, so to have a belief is to express a commitment to the veracity of a proposition or to have a desire is to will the content of a proposition. Thus, intentional states usually come along with what is called *propositional attitudes*.

14. The situation might be more complicated if we accept de Waal's suggestion that apes exhibit what he calls "targeted helping" (de Waal 2009). But, to my knowledge, it is not clear if what Tomasello is discussing here is entirely identical with what de Waal is calling targeted helping; it might be the case, for example, that targeted helping does occur but it does not involve any inference regarding the needs of the other, it might be displayed just when the other is clearly and obviously in need—by feeling pain, for example—but not when the other aims to achieve a goal that is not immediate or self-evident. Anyway, this is merely an educated guess on how to tackle this tension.

15. It must be said that these traces—of past livelihoods, of death bodies, of ancient tools, and so on—serve as the evidentiary basis for paleoanthropology. The act of reading them should be considered deeply *interdiscursive*,

in the sense of necessarily articulating and presupposing several discourses that ground and justify particular readings, discourses coming from other branches of biology and chemistry or physics but also from more entrenched assumptions about how embodied lives are lived. These assumptions might come from our own lives or from the *Geisteswissenschaften*. But this reading is also *intermaterial*, because the act of reading involves reading material traces, empirical evidences, microfossils, and paleoenvironments; in sum, the materiality of these items emerges only by the act of reading how they were materially intertwined with other materialities, or how the embodied subject constructed him/herself by materially engaging in activities that transformed him/her.

16. Here I am not claiming that social sciences and humanities do not possess vices of their own. As has been pointed out, sometimes these disciplines also follow a reasoning in terms of unities and continuities that disregard their internal differences. My point is that this way of "integrating" sciences should be better scrutinized. I thank Daniel Liu for this observation.

17. It might seem perplexing to claim that this tension between the natural and the social sciences made possible to study humanity as a unified phenomenon while, at the same time, gave rise to racial discourses in which humanity was broken down into subgroups; I believe there is no contradiction in these different developments. Both have to do with tracing the boundaries between Ego and Alter, demarcating humanity from nature. I thank Adam Hochman for pointing out this.

18. This is very much in line with the account developed by Winther (2014) when he discusses reification, the philosophic fallacy, and how to overcome these problems.

19. Again, it would seem that William Wimsatt (2013) recognizes this when he claims that we need to avoid any individualistic account on cultural evolution; in fact, he is now trying to introduce a more collectivistic dimension without giving up the individual. Sadly, he still exemplifies an attempt to create an architectonics of knowledge in which biology reigns as the discipline in which integration occurs. Nonetheless, it is certainly an important advance from the models originally developed by Cavalli-Sforza and Feldman (1973) that were centered on the family and the individual as the central core of analysis. These ones illustrated a bias toward a particular form of social arrangement that can lead to Westernized, heterosexist, and patriarchal accounts on the social.

References

Bartra, Roger. 2007. *Antropología del Cerebro. La Conciencia y los sistemas simbólicos*. Mexico: Fondo de Cultura Económica-Pretextos.

Boas, Franz. 1940. *Race, Language, and Culture*. Chicago: University of Chicago Press.

Carroll, Sean. 2005. *Endless Forms Most Beautiful. The New Science of EvoDevo and the Making of the Animal Kingdom*. New York: W.W. Norton & Company.

Cavalli Sforza, Luigi and Marcus Feldman. 1973. "Cultural Versus Biological Inheritance: Phenotypic Transmission from Parents to Children (A Theory of the Effect of Parental Phenotypes on Children's Phenotypes)." *American Journal of Human Genetics* 25: 618–637.

Delisle, Richard G. 2012. "The Disciplinary and Epistemological Structure of Paleoanthropology: One Hundred and Fifty Years of Development." *History & Philosophy of the Life Sciences,* 34(1–2): 283–329.

Derrida, Jacques. 2008. *The Animal that Therefore I Am*. New York: Fordham University Press.

de Waal, Frans. 2009. *The Age of Empathy. Nature's Lessons for a Kinder Society*. New York: Three Rivers Press.

Dilthey, Wilhem. 1989. *Selected Works: I. Introduction to the Human Sciences*. Princeton, NJ: Princeton University Press.

Echeverría, Bolivar. 2000. *La Modernidad de lo Barroco*. Mexico: Biblioteca Era.

Engels, Frederich. 1934 [1876]. *The Part Played by Labour in the Transition from Ape to Man*. Moscow: Progress Publishers.

Feyerabend, Paul. 1993. *Against Method*. New York: Verso.

Foucault, Michel. 1970. *The Order of Things. An Archaeology of the Human Sciences*. New York: Random House Inc.

Foucault, Michel. 1972. *The Archaeology of Knowledge*. New York: Tavistock Publication Limited.

Friedman, Michael. 1974. "Explanation and Scientific Understanding." *Journal of Philosophy*, 71: 5–19.

Giffney, Noreen and Myra Hird. 2008. *Queering the Non/Human*. Burlington: Ashgate Publishing Limited.

Godfrey-Smith, Peter. 2009. *Theory and Reality: An Introduction to the Philosophy of Science*. Chicago: University of Chicago Press.

Haraway, Donna. 2007. *When Species Meet*. Minneapolis: University of Minnesota Press.

Heidegger, Martin. 1962. *Being and Time*. New York: Harper and Row, Publishers, Incorporated.

Hodge, Jon. 2000. "Knowing about Evolution: Darwin and His Theory of Natural Selection." In *Biology and Epistemology*, edited by Richard Creath and Jane Maienschein, 27–47. Cambridge: Cambridge University Press.

Hull, David. 1988. *Science as a Process*. Chicago: University of Chicago Press.

Jasanoff, Sheila. 2004. *States of Knowledge: The Co-Production of Science and Social Order*. London & New York: Routledge.

Kitcher, Philip. 1989. "Explanatory Unification and the Causal Structure of the World." In *Scientific Explanation*, edited by P. Kitcher and W. Salmon, 410–505. Minneapolis: University of Minnesota Press.

Kuhn, Thomas S. 2012. *The Structure of Scientific Revolutions*. Chicago: University of Chicago Press.

Laland, Kevin, Kim Sterelny, John Odling-Smee, William Hoppitt, and Tobias Uller. 2011. "Cause and Effect in Biology Revisited: Is Mayr's Proximate-Ultimate Dichotomy Still Useful?" *Science*, 334: 313–325.

Latour, Bruno. 1993. *We Have Never Been Modern*. Cambridge: Harvard University Press.

Levy, Pierre. 1995. *Qu'est-ce que le virtuel?* Paris: La Découverte.

Marder, Michael. 2013. *Plant-Thinking: A Philosophy of Vegetable Life*. New York: Columbia University Press.

Marks, Jonathan. 2011. *The Alternative Introduction to Biological Anthropology*. Oxford: Oxford University Press.

Maynard-Smith, John and Eörs Szathmáry. 2000. *The Origins of Life. From the Birth of Life to the Origin of Language*. Oxford: Oxford University Press.

Mc Manus, Fabrizzio. 2009. "Rational Disagreements in Phylogenetics." *Acta Biotheoretica*, *57*(1–2), 99–127.

Mc Manus, Fabrizzio. 2013. *¿Naces o Te Haces? La Ciencia detrás de la Homosexualidad*. Mexico: Paidós de México.

Nelson, Gareth and Norman Platnick. 1981. *Systematics and Biogeography*. New York: Columbia University Press.

Schmitt, Frederick. 2003. *Socializing Metaphysics. The Nature of Social Reality*. Lanham: Rowman & Littlefield Publishing Group.

Schultz, Emily. 2015. "La Construcción de Nicho y el Estudio de los Cambios de Cultura en Antropología: Desafíos y Perspectivas." *Interdisciplina* 3(5): 131–160.

Searle, John. 1995. *The Construction of Social Reality*. New York: The Free Press.

Sheldon, Rebekah. 2015. "Form/Matter/Chora: Object-Oriented Ontology and Feminist New Materialism." In *The Nonhuman Turn*, edited by R. Grusin, 193–222. Minneapolis: University of Minnesota Press.

Sober, Elliott. 1991. *Reconstructing the Past: Parsimony, Evolution, and Inference*. Cambridge, MA: The MIT Press.

Sterelny, Kim. 2012. *The Evolved Apprentice. How Evolution Made Humans Unique*. Cambridge: The MIT Press.

Striedter, Georg. 2005. *Principles of Brain Evolution*. Sunderland: Sinauer Associates.

Tattersall, Ian. 2012. *Masters of the Planet. The Search for Our Human Origins*. New York: Palgrave Macmillan.

Tomasello, Michael. 2014. *A Natural History of Human Thinking*. Cambridge: Harvard University Press.

Viveiros de Castro, Eduardo. 2004. "Perspectivismo y Multinaturalismo en la América Indígena." In *Tierra Adentro. Territorio Indígena y Percepción del entorno*, edited by Alexandre Surallés and Pedro García Hierro, *Tierra Adentro. Territorio Indígena y Percepción del entorno*, 37–80. Lima: Grupo Internacional de Trabajo sobre Asuntos Indígenas.

Von Wright, Georg H. 2004. *Explanation and Understanding*. Ithaca: Cornell University Press.

Wimsatt, William. 2013. "Articulating Babel: An Approach to Cultural Evolution." *Studies in History and Philosophy of Biological and Biomedical Sciences* 44: 563–571.

Winther, Rasmus. 2014. "James and Dewey on Abstraction." *The Pluralist* 9 (2): 1–28.

3 Human and Mammalian Evolution: Is There a Difference?

John de Vos and Jelle W. F. Reumer

Introduction

When God created the world, he did so in a succession of different steps. The creation of animals was one such step. The creation of mankind was another one. Ever since, mankind has been considered (i.e., has considered itself) not to be part of the animal kingdom. This notion—that *Homo sapiens* is a species next to, above, or outside the mammalian world—has long perverted science. Ernst Haeckel's famous "Stammbaum des Menschen/Pedigree of Man," published in 1874, shows "man" in the highest branch of the tree, above the rest of the living world, although part of the apes.

A favorite saying among mammalian paleontologists is the following: There are three classes of mammal paleontologists. The lowest one consists of those who study herbivores; that is, rodents and other small mammals, artiodactyls (bovids, cervids), and perissodactyls (horses). Although the fossil record of herbivores is vast, scientific output is unimpressive and, in addition, specimens are stored in cheap cardboard boxes. Higher up in the hierarchy are those who study carnivores. Although these paleontologists have less material to work with (because their subjects are higher up in the ecological food chain and thus less numerous), they produce more articles. Their fossils are carefully wrapped and stored in plastic boxes. High above these paleontologists are those who study fossil hominids. Although the human fossil record pales in the face of the fossil record of mammals in general, these paleontologists generate a disproportionate number of publications in high-ranking journals with the highest impact factors. Human fossils are stored in fireproof safes, often have nicknames (e.g., Lucy, Turkana boy, Hobbit, Little Foot), and are treated like icons. It is often easier to gain access to Fort Knox or Buckingham Palace than to see these specimens, let alone touch and study them. Those paleontologists who study or do analyses on fossil hominids and who consider theirs a separate profession are referred to as paleoanthropologists. This distinction reflects the general opinion that paleontology and paleoanthropology are different scientific enterprises. It is this misunderstanding that we discuss here.

We are surprised by the ongoing debate about whether or not the study of human evolution, and of *Homo sapiens* in particular, is part of evolutionary science. To us, as evolutionary biologists, it seems obvious that human evolution and mammalian evolution are inseparable and identical processes. Why should paleoanthropology be a separate science instead of part of mammalian paleontology? If human evolution does not differ from the evolution of other mammalian groups—and we believe it does not differ—then identical environmental factors may lead to identical adaptational responses, that is, to convergences. Can these be found?

We intend to answer this intriguing question by briefly discussing two historical examples: (1) the footprint trails of a tridactyl horse and a hominid from the Tanzanian site of Laetoli G and (2) the discovery of pygmy forms, such as dwarf elephants (from e.g., Malta, Sicily, Crete, and Flores), and dwarf hippopotamuses and other artiodactyls from islands of the Mediterranean and the Indonesian Archipelago, in relation to a small hominin from Flores (*Homo floresiensis*). We conclude with (3) a hypothesis about the pelage of *Homo neanderthalensis* in comparison with the woolly mammoth and woolly rhino as adaptations to the climatic circumstances on the Late Pleistocene mammoth steppe.

Man and Horse

In 1976, Mary Leakey discovered a rich locality with footprints that was excavated in 1978 and published the next year as "the Laetoli Footprints." The presence of prints made by hominids amidst a rich mammalian ichnofauna was groundbreaking. At the site, three hominid trails are intersected by two *Hipparion* trails. Since then, numerous articles have been published on the Laetoli G hominid footprints (see De Vos et al. 1998 for an overview), but only few on the two *Hipparion* trails (Renders 1984; Renders and Sondaar 1987). Nevertheless, in spite of the lopsidedness of scientific and public attention, the footprints of hominid and equid provide information about the locomotion not only of the first bipedal hominids, but also of the tridactyl horse, and, together with preserved footprints of other species, ultimately about the Pliocene ecosystem in Eastern Africa.

The evolution of horses proceeded from the Eocene (ca. 55 Ma), with small, low-crowned, browsing forest-dwelling animals, to modern large, high-crowned grazers living on the grassy plains of the American prairies, the Eurasian steppes, and the African savannahs. Adaptation toward a life beyond the forests and woodlands led to long-legged horses with only one toe per extremity (Franzen 2010). This development accelerated in the Miocene when grassland ecosystems increased. The one-toed genus *Equus* appeared first in North America during the Pliocene, ca. 2–3 Ma ago, and subsequently spread to the Old World. With high-crowned teeth, a digestive system based on caecum fermentation, and long and sturdy legs suited for running, *Equus* is uniquely adapted to living on grassland (Franzen 2010). The Key Evolutionary Innovations (KEIs) involved in horse evolution are (1) lengthening of the legs, (2) an increase in hypsodonty and enamel wrinkling, and (3) development of a larger brain in relation to increasing socialization (Edinger 1948).

Antelopes (Bovidae) underwent a rapid increase in biodiversity in conjunction with climatic cooling and a consequent increase in grassland ecosystems (Vrba 1995). Especially from the Early Pliocene after 4 Ma ago and until the Middle Pleistocene, the number of African antelopes soared. Indeed, the increase in taxic diversity between 2.7 Ma and 2.5 Ma has been counted as 37 FADs (First Appearance Dates, Vrba 1995). Also in antelope evolution, KEIs involve lengthening of the legs, dental evolution, and increasing socialization (i.e., living in a herd structure). Further, this is precisely the period in time in which there was a major increase in hominids (especially within *Australopithecus*) and in which obligate hominid bipedalism emerged. We doubt that this is a mere coincidence.

Hominids are primates. "Primates of modern aspect" appear in the fossil record ca. 55 Ma, and striding (=modern) bipedal hominids (e.g., Turkana boy) after ca. 2 Ma ago in Africa. Current consensus is that modern-type bipedalism is an adaptation for living in an open environment, which resulted from the demise of African forests during the Late Pliocene/Early Pleistocene (De Vos et al. 1998). Recently, Uno et al. (2016) provided new evidence of an increase in grassland ecosystems, from which they concluded that "the biomarker vegetation record suggests that the increase in open, C4 grassland ecosystems over the last 10 Ma may have operated as a selection pressure for traits and behaviors in *Homo* such as bipedalism, flexible diets, and complex social structure" (Uno et al. 2016, p. 6355). Thus, both the evolution of human bipedalism and erect posture on the one hand, and of the long-legged running gait in horses on the other, are the result of Miocene-through-Pleistocene climate change in conjunction with the reduction of forest ecosystems and increase of open habitat. Both evolutionary processes are rooted in similar circumstances. The anatomical differences between hominids and equids can be explained by the different points of departure: that is, from a knuckle-walking plantigrade ape and a quadrupedal digitigrade equid, respectively. Yet, from an evolutionary biological perspective both processes are identical. KEIs involved in human evolution are the upright posture/ bipedalism, dental evolution (in this case, among other, a reduction in canine size), and an increase in brain capacity in relation to greater dexterity and socialization. Humans, antelopes, and horses are mammals that adapted to a new environment, and their evolution reflects their convergences.

Humans and Island Dwarfs

For over a century paleontologists have studied island fossil vertebrates, resulting in a wealth of data from islands in the Mediterranean, the Philippines, Indonesia, the Californian Channel Islands, and many other islands and archipelagos (see Van der Geer et al. 2010 for an extensive overview). Although the mechanisms leading to observed phenomena remain unclear, these studies have given rise to what is called the "Island Rule." That is, in general, small mammals (shrews, hedgehogs, rodents, leporids) become larger when isolated on islands, and large mammals (elephantids, hippopotamids, bovids, cervids) become smaller. A notorious example is the dwarf elephant *Elephas falconeri* from Spinagallo Cave,

Sicily, with a shoulder height of only 90 cm in adult females, and 1.3 m in adult males (Van der Geer et al. 2010). Similar examples abound from islands around the globe. We now know, to list a few examples, of dwarf elephantids (genera *Elephas*, *Mammuthus*, *Stegodon*) from Sicily, Malta, Crete, Santa Rosa (and adjacent Channel Islands), Java, and Flores; of dwarf bovids from Mallorca, Menorca, and the Philippines; of dwarf cervids from Crete, Karpathos, the Ryu-Kyu Islands; and of dwarf hippopotamids from Crete, Malta, and Cyprus. The dwarfing of larger mammals appears to be an adaptation to insular circumstances, especially the absence of large mammalian predators (carnivores) and of the necessity to be large in order to escape from them. An important factor here is the ratio of body surface-to-volume: the smaller a large mammal becomes, the smaller is the risk of metabolic overheating. It is also important to note that size reduction can occur quite rapidly (Van der Geer et al. 2010), which is a fact of which Hawks (2016) was unaware when he wrote: "I'm very surprised to see paleoanthropologists in the press commenting that the dwarfing of *Homo floresiensis* was very rapid." (Hawks 2016, final paragraph). In addition to dwarfism, the development of limb shortening and of low-gear locomotion is another island adaptation seen in larger mammals (Sondaar 1977; Van der Geer et al. 2010; Van Heteren 2012).

Although until fairly recently one might have wondered if humans would be an exception to the Island Rule, the possibility emerged with the discovery of the remains of a Late Pleistocene hominid on the Indonesian island of Flores (Morwood et al. 2004; Morwood and Van Oosterzee 2007; see also Van den Bergh et al. 2016). Claims of microcephaly notwithstanding, the specimens are more reasonably seen as evidence of island dwarfing and of a separate species. More recently, a possible second example of a small hominid was discovered in Callao Cave on the island of Luzon (Philippines; Mijares et al. 2010).

The adaptations seen in larger island mammals can be summarily pigeon-holed as "dwarfism," but that would be too simplistic. Island adaptations comprise a set of characteristics, of which dwarfism is only one. They also include relative limb shortening, development of low-gear locomotion, relative decrease in brain size, and the emergence of specific dental features, such as reduction in number of molars, hypsodonty, hypselodont incisors, and/or changes in occlusal ridge patterning (Van der Geer et al. 2010). Altogether, such changes often blur the picture of the descent of the island taxa from their mainland ancestors.

Homo floresiensis is now considered to belong to a *H. erectus* clade (Zeitoun et al. 2016), and perhaps even a descendant of *H. erectus*, which is known from Indonesia from at least the early Middle Pleistocene (Dubois 1894; Van Heteren 2012; Van den Bergh et al. 2016). Among other features, *H. floresiensis* had short limbs, low-gear locomotion, and a small brain. In short, there is no morphological difference between island elephantids, bovids, cervids and hippopotamids and island hominids (i.e., *H. floresiensis*). Hominids are large mammals, and much like other large mammals, they dwarf and adapt similarly to insular circumstances. It is another example of a convergence.

Woolly "Man"

During the Late Pleistocene, the prevailing ecosystem in much of Eurasia was the so-called Mammoth Steppe (Guthrie 1982), which was a highly productive ecosystem with grasses, herbs, and low shrubs, and where a fauna called the *Mammuthus-Coelodonta* Faunal Complex thrived (hereafter MCFC; Kahlke 1999). The most common elements of this MCFC are the two eponymic taxa, the woolly mammoth (*Mammuthus primigenius*) and the woolly rhinoceros (*Coelodonta antiquitatis*), as well as various other species that sported a woolly pelage, such as the musk oxen (*Ovibos moschatus*) and the cave bear (*Ursus spelaeus*). A woolly pelage was the typical adaptation to the prevailing cold and dry circumstances. Most of the larger fauna went extinct at the end of the Pleistocene as a combined result of climate change towards a warmer and more mesic climate, and the fragmentation of their natural habitat. Among the taxa that disappeared entirely or regionally by the onset of the Holocene, are the woolly mammoth, North American *Mastodon*, Eurasian woolly rhinoceros, musk oxen (extinct in Eurasia only), Eurasian giant deer (*Megaloceros giganteus*), North American *Equus*, and a suite of carnivores, such as *Homotherium*, North American *Smilodon*, the hyena (*Crocuta crocuta*), the cave lion (*Felis spelaea*), and the cave bear (*Ursus spelaeus*).

Another victim of the Late Pleistocene was *Homo neanderthalensis*, which was a large-brained hominid that lived in a broad geographic band across Europe and part of western Asia (e.g., Endicott et al. 2010). Since "Classic" European Neanderthals were adapted to the harsh circumstances of the Late Pleistocene, and, albeit in relatively low numbers, lived on the Mammoth Steppe as part of the MCFC, one may wonder about their physical appearance. Indeed, since so many MCFC species were provided with a woolly pelage (mammoth, rhino, musk oxen, cave bear), we ask why this (rather than hairlessness) should not also have been the case with *H. neanderthalensis*. Hence, we propose that *H. neanderthalensis* was clad in a relatively thick, red or brownish-coloured pelage. We base our hypothesis on the fact that humans are mammals and therefore subject to the same evolutionary processes, including adaptations to a cold and dry climate, as other (large) mammals. For, indeed, why should most (if not all) MCFC mammals have had a thick woolly covering, and *H. neanderthalensis* be the exception? And in this context, we might consider seriously the likelihood that woolly Neanderthals went extinct along with the other woolly mammals.

Conclusion

Homo sapiens is a mammal. The evolution of the genus *Homo*, and of hominids in general, is, therefore, a part of mammalian evolution. Evolutionary processes that shape other large mammals, whether artiodactyl, perissodactyl, proboscidean, or any order of mammal, are in principle the same as, and cannot differ from, the evolutionary processes that shape hominids. The results of evolution within different groups of large mammals reveal interesting convergences. Here we briefly describe two such convergences. First, the adaptation of a long-legged running posture in association with the increase of grass ecosystems in the later part of the Cenozoic. The evolution of horses and African antelopes are prime examples of such adaptations. In this regard, hominid upright stance and bipedalism can be seen as a convergent adaptation. Second, is the dwarfing of larger mammals that become isolated on islands. Not only is this evidenced in the existence of pygmy elephants, pygmy hippopotamuses, dwarfed deer, and dwarfed bovids, but dwarfing is also documented for hominids by *Homo floresiensis*. Finally, based on the fauna that was contemporaneous with Neanderthals, we propose a third convergence: that is, like other Late Pleistocene mammals from the mammoth steppe, *Homo neanderthalensis* was also endowed with a thick, woolly pelage.

Acknowledgments

We'd like to thank Jeffrey Schwartz and the KLI for the invitation and the splendid organization.

References

De Vos, J., P. Y. Sondaar, and J. W. F. Reumer. 1998. "The Evolution of Hominid Bipedalism." *Anthropologie* 36(1–2): 5–16.

Dubois, E. 1894. "*Pithecanthropus erectus*, eine menschenähnliche Uebergangsform aus Java." Batavia: Landsdrukkerij.

Edinger, T. 1948. "Evolution of the Horse Brain." *Memoirs Geological Society of America* 25: 1–177.

Endicott, P., S. Y. W. Ho, and C. B. Stringer. 2010. "Using Genetic Evidence to Evaluate Four Palaeoanthropological Hypotheses for the Timing of Neanderthal and Modern Human Origins." *Journal of Human Evolution* 59(1): 87–95.

Franzen, J. L. 2010. *The Rise of Horses. 55 Million Years of Evolution*. Baltimore, MD: Johns Hopkins University Press.

Guthrie, R. D. 1982. "Mammals of the Mammoth Steppe as Paleoenvironmental Indicators." In *Paleoecology of Beringia*, edited by D. M. Hopkins, C. E. Schweger, and S. B. Young, 307–329. New York, NY: Academic Press.

Hawks, J. 2016. "Hominin Remains from Mata Menge, Flores." Accessed July 14, 2016. http://johnhawks.net /weblog/fossils/flores/mata-menge-van-den-Bergh-remains-2016.html.

Kahlke, R.-D. 1999. *The History of the Origin, Evolution and Dispersal of the Late Pleistocene* Mammuthus-Coelodonta *Faunal Complex in Eurasia (Large Mammals)*. Rapid City, UT: Mammoth Site of Hot Springs.

Leakey, M. D. and R. L. Hay. 1979. "Pliocene Footprints in the Laetolil Beds at Laetoli, Northern Tanzania." *Nature* 278: 317–323.

Mijares, A. S., F. Détroit, P. Piper, R. Grün, P. Bellwood, M. Aubert, G. Champion, N. Cuevas, A. De Leon, and E. Dizon. 2010. "New Evidence for a 67,000-Year-Old Human Presence at Callao Cave, Luzon, Philippines." *Journal of Human Evolution* 59: 123–132.

Morwood, M. J., R. P. Soejono, R. G. Roberts, T. Sutikna, C. S. M. Turney, K. E. Westaway,, W. J. Rink, J.-X. Zhao, G. D. Van den Bergh, D. A. Rokhus, D. R. Hobbs, M. W. Moore, M. I. Bird, and L. K. Fifield. 2004. "Archaeology and Age of *Homo floresiensis*, a New Hominin from Flores in Eastern Indonesia." *Nature* 431: 1087–1091.

Morwood, M. J. and P. Van Oosterzee. 2007. *A New Human: the Discovery of the Hobbits of Flores.* Washington, DC: Smithsonian Books.

Renders, E. 1984. "The Gait of *Hipparion* sp. from Fossil Footprints in Laetoli, Tanzania." *Nature* 308: 179–181.

Renders, E. and P. Y. Sondaar. 1987. *"Hipparion."* In *Laetoli: a Pliocene Site in Northern Tanzania,* edited by M. D. Leakey and J. M. Harris, 471–481 Oxford: Oxford University Press.

Sondaar, P. Y. 1977. "Insularity and Its Effect on Mammal Evolution." *Major Patterns in Vertebrate Evolution,* edited by M. N. Hecht, P. L. Goody, and B. M. Hecht, 671–707. New York, NY: Plenum.

Uno, K. T., P. J. Polissar, K. E. Jackson, and P. B. deMenocal. 2016. "Neogene Biomarker Record of Vegetation Change in Eastern Africa." *Proceedings of the National Academy of Science USA.* Accessed June 15, 2016. http://www.pnas.org/cgi/doi/10.1073/pnas.1521267113.

Van den Bergh, G. D., Y. Kaifu, I. Kurniawan, R. T. Kono, A. Brumm, E. Setiyabudi, F. Aziz, and M. J. Morwood. 2016. "*Homo floresiensis*-like Fossils from the Early Middle Pleistocene of Flores." *Nature* 534: 245–248.

Van der Geer, A., G. Lyras, J. De Vos, and M. Dermitzakis. 2010. *Evolution of Island Mammals. Adaptation and Extinction of Placental Mammals on Islands.* Chichester, UK: Wiley-Blackwell.

Van Heteren, A. 2012. "The Hominins of Flores: Insular Adaptations of the Lower Body." *Comptes Rendus Palévol* 11: 169–179.

Vrba, E. S. 1995. "The Fossil Record of African Antelopes (Mammalia, Bovidae) in Relation to Human Evolution and Paleoclimate." In *Paleoclimate and Evolution, with Emphasis on Human Origins,* edited by E. S. Vrba, G. H. Denton, T. C. Partridge, and L. H. Burckle, 385–424. New Haven, CT: Yale University Press.

Zeitoun, V., V. Barriel, and H. Widianto. 2016. "Phylogenetic Analysis of the Calvaria of *Homo floresiensis.*" *Comptes Rendus Palévol* 15: 555–568.

4 What's Real About Human Evolution? Received Wisdom, Assumptions, and Scenarios

Jeffrey H. Schwartz

Introduction

Articles about human evolution continue to get top billing in the media. Why? Because, for scientists, interested others, and creationists alike, the question of our species' origin continues to capture the imagination. Yet, even a cursory review of the scientific literature reveals that publications dealing with the discovery and systematic analysis of new hominid fossils are far outnumbered by derivative studies, which rely on someone else's allocation of specimens to specific taxa, which, in turn, is too often based on a history of underlying assumption. Consider, for instance, ongoing attempts to define the genus *Homo*.

Although Linnaeus (1735) coined the genus and species *Homo sapiens* in 1735, he defined this taxon not on the basis of morphology, but with the phrase "*nosce te ipsum*" (know thyself). Historically, it was thus left to Blumenbach (1969) who, in 1775 and 1795, provided not only the first anatomical description of the species but also, because creationist belief denied humans a fossil record, the first anatomical description of the genus: for example, large rounded skull; small face, jaws, and teeth (especially incisors and canines), with canines similar to incisors and aligned with them; long thumb; centrally placed foramen magnum; bowl-shaped pelvis; and bipedal stance.

When discovery of the Feldhofer Grotto Neanderthal remains in about 1857 provoked the first public debate about human antiquity, Huxley (1863) declared that this individual was (merely) a brutish representative from the past of a hierarchy of human races that terminated in the most modern, the European. But Huxley referred only to the Feldhofer calotte, which he compared in lateral outline with the skull of an Australian Aborigine, as putative support of his assertion that it would take little to morph one skull shape into the other. King (1864), however, rejected Huxley's claim and its reliance on the skullcap alone. Upon considering all of the Feldhofer Grotto remains, which included various postcranial bones, King concluded that they were sufficiently unlike their counterparts in living humans to warrant their place in a separate species, *H. neanderthalensis*. However, while the features King, and subsequently others, proposed as distinguishing Neanderthals from extant humans (EH) are real, no one has ever considered how their assumptions would impact a diagnosis of the genus *Homo*, regardless of whether Neanderthals were conceived as a variant of the species *sapiens* (Trinkaus 2006, Wolpoff 1989) or as a distinct species (Tattersall and Schwartz 2000).

Although one may excuse nineteenth- and early twentieth-century scholars for placing Neanderthals in the genus *Homo* (what else would one do with the first-recognized fossil hominid, especially in light of the fact that humans are unique in being the only living species of their kind?), one wonders why, at some point, this taxonomic designation was never questioned. If it had been any other mammal, it probably would have. Nevertheless, it was not, which made possible the allocation to *Homo* of specimens that differed from Neanderthals at least as much as Neanderthals do from EH (Schwartz and Tattersall 2002b, 2003).

Returning to human paleontology in the first half of the twentieth century, the ways in which fossils were assessed is telling. Although there were fleeting attempts to refer some African specimens to different supra-specific taxa—*Cyphanthropus* for the Tighenif material (Pycraft et al. 1928) and *Homo* (*Africanthropus*) for the Florisbad partial skull (Dreyer 1935)—the prevalent systematic activity was to allocate specimens to different species of *Homo*. However, with the naming of *Australopithecus* and *Paranthropus* (Broom 1938, Dart 1925), and for a short time *Plesianthropus* (Broom 1947) and the purportedly *Homo*-like *Telanthropus* (Robinson 1953, Broom and Robinson 1949), most south African hominids were regarded as representing species of distinctly different genera. Farther east, and following in the taxonomic footsteps of the earlier named *Pithecanthropus* (Dubois 1894), Asian specimens were also allocated with confidence to different genera: *Sinanthropus* (specimens from Zhoukoudian, China [Black 1927, Weidenreich 1943]); *Javanthropus* (Weidenreich 1937; the Javanese Ngandong specimens Oppenoorth [Oppenoorth 1932] had relegated to *Homo* [*Javanthropus*]); and *Meganthropus* (some specimens from Sangiran, Java [Weidenreich 1945]). As for European fossils, with the singular exception of the Mauer jaw, which served as the type of a new species, *H. heidelbergensis* (Schoetensack 1908), specimens were segregated into *H. sapiens* and *H. neanderthalensis*.

Conspicuously absent from any of these endeavors, however, was serious and morphologically informed consideration of the question, "What constitutes the genus *Homo*?" either in rationalizing the allocation of specimens to old or new species of that genus, or in justifying the erection of new genera. Instead, given the loose application by paleoanthropologists of the rules of zoological nomenclature, new species of *Homo* (e.g., *heidelbergensis*, *florisbadensis* [*helmei*], *rhodesiensis*), as well as new genera, were diagnosed as entities unto themselves.

In 1950, the ornithologist and self-appointed systematist of the modern evolutionary synthesis, Mayr (1950), declared that paleoanthropologists had made a mess of hominid systematics in creating "a bewildering number of names," and he was going to set things straight. However, while his depiction—"bewildering number of names"—may have been true of paleoanthropology, it was no less true of other areas of vertebrate paleontology, wherein, prior to the consistent application of the *International Code of Zoological Nomen-*

clature (*ICZN*), well-meaning systematists strove to reflect taxonomically an ever-expanding picture of taxic diversity. Undaunted by his lack of knowledge of hominid fossils (and, given his sparse bibliography, his unfamiliarity with the literature), Mayr applied his ecological niche/adaptation-based idea of defining a genus to the human fossil record: Since all hominids were bipedal, and therefore presumed to have exploited their surrounding circumstances similarly, they should all be relegated to the same genus, *Homo*. Within this genus, Mayr envisioned three chronologically transforming species, *transvaalensis->erectus->sapiens*.

The embrace of Mayr's scenario by the dominant American physical anthropologist Washburn (1951) cemented paleoanthropological practice: Define species not on the basis of their morphology, but their geological age. Although Mayr (1963) later partly acceded to Robinson's (1954) case for recognizing the australopith genera *Australopithecus* and *Paranthropus*—Mayr chose *Australopithecus* to represent all specimens—the use of chronology as the basis for allocating specimens to the default alternative genus, *Homo*, and to either *H. erectus* or *H. sapiens*, remained intact. Unacknowledged, however, was the fact that, with "bipedalism" no longer a diagnostic feature of the genus *Homo*, this taxon remained undiagnosed. Nevertheless, the Gestalt of *Homo* persisted, such that, in spite of the circularity of the argument, once a site was declared the bearer of representatives of the genus, any morphology of any so-assigned specimen was interpreted in light of this scenario.

Throughout this history, only Le Gros Clark (1955) ventured to define the genus *Homo*, although he did so in very general terms: for example, a bipedal gait dissimilar to that of australopiths, erect posture (in spite of this also being characteristic of australopiths), cranial capacity $\geq$800 cm^3 (based on the smallest *H. erectus* skull), and the wherewithal to make tools. Soon thereafter, Leakey and colleagues (1964) referred to *Homo*, and to their newly named species *habilis*, a partial mandible, hand, and two parietals (OH 7, holotype), as well as a partial foot (OH 8), on the belief that this species, and not the contemporaneous individual represented by the *Zinjanthropus boisei* partial cranium OH 5, had been bipedal and the creator of the crude Olduwan stone tools also found in Upper Bed I. However, in order to include holotype OH 7 in *Homo*, Leakey and colleagues had to redefine the genus. They did so by lowering its threshold cranial capacity to 600 cm^3, which, on the assumption that the two partial, reconstructed parietals represented the same individual (in spite of their distinctly different profiles [Schwartz and Tattersall 2003]), co-author Tobias had estimated.

Nevertheless, while a cranial capacity of 600 cm^3 subsequently became the minimum requirement for membership in *Homo*, its utility was limited. For, although some specimens were crania (e.g., KNM-ER 1470, *H. rudolfensis* [Alexeev 1986]), others were not (e.g., the type specimen of *H. ergaster* is mandible KNM-ER 992 [Groves and Mazák 1975]). Consequently, and without morphological basis for assigning specimens representing different parts of the skeleton to the same species (e.g., *erectus*), much less to the genus

Homo, the disparity between specimens within a species, and thus within the genus, had to be explained as the result of morphological responses to differing adaptive circumstances (Antón, Potts, and Aiello 2014, Spoor et al. 2015, Villmoore et al. 2015, Wood and Collard 1999a, b). Even attempts to make the genus *Homo* a more workable taxon by excluding from it the species *rudolfensis* and *habilis* (albeit the constitution of these species was not questioned; Wood 1992, Wood and Collard 1999a, b), or just plain ignoring or dismissing as relevant the type specimens of presumed species of *Homo* (Antón, Potts, and Aiello 2014), serves only to exacerbate the undiagnosability of the genus *Homo*.[1]

Thus, despite creating the taxon *Kenyanthropus platyops* for a 3.5 ma flat-faced cranium (KNM-WT 4000), in which genus they suggested the flat-faced KNM-ER 1470 might also be subsumed, M. Leakey and colleagues' (2001) primary comparisons were with specimens already branded with generic identity—*Australopithecus*, *Paranthropus*, and *Ardipithecus*—while, when contrasting KNM-WT 4000 with *Homo*, they turned to specimens that were generally accepted as representing *H. habilis sensu lato* (for the latter, see Wood [1992]), Thus, while reevaluating *Homo* was not their goal, Leakey and colleagues followed the paleoanthropological practice of accepting prior designations of specimens to species that were also accepted as belonging to that genus. Clearly, these endeavors are circular at best.

Is *Sensu Lato* Systematically Meaningful?

The manner in which post-Mayr paleoanthropologists continue to allocate specimens to species of *Homo* has led to the widespread embrace of the concept *sensu lato* (*s.l.*), especially in conceiving *Homo habilis* (e.g., Leakey et al. [2001], Strait and Grine [2004], and Wood [1992] and *H. erectus* (e.g., Antón [2003], Lordkipanidze and colleagues [2013], and Rightmire [1990]). Further, even though the term *sensu lato* has not been used formally in association with *H. sapiens*, referring to it specimens that are assumed to belong to the species (and thus to the genus) as "archaic" and "anatomically modern" tacitly employs the designation *H. sapiens s.l.*

By definition, *sensu lato* means "in the wide or broad sense." In stark contrast, *sensu stricto* means "in the strict sense." The question, however, is: Is *sensu lato* systematically meaningful? Would one, for instance, conceive of an "archaic" versus "anatomically modern" *T. rex*, and thus of a *T. rex s.l.*? Although schooled systematists would immediately respond "no," the fact that paleoanthropologists embrace the concept *sensu lato* as valid serves to highlight just how far human paleontological practice deviates from the rest of paleontology.

Regardless of *sensu lato* being accepted in paleoanthropological practice, there remains the question of how, in the first place, any species (especially *habilis*, *erectus*, *sapiens*) was originally defined. This is not a trivial concern because, if a proper diagnosis had been attempted, the proposed species must have been conceived as *sensu stricto*. Only afterward, by the accretion of specimens that differ significantly from the holotype, does it become necessary to invoke the term *sensu lato*.

Unappreciated by those who embrace this systematically uninformative concept is that they are abandoning—indeed, rejecting—"species" as a systematically meaningful taxonomic unit. For, how valid is it to diagnose a species on the basis of features that are presumably unique to its members, and then do an about face and allow any specimen, which anyone, for whatever reason—geography, chronology, or personal bent—thinks should be placed in that species, to be part of it? Unfortunately, since this modus operandi is pro forma in paleoanthropology, the result is inevitable: As the range of dissimilarity increases between the holotype and specimens assigned to its hypodigm, and then increasingly between these specimens as well, so, too, does the likelihood of finding a specimen within the continually enlarging hypodigm that, in general shape or some feature, compares favorably to a newly unearthed fossil. And once that specimen is included in the hypodigm, the now-expanded web of general similarity makes possible the inclusion in it of additional specimens whose dissimilarity with the holotype—if compared directly—is striking: for example, the type of *Homo erectus*, Trinil 2, and KNM-WT 15000.

Yet, as inconceivable as this practice may be to the work-a-day vertebrate systematist, not only is any concept of taxon-as-hypothesis not embodied in paleoanthropological practice, as discussed above, neither is the systematic importance of the holotype (Potts et al. 2004, Antón, Potts, and Aiello 2014, Walker and Leakey 1993). Indeed, albeit schizophrenically, while Mayr (1950) dismissed out of hand the validity of virtually all previously named hominid genera and species, and in the process paid no heed to named type specimens, he (Mayr 1969) was also cognizant of the fact, as clearly stated in the *ICZN*, that only the specimen or specimens representing the holotype can be the rightful name-bearers of that species. Further, while "species" is a hypothesis to be tested, so too is the construction of its hypodigm.

In practice, vertebrate, mammalian, and even nonhominid primate systematists and paleontologists, construct hypodigms on the basis of detailed, favorable comparisons between its potential members and the holotype (Gazin 1958, Russell, Louis, and Savage 1967, Simpson 1945, 1961, Szalay 1976). Consequently, if presented with a *sensu lato* "species," they most certainly would suspect that this was an undiagnosed taxon that likely subsumed more than one taxon, and seek to test that possibility. In contrast, by embracing a *sensu lato* grouping as real and systematically valid, paleoanthropologists accept *a priori*, and therefore must describe, the morphological differences between specimens as individual variation (Antón 2003, Bae 2010, Curnoe 2010, Harvati et al. 2011, Lordkipanidze et al. 2013, Mayr 1950, Domínguez-Rodrigo et al. 2015). How Zollikofer and colleagues (2014) have interpreted specimens from Dmanisi, Republic of Georgia, serves to illustrate.

With the discovery of Skull 5, Zollikofer et al. restated the assumption underlying their analysis: Since all specimens were found in the same ~1.8 Ma stratum, they had to be variants of the same species (Lordkipanidze et al. 2013). Further, since the Dmanisi fossils were supposedly not australopith- or *Homo habilis*-like, but were in some metrical dimensions similar to specimens attributed to an accepted but undefined, geographically far-flung, and wildly variable *H. erectus s.l.* (Rightmire, Lordkipanidze, and Vekua 2006), these specimens must belong to that species.

In the Skull 5 publication, they more clearly stated their preconception by asserting that the Dmanisi specimens (four mandibles that might be associated with four of the five skulls) actually represented a regional variant (paleodeme) of *H. erectus* (*s.l.*), which they classified as the sub-subspecies *H. erectus ergaster georgicus*. They justified (p. 329) their conclusion by quoting Darwin (the fifth edition of *The Origin* [Darwin 1869]): "As already noted by Darwin, recognizing species diversity comes 'at the expense of admitting much variation' within species."

In response to Lordkipanidze and colleagues, Spoor (2013) noted that their geometric-morphometric analysis did not capture the morphological detail necessary to demonstrate the taxic unity of the Dmanisi sample, and that the *ICZN* does not sanction quadrinomial taxa. Thus, *Homo er. erg. georgicus* was taxonomically invalid. Schwartz et al. (2014) then pointed out that, by including in *H. erectus* all Asian, as well as east and a few south African specimens that were not deemed *H. habilis* (*s.l.*) or australopith, Lordkipanidze and colleagues ignored not only the distinctive morphology of the type specimen of *H. erectus* (the Trinil 2 calvaria), but also the clear-cut morphological differences between specimens that, via Mayrian dogma, had come to be subsumed in *H. erectus*. Rather, when morphology is taken into account, one finds that: (1) among the Asian specimens, only those from Sangiran share the derived features that are distinctive of Trinil 2; (2) the Zhoukoudian and Ngandong specimens are sensibly hypothesized as representing different morphs; (3) none of the African specimens, represented in east Africa by crania KNM-WT 15000, ER 3883, ER 3733, and OH 9 share derived features with any Asian specimen (see also Andrews [1984]); and (4), the East African crania themselves do not constitute a morphologically unified group (Schwartz and Tattersall 2000, Schwartz 2004, Schwartz and Tattersall 2003).

In their rejoinder, Zollikofer and colleagues (2014) again sought support in citing Darwin (1859; the first edition of *The Origin*): "It should be remembered that systematists are far from pleased at finding variability in important characters." They then proceeded to defend their use of a quadrinomial on the grounds that while the *ICZN* does not regulate its use, the *International Code of Botanical Nomenclature* does, such that plant systematists frequently use it to denote infraspecific groups (i.e., demes, local populations). Turning from plants to animals, Zollikofer and colleagues stated: "Demic differences are also present in animal species, such that we maintain the long-held argument that nomenclature has to serve evolutionary and (paleo-) population biology, not vice versa *(13)*" (p. 360–b).

In truth, however, their reference *(13)* is an editorial comment in *Nature* (1885, p. 316) on Garman's (1884) proposal of a system of animal classification that provided for infraspecific names. Further, while Zollikofer and colleagues submit that Garman claimed that the "advantage" of his system was "apparent," the editors who drafted the cited commentary actually concluded (p. 316): "But is this not the case in which it may be said that the proposed remedy is as bad as the disease?" In addition, then, to the use of quadrinomials not being regarded as serving the needs of evolutionary biology, it seems the answer to the *Nature* commentary is "yes," as the case of *H. erectus ergaster georgicus* demonstrates. Nevertheless, it remains the case that animal taxonomy is governed not by the *ICBN*, but the *ICZN*.

With regard to the *ICZN*, Zollikofer and colleagues (2014) are correct in stating that it does not regulate infrasubspecific names. However, this *Code* clearly provides rules for the use of an infrasubspecific name, as well as for assigning an infrasubspecific name as an addition to a trinomen (Articles 45.5 and 45.6). Specifically, Article 45.5.1 states that an infraspecific name "cannot be made available from its original publication by any subsequent action . . . except by a ruling of the Commission." Further, since Article 45.6.1 recognizes an infraspecific name "if its author expressly gave it infrasubspecific rank, or if the content of the work unambiguously reveals that the name was proposed for an infrasubspecific entity," the *Code* in effect does regulate naming an infrasubspecific group by providing specific rules for designating the taxa that would subsume it. Indeed, the protocol for naming a genus, species, and subspecies must be satisfied prior to other considerations (Articles 10.2, 11, 19, 20, 23–to-34, and 45.1–45.4). Consequently, while the *Code* does not explicitly prevent assigning an infrasubspecies *georgicus* to *Homo erectus ergaster*, it certainly governs the designations *Homo*, *erectus*, and *ergaster*. And, of these hierarchically nested taxa, only one—*ergaster*—can claim to have been diagnosed in accordance with the *Code*.

More broadly, as is also exemplified by the general paleoanthropological disregard for the significance of the type/holotype, were the *Code* that is used by every other zoologist and vertebrate paleontologist applied to human fossils, paleoanthropological practice would move from the largely subjective to the more objective.

Multiregionalism and the Submersion of Species *Erectus*

Mayr's notion of a wildly variable species *erectus* (which encompassed all Asian and a few non-Asian hominids) that had morphed into species *sapiens,* was consistent with Weidenreich's (1946) "candelabra" model of *Homo sapiens'* origins. There Weidenreich depicted Old World regional variants of a panmictic pre-*sapiens* species that, while maintaining its genetic integrity through interbreeding, transformed seamlessly and gradually into regional variants of the species *sapiens*, that were also prevented from becoming distinct species because of interbreeding.

In their formulation of a "multiregional model" to account for the emergence of modern human populational variation, Wolpoff and colleagues (2000) embraced Weidenreich's conception, but with the proviso: Their (preconceived) notion of slow, gradual morphological change from "early" to recent *Homo* reflected an accretion of features that, depending on the specific morphology of any given fossil, were expressed differentially in populations in different regions of the Old World. Hence, for example, the specimens from ~100 ka Skhul Cave, Israel, which have a superoinferiorly (s/i) short, but continuous and rounded brow, and, as in Skhul IV and V, a somewhat s/i tall and rounded cranium, can be seen as presenting a mixture of Neanderthal and recent human features, while, in their robusticity, some specimens from the Upper Paleolithic site Mladeč express a Neanderthal heritage (Frayer et al. 2006, Wolpoff, Frayer, and Jelínek 2006).

In this and other versions of the multiregional scenario, "modern" *sapiens* emerged through a process of transformation of "archaic" members of the species that, in addition to those regarded as Neanderthal, had been the basis at one time or another of notions such as "pre-Neanderthaloid" and "Neanderthaloid" (Hrdlicka 1930, McCown and Keith 1939). Given the precedence of chronology over morphology in relegating specimens either to *erectus* or to *sapiens*, it is not surprising that multiregionalists championed an accretional version of Weidenreich's "candelabra" model of modern human origins, or that Pilbeam (1972) could suggest that the younger-than-*erectus* Ngandong specimens, which look nothing like Neanderthals, represented an Asian variant of the latter hominid. Pushing multiregional evolution to the extreme, Wolpoff and colleagues (1994) argued that, since (their conception of) *H. erectus* had continuously transformed into (their conception of) *H. sapiens*, recognizing two species was meaningless because there had been no genetic discontinuity between them. Consequently, the species name *erectus* should be sunk, and its presumed members subsumed in species *sapiens*.

A question one may ask is: Is the multiregional explanation, which, as defined by Wolpoff et al. (2000), promotes the scenario of continually semi-subdividing and reticulating populations within the framework of a smoothly transformation species, testable? According to them (Wolpoff et al. 2001), it is. But it is not, because it is first assumed that (1) evolutionary change is gradual and continuous, (2) *H. erectus* and *H. sapiens* are the only taxa involved, (3) the specimens they accept as constituting *H. erectus* and *H. sapiens* can only belong to those species, (4) these species are linked phylogenetically via their chronological relationship, (5) differences between specimens regarded as *H. sapiens* represent a subset of such a lineage of change such that (as Huxley [1863] proffered for the Feldhofer Grotto Neanderthal) they represent a continuum of change from the "archaic" to the "anatomically modern," and (6), there are distinctly *erectus*, AS, and AMS features that are heritable as such, and, therefore, potentially identifiable in any given specimen. The circularity is apparent: Since evolutionary change is gradual and continual, and features are discrete entities that can be identified individually in any given specimen, the differential expression of features among specimens deemed *H. erectus*, AS, and AMS demonstrates the gradual accretion of features, from the archaic to the modern.

On the Nature of Morphological Difference

Since mainstream scenarios about human evolution rely on notions of the tempo and mode of evolutionary change, it seems appropriate to address them, especially the belief that while evolutionary change is continual and gradual, morphological features remain discrete entities (with distinct genetic bases, i.e., there are "genes for" features), whose combination reflects a "reticulated" relationship between populations whose genetic interactions wax and wane over time. This conception has been referred to as "mosaic evolution" or "total morphological pattern," and its underpinnings constitute a fusion of Darwinian "evolution" and population genetics. The irony of this widely accepted belief, however, is that the elements of it—gradual and continuous transformation and the discreteness of traits and therefore these genetic underpinnings—are incompatible.

Although the popular received history of evolutionary thought portrays Darwin's model of gradual and continual evolutionary change as being swiftly accepted by a sympathetic intellectual community, with Thomas Huxley its public champion, this cannot be farther from the truth (see reviews by Bowler [1989] and Schwartz [1999]). Rather, while Darwin was one of a coterie of intellectuals from various disciplines (including Huxley and the physicist Lord Kelvin) who supported one's right to think beyond the creationist teachings of the Church of England, Huxley embraced neither a model of gradual change, nor one of natural selection as the agent of evolutionary change. To the contrary, he, and more so the comparative morphologist St. George Mivart (1871), was a saltationist, who, in trying to understand the evolution of complexity, could not conceive how a structure that was crucial for survival—such as a reproductive organ, brain, or lung—could gradually change from an infinitesimally small version of itself into the version required for it to be functional. Rather, these saltationists argued, evolutionarily significant change must be enacted sometime during an individual's development, such that a novel feature would appear abruptly and suddenly and, if it did not kill its bearer, would persist from one generation to the next. From this perspective, Huxley, Mivart, and others (including Darwin's cousin Francis Galton, and later also Hugo de Vries, William Bateson, and Thomas Hunt Morgan early in his career) conceived the processes underlying the emergence of morphological novelty (e.g., the origin of species) as distinct from those that contributed to the persistence of novelty (e.g., adaptation via natural selection, and the survival of species). And it was in the latter consideration they believed Darwin's model of natural selection was best situated: that is, one could re-title Darwin's opus *On the Origin of Adaptation by Means of Natural Selection* (Schwartz 1999).

With the rediscovery of Mendel's paper of 1866 and the concept of traits being the products of units of inheritance, and subsequently Bateson and Saunders's (1902) application of Mendelian principles to animals, the dichotomy between a Darwinian view of evolution (in which features from parents blended in offspring), and a Mendelian perspective (in which features, like their underlying units of inheritance, were discrete entities), led to an intellectual divide between Darwinians and Mendelians (Bowler 1989, Schwartz 1999). Thus, while Darwinians maintained that evolutionary change was gradual and continuous (and thus emphasized continuous variation), Mendelians promoted discontinuous variation, and advocated a conceptual and practical separation between the "origin of species," in which morphological novelty appeared suddenly and in full force, and the "survival of species." Indeed, the advent of fruit fly population genetics was predicated on discovering the specific units of inheritance underlying discrete and specific features (Morgan 1932).

How Darwinism and Mendelism came to be merged into the framework of the "modern evolutionary synthesis" can be traced to Morgan. Although the "father" of fruit fly population genetics, Morgan (1916) initially believed in the discreteness of units of heredity and their morphological counterparts. However, as he became increasingly interested in the problem of reconciling continuous variation with the discreteness of hereditary units, he contrived how to meld Mendelism with Darwinism. His reasoning: Although units of inheritance are discrete entities, but only small-scale changes are evolutionary viable, the resultant and also infinitesimally small phenotypic differences between individuals can be regarded as continuous variation (*nota bene*: Morgan never justified why he rejected as evolutionarily relevant the large-scale changes he also constantly observed in his fruit fly colonies). Thus, since Darwinian evolutionary change is the long-term result of the accumulation of infinitesimally small changes, there was now a genetic explanation for it (Morgan 1922).

The architects of the "modern evolutionary synthesis" embraced the melding of Darwinism and Mendelism, but with the added tenet that only separation of a parental population into subpopulations, followed by selection gradually "pulling" apart descendant populations genetically and thus also morphologically, will lead to the origin of species. But theirs was not the only conception of evolutionary change. Witness the saltationist-like models of the mathematical geneticists Wright (1929) and especially Haldane (1932), physiological geneticist and epigeneticist Waddington (1940), paleontologist Schindewolf (1936), and theoretical developmental geneticist Goldschmidt (1940). Further, the British school, headed by Julian Huxley (1940a), was sympathetic toward non-Darwinian models (e.g., Goldschmidt's), and tried to incorporate a panoply of organisms—not just Darwin's focus on animals—into a model of evolutionary change that gave priority, not to geographic isolation, but to natural selection in generating new species (Huxley 1940b).

In defending their "synthesis" as the only viable explanation of evolutionary change, its architects focused primarily on Goldschmidt (1940). Dobzhansky (1941) rewrote *Genetics and the Origin of Species* in large part to attack Goldschmidt for using his (Dobzhansky's) chromosomal work in a non-Darwinian model of species' origin. Mayr (1942) outright denigrated Goldschmidt with ad hominem slurs. And Simpson (1944) attacked Goldschmidt without reason, and then damned Schindewolf for being typological (the systematic kiss of death) and anachronistic (Simpson 1952). As for the British school, Dobzhansky simply omitted it from discussion, likely because Julian Huxley had not invited him, but

Wright, to represent genetics in the workshop that resulted in the volume, *The New Systematics* (Huxley 1940a). In the end, the success of the "synthesis" lay more in rejecting viable alternative hypotheses than in articulating its own. Additionally, as we now know, while Waddington, Goldschmidt, and at times Wright, were on the right track in thinking "outside the gene," in the 1940s, their ideas could not be demonstrated in the way that fruit fly population genetics was believed to demonstrate Darwinian change.

Although the British and German approaches to deciphering evolution were more synthetic than the Synthesis—which was situated firmly in animals and population genetics—its mantra, "Nothing in biology makes sense except in the light of (this particular way of conceiving) evolution," echoed Morgan's (1910) dismissal of paleontology, comparative and developmental anatomy, ecology, and other biological endeavors, as being evolutionarily informative. Indeed, as Morgan prophesized, the belief was that only population genetics could shed light on evolutionary processes. In this regard, witness the difference between the titles of Dobzhansky's (1937, 1941) *Genetics and the Origin of Species* and Mayr's (1942) *Systematics and the Origin of Species* versus Simpson's (1944) *Tempo and Mode in Evolution*, in which the paleontologist had to be content addressing only large-scale evolutionary patterns, not the advent of species. It is thus in this historical context that one might understand how, when the systematist of the Synthesis Mayr (1950) turned his sights on human evolution, and declared that paleoanthropologists had been getting it wrong, Washburn (1951) strove to make his discipline of physical anthropology compatible with the Synthesis by embracing the centrality of variation and natural selection in conceiving evolution, as well as by endorsing Mayr's collapsing of the human fossil record into a single, linearly and gradually changing monolith.

That this mindset has for so long dictated how paleoanthropologists approach the human fossil record is both a reality and a mystery. Thus, while the rest of paleontology, and systematics in general, moved into the realm of generating testable theories of relationship based on discriminating between (equally testable, hypothesized) primitive (plesiomorphic) retentions that many individuals/morphs/taxa possess, versus derived (apomorphic) features that, by their more restricted distributions, may hypothetically distinguish morph/monophyletic groups, paleoanthropology remained mired in Mayr's legacy of treating morphological difference, not as potentially revealing of taxic diversity, but as variation, and only variation.

But why would Mayr, a bird taxonomist who dealt with detailed morphology and the identification of diverse taxa, approach the human fossil record so differently? Why lump all human fossils into a single, transforming lineage, with myriad specimens presenting myriad morphologies? The answer appears to lie in the following (paraphrased) remark: Were we to find together the bones of a small San bushman ("pygmy") and of a tall Bantu, they would be allocated to separate species, and we would be wrong.

Although this may seem a reasonable remonstration, it is not. For, as any comparative morphologist would immediately observe, the morphological differences between EH,

Neanderthals, and specimens sometimes attributed to *H. heidelbergensis* are real. Were this not the case, there would be no need to invoke the taxonomically invalid and systematically meaningless referents "archaic" and "anatomically modern" to describe what all paleoanthropologists agree are not the same version of hominid. Further, even if one were to initially hypothesize that the San and Bantus represent different species (which, in reality, would be predicated on size, not morphology—for it is unlikely that anyone would suggest that these humans differ in systematically relevant craniodental and postcranial features), this hypothesis would be tested (and falsified) in the context of a broader comparison between fossil hominids as well as between them, extant apes, and humans.

Yet, again, why might Mayr have abandoned the morphologically based approach of bird taxonomy in his pronouncement about hominids? We might interpret this in a historical context, steeped in the Nazi atrocities of World War II (Schwartz 2006). For, given their marked morphological differences, if EH, Neanderthals, and other hominids belong to the same species, how could anyone pretend to perceive races of living humans as distinct entities? Thus, if the morphological differences between hominid fossils, no matter how large or small, represent variation, the differences between extant humans pale in comparison.

But, to return to the matter of the coexisting scenarios "continuous transformation" and "inheritance of markedly differing morphologies lock-stock-and-barrel," the conundrum is how one can embrace the notion that only small changes over long periods of time produce evolutionary difference. For, in truth, the evidence continually demonstrates that the only small differences that can be documented between specimens are those that reflect the differential expression of a feature that is already present.

For instance, whether large or small, bulky or gracile, the fundamental configurations of the uniformly thick brows, suprainiac fossae, anteriorly tapering and puffy faces, and nasal-cavity medial projections of different specimens of Neanderthal are the same (e.g., Schwartz and Tattersall 1996, Trinkaus 2006). In contrast, the configurations of the brow, occipital region, lower face, and nasal cavity of Neanderthals and such specimens as Kabwe, Arago, Bodo, and Petralona are not similar in detail: that is, the latter have superoinferiorly tall brows with flat surfaces that twist laterally and are defined by a superior margin, (when preserved) steeply undercut occipital planes, large lower face, and (as seen at least in Petralona) a series of tiny bumps where, in virtually all mammals, there is a distinct conchal crest (Schwartz and Tattersall 2002a, Schwartz and Tattersall 2002b). On the other hand, as with specimens of Neanderthal, the *degree* to which these features (especially in the brow and face) are expressed in Kabwe, Arago, Bodo, and Petralona differs from specimen to specimen (Schwartz and Tattersall 2002b).

As illustrated, the problem lies in conflating differences *between* kind (e.g., taxically relevant features, as between Neanderthals and specimens currently deemed *H. heidelbergensis*), with differences *within* kind (differential expression of these features, as among specimens of Neanderthal; Schwartz 2006). Thus, for example, although the brow of

Mladeč 2 is more robust than the similarly configured brow of Mladeč 1, this is not demonstration of continuity between Neanderthals and AMS, because, whether bulky or gracile, a Neanderthal brow is always a Neanderthal brow.

A Developmental Understanding of Morphology

The contradiction between the notions "continual transformation" and "identifiable morphology" is exacerbated, and contradicted, by the unfounded assumption underlying the conception "continual transformation"; that is, the belief that point mutations accumulate over time, such that shapes and species morph one into another. Yet, since no paleoanthropologist would mistake a Neanderthal for an EH, and if the authors of DNA analyses are correct in insisting there are identifiable Neanderthal and EH genes, the scenario cannot be correct. But do current DNA analyses of Neanderthals and EH provide insight into the development of these morphological differences, or even morphology in general? At present, the answer is "no."

The analysis of hominid nDNA and mtDNA sequences—of EH and so far of Neanderthals as well as specimens from Sima de los Huesos, Atapuerca, and Denisova Cave (Meyer et al. 2013, Meyer et al. 2012, Prüfer et al. 2013)—is remarkable for its contradictions; that is, the simultaneous assumptions of (ongoing) change and the absence of change, and the assumption that nuclear (n) DNA and mitochondrial (mt) DNA can be treated equally in phylogenetic analysis (Schwartz 2016). In the broad picture, the chimpanzee sequence, whether n- or mtDNA, is taken as representing in its entirety the primitive substrate from which the lineage leading to a Neanderthal-EH ancestor continuously and accumulatively changed, and, after diverging from this common ancestor, EH and Neanderthal sequences continued to change. However, if one embraces the "molecular assumption"—that molecules constantly change such that the accrual of molecular change means that earlier diverging taxa are more genetically dissimilar than more recently diverging taxa—the chimpanzee sequence cannot represent the static, primitive baseline from which to evaluate hominids. Further, in the context of distinguishing between primitive and derived features, sequence similarity may actually reflect non-change (Schwartz 2016).

Upon scrutinizing sequences in the context of an accepted *Pan*-Neanderthal/EH relationship, SNPs (not genes, or alleles, as these single nucleotide differences are deemed and reiterated in the literature) found in Neanderthals and EH and not in *Pan* are interpreted as changes accrued in the lineage leading to a presumed Neanderthals-EH common ancestor. Single nucleotide polymorphisms (SNPs) found in EH and not in Neanderthals, and vice versa, are then identified as Neanderthal or EH "genes," respectively. But, the validity of this interpretation aside—different nucleotides are not alleles—these analyses do not demonstrate a close relationship between *Pan* and hominids, or between Neanderthals and humans. Rather, these relationships are assumed from the beginning, and similarity and dissimilarity evaluated in the context of these presumed relationships.

With regard to n- versus mtDNA, as molecular entities, they could not be more different. For instance, metazoan nDNA sequence length is in the millions of nucleotides; nDNA presents as paired, linear, and discontinuous chromosomes; replication is concurrent with cell division; and, since most of nDNA is noncoding and involved in development, it does not consist of genes in the sense that the (2–3% of the genome) coding region does, wherein nucleotide sequence is translated directly into gene products (i.e., metabolically active proteins and enzymes). On the other hand, mtDNA sequence length is in the thousands of nucleotides; mtDNA presents as a single, circular chromosome (like bacteria) of which there are multitudes more in a cell than nDNA chromosomes; mtDNA of different sequences can coexist in the same cell; mtDNA does not replicate during cell division; mtDNA can mutate freely (while the noncoding region of nDNA cannot) because it lacks repair mechanisms; and, since mtDNA has an aerobic (i.e., enzymatic) function, it is analogous to the coding region of bacteria and metazoans in which nucleotide sequence is of importance, and in which mutation alters metabolic function, not morphology. Further, significantly, and *contra* received wisdom, mtDNA is not exclusively clonal or maternally inherited. Indeed, there is growing evidence that paternal leakage and recombination occurs in invertebrates and vertebrates (see review in Schwartz [2016]). Consequently, not only should applications of mtDNA to hominid evolution be approached with caution, so, too, should analyses that include n- and mtDNA and interpret these molecules as the same thing.

Having said this, one thing that anyone should be able to (but generally does not) understand, is that these kinds of molecular analyses address and take into consideration neither the development of an organism, nor the emergence of morphology. Indeed, as King and Wilson (1975) recognized decades ago when reviewing comparisons between *Pan* and EH based on blood serum proteins and DNA hybridization, in spite of a proposed close relationship between the two hominoids, they are extremely different morphologically. Thus, as King and Wilson pointed out then, and is no less relevant now, the matter of molecular sequence identity is distinctly different from that of morphological identity. Unfortunately, these days, the molecular similarities between *Pan* and EH have been translated into *Pan* being the *de facto* morphological comparison with hominids, and, like its sequences, its features considered to constitute the primitive conditions from which hominids diverged (Lockwood, Kimbel, and Lynch 2004, Ward and Kimbel 1983, Pilbeam 2000, Wood and Harrison 2011, Lovejoy et al. 2009, Kimbel et al. 2013). In further disregard of King and Wilson's epiphany, and as illustrated in molecular analyses that claim demonstration of Neanderthal and EH "genes," much inference about human evolution continues to promulgate the notion that demonstration of sequence similarity also somehow reflects morphological identity, with the unspoken expectation that demonstration of sequence similarity embodies comparisons of "genes for" specific structures. Clearly, this conflation is without basis.

Historically, the conception of there being "genes for" structure, which is inherent in Mendel's sweet pea experiments, became entrenched as a result of Morgan's work on fruit fly chromosomes, in which genes *for* features, such as thoracic bristle number, eye color, and wing length, were conceived as beads of a necklace (Morgan et al. 1926). Subsequently, the "genes for" idea was considered affirmed with the identification of DNA in bacteria, in which nucleotide triplets (codons) are linked to specific amino acids, with a specific sequence of codons constituting a gene that produces a specific sequence of amino acids that yields a specific protein (Watson and Crick 1953). From this emerged the notion of "DNA as the blueprint of life" and the concept "one gene-one protein." Consequently, since DNA was supposed to be the same in all organisms, any DNA sequence was considered significant and inextricably tied to gene products, which, in metazoans and unlike bacteria, is topological and three-dimensional morphology. Hence, comparing DNA sequences to infer phylogenetic relationships became the goal of molecular systematists (see historical review in Schwartz 2016). However, while the details of bacterial versus metazoan genomes, and their marked differences, are now well understood, the notion of their equivalence persists, especially in the paleo- and molecular anthropological literature. Here are the facts.

In bacteria, virtually 98% of the genome is coding, consisting of genes that produce metabolically active proteins and enzymes that allow them to adapt to changes in their environment; only 2–3% of the genome is noncoding, being represented by control or promoter regions (operons), in which change is reversible (Eisen 2000, Jacob and Monod 1959). In the world of bacterial DNA, a change (point mutation, nucleotide substitution) in a codon can yield a different amino acid that then yields a different protein. Hence, a nucleotide substitution changes the gene that encodes a protein. However, since these proteins are metabolically functional, bacteria do not change in their topology (in fact, they do not express topology). But such a change could allow a bacterial population with it to exist in (adapt to) different environmental conditions. With changes in their surrounding circumstances, some bacteria will bear a mutation that allows them to survive in it, and on it goes: from generation to generation, mutations that affect amino acids and their proteins. Consequently, it would seem, a small mutation can, over time, produce something different. However, in bacteria, only proteins change, not morphology (which they do not express).

In contrast to bacterial genomes, in metazoans only ~2–3% of the genome encodes metabolically active proteins. About 97–98% is noncoding, being composed primarily of transcription factors, developmentally regulated genes, and introns (essentially spacers between exons) that, in a hierarchy of expression, underlie the emergence of topology and three-dimensional organismal form, from the gene-regulatory networks (GRNs) that generate basic body plans, to developmental gene batteries (DGBs), that are involved in the differentiation of tissues and the emergence of the details of topology and structure (Davidson and Erwin 2006, Davidson 2010). Thus, while a point mutation in a metazoan's coding region would merely affect metabolic function, without morphological consequence, changes affecting the noncoding region would undermine proper development. The *RUNX2* signaling pathway serves to illustrate.

Were it not for the signaling pathway that involves the *RUNX2* transcription factor, vertebrates would not develop teeth, cartilage, or bone (D'Allisandro, Tagariello, and Piana 2010, Åberg et al. 2004, Kuhlwilm, Davierwala, and Pääbo 2013). As is noted in humans, and replicated in experimentally manipulated mice, dysregulation of the *RUNX2* signaling pathway leads to cleidocranial dysplasia (CCD) and its myriad abnormalities: for example, persistence of the frontal/metopic suture and anterior fontanelle, a disproportionately swollen frontal, a truncated lower face and mandible, plagiocephaly, incomplete ossification of the auditory tube, delayed tooth development and eruption, amelogenesis imperfecta, shortened or absent clavicles and changes in the thorax (with consequent reoriented of the shoulders), and thickened postcranial bones. In addition to these malformations, CCD is associated with dermal lesions, imperforate peritonea, deafness, mental retardation, and improper insertion of the tongue (leading to difficulty in swallowing and, in humans, speech [ibid.]).

Although Green et al. (2010) asserted that only a slight change in the *RUNX2* transcription factor would be necessary to convert normal Neanderthal morphology into normal EH morphology, this is impossible, because every case of dysregulation of the *RUNX2* signaling pathway, whether in humans or in experimentally manipulated mice, demonstrates the opposite: change does not produce normal morphology. That this is true is revealed in the preserved skull and mandible of the Pech de l'Azé Neanderthal child (41–51 ka): that is, rather than EH morphology, this individual presents features diagnostic of CCD in general (Schwartz 2016). Clearly, the emergence of Neanderthal- versus EH-specific morphology is more complex than merely "tweaking" a gene or transcription factor.

Although this is an example of a pathological syndrome, the striking differences between normal parents and offspring with CCD provide a window on mechanisms that could produce the kind of change one might equate with "evolution." This example also illustrates the fundamental interpretive differences between Darwin and Mivart and colleagues. Namely, while Darwin (1859, 1868) rejected as having evolutionary relevance the myriad documented examples from animal husbandry in which offspring that differed markedly from their parents ("monstrosities" and "sports of nature") served as the basis of a new breed of animal or plant, saltationists saw these cases as presenting the possibility that evolutionarily significant novel morphology could originate abruptly, and persist thereafter in generation after generation of offspring. One instance of disagreement between Darwin and saltationists involved the unexpected appearance of an offspring of normal parents that would become the progenitor of the distinct Niata breed of cattle. As Morgan (1903) later remarked:

In Paraguay, during the last century (1770), a bull was born without horns, although his ancestry was well provided with these appendages, and his progeny was also hornless, although at first he was mated with horned cows. If the horned and the hornless were met in a fossil state, we would certainly wonder at not finding specimens provided with semidegenerate horns, and representing the link between both, and if we were told that the hornless variety may have arisen suddenly, we should not believe it and we should be wrong. (p. 315)

The *RUNX2* example also demonstrates how what is considered evolutionary change—the advent of morphological novelty—does not result from the accrual of small genetic changes (mutations, and specifically point mutations). To the contrary, and since functioning signaling pathways are intricately orchestrated, as they must be in order for generations of offspring to recapitulate being morphologically "normal," the advent of morphological novelty must be the consequence of shifting from one functioning signaling pathway to another (Maresca and Schwartz 2006). Furthermore: (1) evolutionarily significant morphological change cannot result from accumulated point mutations because they are infrequent and random with regard to the affected cells and genomes (Drake et al. 1998), that is mutations are not cumulative relative to the same structure (*contra* Morgan, Dobzhansky, Mayr et al., and subsequent received wisdom); and (2) since the vast majority of an individual's cells are somatic, for a mutation to have evolutionary potential, it must affect sex cells. Thus, the likelihood of any mutation being a major source of evolutionarily relevant morphological change decreases significantly (Maresca and Schwartz 2006).

Adding to the unlikelihood of small mutations accruing to produce significant morphological change are not only the facts that mutation is infrequent, but also that the only constant physical source of potential mutation is UV radiation (Maresca and Schwartz 2006). Since cells have myriad repair mechanisms (e.g., heat shock/stress proteins) that function to eliminate the effects of mutation, the question is: How can the *effective* mutation rate increase? One answer: Derail repair mechanisms, and, therefore, disrupt signaling pathways involving developmentally regulated genes. In the case of heat shock proteins, this would mean stressing organisms beyond their cells' capacity both to adjust the window of stress response, and to generate enough heat shock proteins to "correct" possible sources of "mutation," especially reading errors introduced during replication and translation. Furthermore, as noted above, these perturbations must affect sex cells and, specifically, their noncoding regions. For example, while change in the DNA of a somatic cell may lead to cancer, this change is irrelevant to offspring and thus of no potential evolutionary consequence.

The next question is: How can this potential source of morphological change be enacted in successive generations of offspring? Interestingly, an answer lies in the competing evolutionary theories of Haldane (1932), Wright (1931), and Fisher (1930). For, although differing in the evolutionary significance of homozygosity versus heterozygosity, they all began with the realization first noted by Bateson (1913) that most potentially evolutionarily significant mutations occur in the "recessive" (unexpressed) state—which is a logical conclusion based on the fact that most immediately expressed mutations are deleterious or lethal. Thus, from generation to generation, this ("recessive," unexpressed) potential for change will spread throughout the population until heterozygote saturation is reached, and mating between heterozygotes yields individuals that express the novel feature by dint of being homozygous for the "mutation." The process proceeds, with homozygotes producing offspring like themselves, and with heterozygotes adding to the population of homozygous offspring, until the "mutation" shifts—by a mechanism that is still not understood—from the "recessive"/unexpressed to the "dominant"/expressed state (Schwartz 1999). Of importance is recognizing that, while the spread of the "mutation" from generation to generation will be continuous and, depending on generation time, relatively "gradual," when homozygosity is reached, the novel feature/s it underlies will be expressed "suddenly" and (if not causing death) functionally. Importantly, this novelty will be expressed not in only one, but in many individuals. The latter consequence is profound, in that it addresses a long-standing criticism of saltational models of evolutionary change, from Darwin of Mivart to Dobzhansky, Mayr, and Simpson of Goldschmidt: How can more than one individual be the bearer of a feature that was not the result of gradual change, which, through the accrual of infinitesimally small changes, allows individuals to interbreed as the incipient novelty is passed from one generation to the next until it reaches it fully functional state?

The possibility of the above-proposed "sudden origins" model—which incorporates the spread of a "mutation" and its "sudden" appearance in multiple individuals—is demonstrated by a well-documented case in which normal mice in a colony with known pedigree required for experimental use, unexpectedly gave birth to synpolydactylous offspring. In reconstructing the process that led to this unexpected event, in what was supposed to be a genetically known, pure-bred population, Johnson and colleagues (1998) explained it as a stochastic "mutation" involving the *Hoxd-13* signaling complex that had to have emerged in the "recessive"/unexpressed/undetectable state in one individual, and then spread "silently" through the colony until heterozygote saturation made possible its expression in homozygous offspring (which Johnson et al. determined were homozygous for the mutation).

Consequently, this and the *RUNX2*-CCD examples demonstrate that the advent of marked morphological novelty, perhaps of the kind a systematist would consider evolutionarily significant, is less likely the result of the accumulation of small mutations than the result of altering developmental signaling pathways whose downstream effects are not realized until generations after the event or insult occurred (Maresca and Schwartz 2006, Schwartz 1999).

The importance of the differential recruitment of the same genes and transcription factors in the development of different morphologies—rather than thinking there are "genes for" specific structure—is further exemplified by experiments involving the *Rx* gene and bone-modifying transcription factor 4 (*BMP4*) and the alternative splicing of the *Slo* gene in chickens. In the former case, *Rx* is recruited along with *distal-less* and *orthodentical* in the conversion of a bilaterally symmetrical echinoderm larva into a radially symmetrical adult (Lowe and Wray 1997). In vertebrates, *Rx* participates in eye development (Mathers et al. 1997). Further, in knock-out experiments, mice that are manipulated for the null expression of the *Rx* gene develop neither an eye, nor its associated bony elements (Mathers et al. 1997)—which indicates that these structures developed ("evolved") simultaneously (not, for example, bone developing to protect the eye as it enlarged). A similar phenomenon occurs when *BMP4* expression is manipulated in Darwin's finches: Not only does the morphology of the beak, but also its attendant soft and hard tissue structures, change (Abzhanov et al. 2004, Wu et al. 2004).

The matter of there not being specific genes for specific morphologies is also demonstrated by a study of the chick cochlea, in which ~500 proteins were identified (Rosenblatt et al. 1997). According to the central dogma of one gene-one protein, there must be ~500 genes. But there are not. How, then, can we account for this disparity, which is similar to the demonstration that the human genome does not embody enough genes to account for the number of gene products (Consortium 2001)? As Ast (Ast 2005) has summarized, the answer lies not in a DNA sequence itself, but in the myriad ways it can be alternatively spliced by RNA. In the case of the chick, there is a *Slo* "gene"—a sequence of nucleotides comprising various exons and introns—that is defined not by what it "does" or a morphology it "produces," but by "start" and "stop" TATA sequences that are "read" by RNA as it proceeds to alternatively splice the contained exons and introns in ~500 different ways.

The preceding message—morphological change is not a matter of the accrual of slight genetics changes, and that there are not "genes for" specific structure (Pearson 2006, Stotz 2006)—clearly impacts how one approaches not only the hominid, but any fossil record, wherein analysis of preserved morphology is the primary source for hypothesizing taxa and their evolutionary relationships.

Back to Hominids

Although the preceding discussion should provoke a rethinking of the systematic relevance of comparing nDNA sequences without knowing how they are "used" (recall, mtDNA has nothing to do with morphology), the matter at hand is how this element of an organism's biology impacts the interpretation of morphology, and thus of fossils. Witness the fact that virtually all paleoanthropologists embrace the notion that evolutionary change is slow and continuous. The consequence of this belief, however, is that neither morphology nor taxon can be defined because they are constant states of flux. Yet, the paleoanthropological literature is rife with claims that one can readily identify, in the same individual, distinctly different morphologies: for example, features characteristic of Neanderthals and of EH.

However, truth be told, although these assumptions and perceptions inform the interpretation of fossil hominids, one cannot have it both ways. Either features morph from one into another without distinction, or they remain static and identifiable. Furthermore, the realities of the differential expression of signaling pathways in developmental time and space underscores how incompatible with biology are popular scenarios such as "form follows function" and "adaptation molds morphology."

From a developmental perspective, similarities between an array of specimens can be appreciated as due to fundamental, common, and thus primitively retained, developmental processes: that is, GRNs, such as the *RUNX2* signaling pathway and the development of bone, tooth, and cartilage. Features that distinguish one group or clade from another within this taxic array (apomorphies, such as diphyodonty and heterodonty) would result from signaling pathways higher up in the developmental hierarchy (from GRNs to DGBs). Smaller and smaller clades possess their defining features as a result of more specific signaling pathways at higher levels in the developmental hierarchy (DGBs), and so on, to the distinguishing characters of species (i.e., differences in "kind"). Intraspecific, individual variation (i.e., differences within "kind," such as superoinferiorly taller versus shorter supraorbital margins) is the result of differential expression of transcription factors and other threshold effects involved in the signaling pathways that produce the feature itself. Were it not for these developmental underpinnings, one could not identify features that might unite clades and clades within clades, and that, ultimately, lend themselves to the delineation of species.

Informed by this perspective, let us turn to a specific case in hominid systematics: the allocation to *H. erectus* of specimens that span the gamut from the holotype, the Trinil 2 skull cap, with its long, low lateral profile, long, continuous slope into a non-three-dimensional supraorbital region, and rectangular rear profile; to those from Zhoukoudian, with rounded lateral profiles, an anteroposteriorly short posttoral sulcus, superiorly distended, moderately superorinferiorly tall brows, and sub-round rear profile; to OH 9, with its long and moderately curved lateral profile, long posttoral plane that terminates in a posttoral sulcus behind massive, superoinferiorly tall, and anteriorly protruding brows, and roundedly triangular rear profile (see also discussion in Andrews [1984]).

In light of the preceding discussion, we can appreciate that the different features we observe in different specimens are the result of different developmental histories, from their embryological beginnings to their fully expressed adult states. Indeed, although commonly asserted in the paleoanthropological literature (Bastir et al. 2008, Lieberman, McBratney, and Krovitz 2002), these differences are not the result of the ongoing morphing of one specimen's *Gestalt* into another. Rather, the notion that "this specimen's morphology morphed into another specimen's morphology" is predicated on the belief that one can align adult specimens in a transformation series that is developmentally and evolutionarily accurate. In truth, however, no adult specimen is de facto morphologically intermediate between others. "Intermediacy" can be invoked only after specimens have been arranged in a sequence that reflects a paleoanthropologist's basis for doing so, for example, chronology and/or geography. Curiously, although evolutionary biologists were quick to dismiss Haeckel's "biogenetic law" (ontogeny recapitulates phylogeny) because it was based on lining up adult specimens as evidence of both developmental and evolutionary change (see review in Schwartz 1999), the practice is alive and well in paleoanthropology. Thus, in the case of *H. erectus* among many others, discussions of the species' "evolution" derive primarily from seriating specimens chronologically, explaining differences between far-flung specimens samples as "demic" and having been molded to meet different adaptive needs, and justifying how one specimen could have morphed into another (Antón 2003, Lordkipanidze et al. 2013, Rightmire, Lordkipanidze, and Vekua 2006, Rightmire 1990, Zollikofer et al. 2014). No matter how seductive, these scenarios are not, as they should be, based first on morphology.

Conclusion

The human fossil record is now sufficiently represented to invite—indeed, to warrant—systematic and phylogenetic analyses that are not constrained by received wisdom, which continues to dictate how specimens "fit into" traditional scenarios of human evolution. Yet, beginning in the 1970s with the acceptance of *Australopithecus afarensis*, it is often only specimens of some antiquity have been afforded taxonomic recognition: for example, for genera, *Ardipithecus ramidus, Orrorin tugenensis, Kenyanthropus platyops, Sahelanthropus tchadensis,* and for species, *Australopithecus anamensis,* and *Au. sediba*. Furthermore, while there have been attempts to resurrect *H. heidelbergensis* and *H. neanderthalensis*, and to use *H. ergaster* to receive African "*erectus,*" these species remain housed in a genus that has become a default taxon: that is, a specimen is placed in *Homo* if it is deemed not "australopith" and/or too ancient. The naming of *H. naledi* serves to illustrate. Yet, not only has the genus *Homo* never been properly diagnosed (Schwartz and Tattersall 2015), the arbitrary way in which specimens are referred to it compounds and exacerbates the problem. For, as the array of morphologically disparate specimens assigned to *Homo* increases, so, too, does the impossibility of defining the genus in the only systematically meaningful way: through detailed morphological comparisons and the testing both of named species and of their potential phylogenetic relationship, which, after all, are not facts, but hypotheses. Granted, the hypothesis "genus" cannot be tested as can the hypothesis "species" but, since the former would denote a clade (i.e., minimally three taxa), its use can be applied only after a theory of relatedness of nested sets of species has been generated and its constituent hypotheses tested. Consequently, the current paleoanthropological practice of erecting a new species and relegating it to a genus, on the belief that it does not belong to another, is the wrong way 'round.

Perhaps the relative ease with which newly discovered fossils are allocated to new species reflects a psychological predisposition toward recognizing something new as being something potentially different. Curiously, however, considering the possibility that taxic diversity may also be represented in previously pigeon-holed and especially iconic specimens remains off limits (and as Andrews [1984] commented for *H. erectus*, heretical). Perhaps the reluctance of paleoanthropologists to revisit already-interpreted specimens is because that would mean questioning tradition and received wisdom—which is precisely what tradition and received wisdom prevent—as well as questioning what one was taught, has built a career upon, and will teach the next generation.

In curious contradiction to recapitulating received wisdom is the rapid and widespread embrace of subsuming humans and their fossil relatives, as well as *Pan*, within the taxonomic tribe Hominini. For doing so limits the systematic wiggle room necessary, at least psychologically, to reassess mainstream interpretations of human evolution: for example, "If it's not an australopith, it must be *Homo*." For this reason, I use "hominid" to designate the clade that subsumes humans and their closest fossil relatives. For those who would reject this use of "hominid" because it does not include *Pan*, I must point out that adopting "hominid" for humans and their closest relatives does not violate one's preferred extant human-extant ape relationship. Furthermore, collapsing the entire human fossil record into one hominin subtribe promulgates the conflation of "closest *living* relative" with "closest relative." Indeed, no matter how much (or little) taxic diversity any one of us recognizes in the human fossil record, the fact is that no ape, *Pan* included, is our closest relative.

Historically, the reason that humans were removed from their own family Hominidae and, with *Pan*, relegated to tribe Hominini is primarily due to Goodman's (Goldberg et al. 2003, Goodman et al. 1983, Goodman et al. 1998) insisting that molecular similarity between humans and African apes, and chimpanzees specifically, should be reflected in classification. At one point in time, he and his colleagues even suggested putting chimpanzees and all hominids in the same genus, *Homo* (Goodman et al. 1998). Goodman justified his taxonomic deflation of hominids on the grounds that he was following strict Hennigian cladistics, in which a theory of relationship is translated directly into classification (Hennig 1966). It remains a mystery, however, how Goodman's suggestion took hold, especially because stalwart synthesis Darwinians were opposed not only to Hennig's approach to systematics, but to the direct translation of hypothesized cladistic relationships into a classification (Mayr 1969).

Unfortunately, once bestowed, taxonomic names are difficult to dislodge, especially in human evolutionary studies, and, as mentioned above, most paleoanthropological endeavors begin with the assumption that named specimens actually represent the taxa into which they were shoehorned. Thus, if one wants to study, say, australopith endocast or postcranial morphology, one uses specimens already deemed *Au. africanus* or *Paranthropus robustus*. And so the literature grows, in apparent affirmation of the supposed reality of these taxa and the specimens they have come to subsume. Also unfortunate is the fact that the ascension of paleoanthropology as a discipline coincided with the decline in the teaching of systematics. This is demonstrated in the general paleoanthropological disregard for the type specimen (e.g., Antón et al. [Antón 2003, Antón, Potts, and Aiello 2014] did not consider mandible OH 7 in their attempt to define the genus *Homo* because they thought it too distorted for their purposes), and the widespread use—both tacitly and explicitly—of *sensu lato*. For, once one undertakes a study of, for example, *H. erectus sensu lato* (Antón 2003, Antón, Potts, and Aiello 2014), the holotype becomes lost in a forest of remarkable "variation" such that any specimen can be used to represent the species. Further, lack of familiarity with the constraints of systematic practice and phylogenetic reconstruction has led to assertions by paleoanthropologists that they can actually identify homoplasy or homoplastic features in the absence of a theory of relatedness (Lockwood and Fleagle 1999, Gibbs, Collard, and Wood 2002, Gibbs, Collard, and Wood 2000). In truth, however, homoplasy is a hypothesis that can only be articulated after one has chosen one of alternative theories of relationship, wherein the shared similarities supporting alternatives to one's preferred phylogeny must be homoplastic (Schwartz 2008). And, indeed, in the case of Lockwood and Fleagle, and Wood and colleagues, the identification of homoplasy was predicated not after generating a theory of relationship, but in the context of one that was assumed from the start.

What to do? A simple response would be to abandon taxonomic names, and start from scratch, evaluating every specimen as if it had been newly discovered. From there, hypothesize morphs, sister morphs, and clades of morphs. And then, and only then, turn to the task of assigning names. If one ends up with the same scheme as one currently in favor, so be it. It has been tested. If not? That would be interesting.

Acknowledgments

Thanks to the KLI and Gerd Müller for sponsoring this workshop, the participants for challenging discussion, and especially Eva Lackner and Isabella Sarto-Jackson, without whose efforts this event would not have been as intellectually comfortable as it was. Thanks also to Peter Andrews for insightful and helpful criticisms of this contribution.

Note

1. Paleoanthropologists' rejection of, and disregard for the taxonomic and systematic relevance of type specimens is baffling. True, sometimes—although, in reality, infrequently enough to be a non-issue—a specimen chosen as the holotype of a new taxon is morphologically uninformative, as is the weathered, cracked, and dentally undecipherable mandibular corpus Omo 18-1967-18, holotype of *Paraustralopithecus aethiopicus*. However, just because someone created the name, does not mean that others have to use it and refer other specimens to its hypodigm (unless, if ever, it can be linked to specimens with usable morphology). In the meantime, proceed with rigor, and deal with specimens that are morphologically intact and informative. If a specimen is sufficiently different from others to warrant special attention, perhaps then create a name for it. It only further muddies the

waters, and makes phylogenetically and systematically meaningful comparison impossible, if, as Ánton et al. (2003) and Ánton, Potts and Aiello (2014) advocate, one ignores a type specimen (either because it lacks usable morphology or, as in the case of Ánton and colleagues and the OH 7 mandible, one basically "doesn't like it") and uses a more "acceptable" specimen from an inappropriately constituted hypodigm as the comparative base from which to assign other specimens to that taxon. In further response to paleoanthropologists who *de facto* eschew the use of type specimens, one may ask: If not by comparison with a specimen that was given a species or genus and species name, how does one decide which other specimens might also be so recognized? Should one make up names—but on what basis, or, as Haeckel did when coining "Pithecanthropus," anticipate the discovery of a missing link?—and then wait for specimens to come along that one thinks could be given it? By dismissing the OH 7 mandible as relevant to defining the species *habilis*, and choosing another specimen as the referent name bearer, one is still using a specific specimen as the basis for including other specimens in that taxon. The *International Code of Zoological Nomenclature* exists for a practical reason, from which paleoanthropology is not exempt, unless, of course, one believes that paleoanthropologists are more informed than every other systematist, and that humans should be treated differently than the rest of the biological world.

References

Åberg, T., A. Cavender, J. S. Gaikwad, A. Bronckers, W. Xiuping, J. Waltimo-Sirén, I. Thesleff, and R. N. D'Souza. 2004. "Phenotypic Changes in Dentition of Runx2 Homozygote-Null Mutant Mice." *Journal of Histochemistry & Cytochemistry* 52:131–139.

Abzhanov, A., M. Protas, B. R. Grant, P. R. Grant, and C. J. Tabin. 2004. "*Bmp4* and Morphological Variation in Beaks in Darwin's Finches." *Science* 305:1462–1465.

Alexeev, V. P. 1986. *The Origin of the Human Race*. Moscow: Progress Publishers.

Andrews, P. 1984. "On the Characters that Define *Homo erectus*." *Courier Forschungs Senckenberg* 69:167–175.

Antón, S. C. 2003. "Natural History of *Homo erectus*." *Yearbook of Physical Anthropology* 46:126–170.

Antón, S. C., R. Potts, and L. Aiello. 2014. "Evolution of Early *Homo*: an Integrated Biological Perspective." *Science* 345: 45–58. doi: 10.1126/science.1236828.

Ast, G. 2005. "The Alternative Genome." *Scientific American* 292:59–65.

Bae, C. J. 2010. "The Late Middle Pleistocene Hominin Fossil Record of Eastern Asia: Synthesis and Review." *Yearbook of Physical Anthropology* 53:75–93.

Bastir, M., A. Rosas, D. E. Lieberman, and P. O-Higgins. 2008. "Middle Cranial Fossa Anatomy and the Origin of Modern Humans." *The Anatomical Record* 291:130–140.

Bateson, W., and E. R. Saunders. 1902. *Reports to the Evolution Committee of the Royal Society. Report 1. Experiments Undertaken by W. Bateson, F. R. S., and Miss E. R. Saunders*. London: Harrison & Sons.

Bateson, William. 1913. *Problems of Genetics*. New Haven: Yale University Press.

Black, D. 1927. "On a Lower Molar Hominid Tooth from the Chou Kou Tien Deposit." *Palaeontologia Sinica* 7:1–28.

Blumenbach, J. F. 1969. *On the Natural Varieties of Mankind (De Generis Humani Varietate Nativa)*. Translated by T. Bendyshe. Vol. 1775 and 1795 treatises. New York: Bergman Publishers. Original edition, 1865. Anthropological Society of London.

Bowler, P. 1989. *Evolution: The History of an Idea*. Berkeley: University of California Press.

Broom, R., and J. T. Robinson. 1949. "A New Type of Fossil Man." *Nature* 164:322–323.

Broom, Robert. 1938. "The Pleistocene Anthropoid Apes of South Africa." *Nature* 142:377–379.

Broom, Robert. 1947. "Discovery of a New Skull of the South African Ape-Man, Plesianthropus." *Nature* 159:672.

Clark, W. E. Le Gros. 1955. *The Fossil Evidence for Human Evolution*. Chicago: University of Chicago Press.

Consortium, International Human Genome Sequencing. 2001. "Initial Sequencing and Analysis of the Human Genome." *Nature* 409:860–925.

Curnoe, D. 2010. "A Review of Early *Homo* in Southern Africa Focusing on Cranial, Mandibular and Dental Remains, with the Description of a New Species (*Homo Gautengensis* sp. nov.)." *HOMO - Journal of Comparative Human Biology* 61:151–177.

D'Allisandro, G., T. Tagariello, and G. Piana. 2010. "Cleidocranial Dysplasia: Etiology and Stomotognathic and Craniofacial Abnormalities." *Minvera Stomatologica* 59:117–127.

Dart, Ray. 1925. "*Australopithecus africanus*: The Man-Ape of South Africa." *Nature* 115:195–199.

Darwin, C. 1869. *On the Origin of Species by Means of Natural Selection, or the Preservation of Favoured Races in the Struggle for Life*. 5th ed. London: John Murray.

Darwin, C. 1872. *On the Origin of Species by Means of Natural Selection, or the Preservation of Favored Races in the Struggle for Life*. 6th ed. London: John Murray.

Darwin, Charles. 1859. *On the Origin of Species by Means of Natural Selection, or the Preservation of Favored Races in the Struggle for Life*. London: John Murray. Original edition. Reprint, 1964.

Darwin, Charles. 1868. *The Variation of Animals and Plants under Domestication*. London: John Murray.

Davidson, E. H. 2010. "Emerging Properties of Animal Gene Regulatory Networks." *Nature* 468:911–920.

Davidson, E. H., and D. H. Erwin. 2006. "Gene Regulatory Networks and the Evolution of Animal Body Plans." *Science* 311:796–800.

Dobzhansky, T. H. 1941. *Genetics and the Origin of Species*. 2nd ed. New York: Columbia University Press.

Dobzhansky, Theodosius. 1937. *Genetics and the Origin of Species*. New York: Columbia University Press.

Domínguez-Rodrigo, M., T. R. Pickering, S. Almécija, J. L. Heaton, E. Baquedano, A. Mabulla, and D. Uribelarrea. 2015. "Earliest Modern Human-like Hand Bone from a New >1.84-Million-Year-Old Site at Olduvai in Tanzania." *Nature Communications* 6: 7987. doi: 10.1038/ncomms898.

Drake, J. W., B. Charlesworth, D. Charlesworth, and J. F. Crow. 1998. "Rates of Spontaneous Mutation." *Genetics* 148:1667–1686.

Dreyer, T. 1935. "A Human Skull from Florisbad." *Proceedings of the Academy of Science Amsterdam* 38:119–128.

Dubois, Eugene. 1894. *Pithecanthropus erectus, eine Mesnchenähnliche Übergangsform aus Java*. Batavia: Landesdruckerei.

Eisen, J. A. 2000. "Assessing Evolutionary Relationships among Microbes from Whole-Genome Analysis." *Current Opinion in Microbiology* 3:475–480.

Fisher, R. A. 1930. *The Genetical Theory of Natural Selection*. Oxford: Oxford University Press.

Frayer, D. W., J. Jelínek, M. Oliva, and M. H. Wolpoff. 2006. "Aurignacian Male Crania, Jaws and Teeth from the Mladeč Caves, Moravia, Czech Republic." In *Early Modern Humans at the Moravian Gate*, edited by M. Teschler-Nicola, 185–272. Vienna: Springer.

Garman, S. 1884. "On the Use of Polynomials as Names in Zoology." *Proceedings of the Boston Society of Natural History* 1884.

Gazin, C. L. 1958. "A Review of the Middle and Upper Eocene Primates of North America." *Smithsonian Miscellaneous Collections* 144:1–112.

Gibbs, S., M. Collard, and B. Wood. 2000. "Soft-Tissue Characters in Higher Primate Phylogenetics." *Proceedings of the National Academy of Science* 97:11130–11132.

Gibbs, S., M. Collard, and B. A. Wood. 2002. "Soft-Tissue Anatomy of the Extant Hominoids: a Review and Phylogenetic Analysis." *Journal of Anatomy* 200:3–49.

Goldberg, A., D. E. Wildman, T. R. Schmidt, M. Hüttemann, M. Goodman, M. L. Weiss, and L. I. Grossman. 2003. "Sister Grouping of Chimpanzees and Humans as Revealed by Genome-Wide Phylogenetic Analysis of Brain Gene Expression Profiles." *Proceedings of the National Academy of Science* 100:5873–5878.

Goldschmidt, R. B. 1940. *The Material Basis of Evolution*. New Haven: Yale University Press. Original edition. Reprint, 1982.

Goodman, M., G. Braunitzer, A. Stangl, and B. Schrank. 1983. "Evidence on Human Origins from Haemoglobins of African Apes." *Nature* 303:546–548.

Goodman, M., C. A. Porter, J. Czelusniak, S. L. Page, H. Schneider, J. Shoshani, G. Gunnell, and C. P. Groves. 1998. "Toward a Phylogenetic Classification of Primates Based on DNA Evidence Complemented by Fossil Evidence." *Molecular Phylogenetics and Evolution* 9:585–598.

Green, R. E., J. Krause, A. W. Briggs, T. Maricic, U. Stenzel, M. Kircher, N. Patterson, L. Heng, Z. Weiwei, M. H.-Y. Fritz, N. F. Hansen, E. Y. Durand, A.-S. Malaspinas, J. D. Jensen, T. Marques-Bonet, C. Alkan, K. Prüfer, M. C. Meyer, H. A. Burbano, J. M. Good, R. Schultz, A. Aximu-Petri, A. Butthof, B. Höber, B. Höffner, M. Siegemund, A. Weihmann, C. Nusbaum, E. S. Lander, C. Russ, N. Novod, J. Affourtit, M. Egholm, C. Verna, P. Rudan, D. Brajkovic, Z. Kucan, I. Gušic, V. B. Doronichev, L. V. Golovanova, C. Lalueza-Fox, M. de la Rasilla, J. Fortea, A. Rosas, R. W. Schmitz, P. L. F. Johnson, E. E. Eichler, D. Falush, E. Birney, J. C. Mullikin, M. Slatkin, R. Nielsen, J. Kelso, M. Lachmann, D. Reich, and S. Pääbo. 2010. "A Draft Sequence of the Neandertal Genome." *Science* 238:710–722.

Groves, C., and V. Mazák. 1975. "An Approach to the Taxonomy of the Hominidae: Gracile Villafranchian Hominids of Africa." *Casopis pro mineralogii a geologii* 20 (225–246).

Haldane, J. B. S. 1932. *The Causes of Evolution*. New York: Harper & Brothers.

Harvati, K., C. Stringer, R. Grün, M. Aubert, P. Allsworth-Jones, and C. A. Folorunso. 2011. "The Later Stone Age Calvria from Iwo Eleru, Nigeria: Morphology and Chronology." *PLoS ONE* 6/e24024:1–8.

Hennig, W. 1966. *Phylogenetic Systematics*. Chicago: University of Chicago Press.

Hrdlicka, A. 1930. "The Skeletal Remains of Early Man." *Smithsonian Miscellaneous Collections* 83:1–379.

Huxley, J. 1940a. "Introductory: Towards the New Systematics." In *The New Systematics*, edited by J. Huxley, 1–46. Oxford: Clarendon Press.

Huxley, J., ed. 1940b. *The New Systematics*. Oxford: Clarendon Press.

Huxley, Thomas Henry. 1863. "On Some Fossil Remains of Man." In *Man's Place in Nature*. New York: D. Appleton.

Jacob, F., and J. Monod. 1959. "Genes of Structure and Genes of Regulation in the Biosynthesis of Proteins." *Comptes Rendus Hebdomadaires des Seances de l'Academie des Sciences* 249:1282–1284.

Johnson, K. R., O. S. Hope, L. R. Donahue, P. Ward-Bailey, R. T. Bronson, and M. T. Davisson. 1998. "A New Spontaneous Mouse Mutation of *Hoxd13* with a Polyalanine Expansion and Phenotype Similar to Human Synpolydactyly." *Human Molecular Genetics* 7:1033–1038.

Kimbel, W., G. Suwa, B. Asfaw, and T. D. White. 2013. *"Aridipithecus ramidus* and the Evolution of the Human Cranial Base." *American Journal of Physical Anthropology* 150(S56):166–167.

King, M.-C., and A. C. Wilson. 1975. "Evolution at Two Levels in Humans and Chimpanzees." *Science* 188:107–116.

King, W. 1864. "The Reputed Fossil Man of the Neanderthal." *Quarterly Journal of Science* 1:88–97.

Kuhlwilm, M., A. Davierwala, and S. Pääbo. 2013. "Identification of Putative Target Genes of the Transcription Factor *RUNX2." PLoS ONE* 8. doi: 10.1371/journal.pone.0083218.

Leakey, L. S. B., P. V. T. Tobias, and J. T. Napier. 1964. "A New Species of Genus *Homo* from Olduvai Gorge." *Nature* 202:7–9.

Leakey, M. D., F. Spoor, F. H. Brown, P. N. Gathogo, C. Kiarie, L. N. Leakey, and I. McDougall. 2001. "New Hominin Genus from Eastern Africa Shows Diverse Middle Pliocene Lineages." *Nature* 410:433–443.

Lieberman, D. E., B. M. McBratney, and G. Krovitz. 2002. "The Evolution and Development of Cranial Form in *Homo sapiens." Proceedings of the National Academy of Sciences (USA)* 99:1134–1139.

Linnaeus, C. 1735. *Systema Naturae per Regna Tria Naturae, Secundum Classes, Ordines, Genera, Species cum Characteribus, Differentiis, Synonymis, Locis.* Stockholm: Laurentii Salvii.

Lockwood, C. A., and J. Fleagle. 1999. "The Recognition and Evaluation of Homoplasy in Primate and Human Evolution." *Yearbook of Physical Anthropology* 42:189–232.

Lockwood, C. A., W. H. Kimbel, and J. M. Lynch. 2004. "Morphometrics and Hominoid Phylogeny: Support for a Chimpanzee-Human Clade and Differentiation among Great Ape Subspecies." *Proceedings of the National Academy of Science* 101:4356–4360.

Lordkipanidze, D., M. Ponce de León, A. Margvelashvili, Y. Rak, G. P. Rightmire, A. Vekua, and C. Zollikofer. 2013. "A Complete Skull from Dmanisi, Georgia, and the Evolutionary Biology of Early *Homo." Science* 342:326–331. doi: 10.1126/science.1238484.

Lovejoy, A. O., G. Suwa, G. G. Simpson, J. H. Matternes, and T. D. White. 2009. "The Great Divide: *Ardipithecus ramidus* Reveals the Postcrania of our Last Common Ancestors with African Apes." *Science* 326. doi: 10.1126/science.1175833.

Lowe, C. J., and G. A. Wray. 1997. "Radical Alterations in the Roles of Homeobox Genes during Echinoderm Evolution." *Nature* 389:718–721.

Maresca, B., and J. H. Schwartz. 2006. "Sudden Origins: a General Mechanism of Evolution Based on Stress Protein Concentration and Rapid Environmental Change." *The Anatomical Record (Part B: The New Anatomist)* 289:38–46.

Mathers, P., A. Grinberg, K. Mahon, and M. Jamrich. 1997. "The Rx Homeobox Gene Is Essential for Vertebrate Eye Development." *Nature* 387:604–607.

Mayr, E. 1963. "The Taxonomic Evaluation of Fossil Hominids." In *Classification and Human Evolution*, edited by S. L. Washburn, 332–346. Chicago: Aldine.

Mayr, E. 1969. *Principles of Systematic Zoology.* New York: McGraw-Hill.

Mayr, Ernst. 1942. *Systematics and the Origin of Species.* New York: Columbia University Press.

Mayr, Ernst. 1950. "Taxonomic Categories in Fossil Hominids." *Cold Spring Harbor Symposium on Quantitative Biology* 15:109–118.

McCown, T. D., and A. Keith. 1939. *The Stone Age of Mount Carmel: The Fossil Remains from the Levalloiso-Mousterian.* Oxford: Clarendon Press.

Meyer, M., Q. Fu, A. Aximu-Petri, I. Glocke, B. Nickel, J.-L. Arsuaga, I. Martinez, A. Gracia, J. M. Bermúdez de Castro, E. Carbonell, and S. Pääbo. 2013. "A Mitochondrial Genome Sequence of a Hominin from Sima de los Huesos." *Nature* 505:403–406.

Meyer, M., M. Kircher, M. T. Gansauge, H. Li, F. Racimo, S. Mallick, J. G. Schraiber, F. Jay, K. Prüfer, C. de Filippo, P. H. Sudmant, C. Alkan, Q. Fu, R. Do, N. Rohland, A. Tandon, M. Siebauer, R. E. Green, K. Bryc, A. W. Briggs, U. Stenzel, J. Dabney, J. Shendure, J. O. Kitzman, M. F. Hammer, M. V. Shunkov, A. P. Derevianko, B. Patterson, C. Andrés, E. E. Eichler, M. Slatkin, D. Reich, J. Kelso, and S. Pääbo. 2012. "A High-Coverage Genome Sequence from an Archaic Denisovan Individual." *Science* 338:222–226.

Mivart, St. G. 1871. *On the Genesis of Species*. London: John Murray.

Morgan, T. H. 1903. *Evolution and Adaptation*. New York: MacMillan.

Morgan, T. H. 1910. "Chance or Purpose in the Origin and Evolution of Adaptation." *Science* 31:201–210.

Morgan, T. H. 1916. *A Critique of the Theory of Evolution*. Princeton: Princeton University Press.

Morgan, T. H. 1922. "On the Mechanism of Heredity." *Proceedings of the Royal Society (Biology)* 94:162–197.

Morgan, T. H. 1932. "The Rise of Genetics." *Science* 76:261–288.

Morgan, T. H., A. H. Sturtevant, H. J. Muller, and C. B. Bridges. 1926. *The Mechanism of Mendelian Heredity*. New York: Henry Holt. Original edition, 1915.

Oppenoorth, W. 1932. "*Homo (Javanthropus) soloensis*. Een Plistocene Mensch van Java." *Wetenschappelijke Mededeelingen Dienst Mijnbouw Nederlandsch-Indië* 20:49–75.

Pearson, H. 2006. "What Is a Gene?" *Nature* 441:399–401.

Pilbeam, D. 1972. *The Ascent of Man: an Introduction to Human Evolution*. New York: Macmillan.

Pilbeam, D. 2000. "Hominoid Systematics: the Soft Evidence." *Proceedings of the National Academy of Science USA* 97:10684–10686.

Potts, R., A. K. Behrensmeyer, A. Deino, P. Ditchfield, and J. D. Clark. 2004. "Small Mid-Pleistocene Hominin Associated with East African Acheulean Technology." *Science* 305:75–78.

Prüfer, K., F. Racimo, N. Patterson, F. Jay, S. Sankararaman, S. Sawyer, A. Heinze, G. Renaud, P. H. Sudmant, C. de Filippo, L. Heng, S. Mallick, M. Dannemann, Q. Fu, M. Kircher, M. Kuhlwilm, M. Lachmann, M. Meyer, M. Ongyerth, M. Siebauer, C. Theunert, A. Tandon, P. Moorjani, J. Pickrell, J. C. Mullikin, S. H. Vohr, R. E. Green, I. Hellmann, P. L. F. Johnson, H. Blanche, H. Cann, J. O. Kitzman, J. Shendure, E. E. Eichler, E. S. Lein, T. E. Bakken, L. V. Golovanova, V. B. Doronichev, M. V. Shunkov, A. P. Derevianko, B. Viola, M. Slatkin, D. Reich, J. Kelso, and S. Pääbo. 2013. "A Complete Genome Sequence of a Neanderthal from the Altai Mountains." *Nature* 505: 43–49. doi: doi:10.1038/nature12886.

Pycraft, W. P., E. G. Smith, M. Yearsley, R. A. Smith, and A. T. Hopwood. 1928. *Rhodesian Man and Associated Remains*. Edited by F. A. Bather. London: British Museum (Natural History).

Rightmire, G. P., D. Lordkipanidze, and A. Vekua. 2006. "Anatomical Descriptions, Comparative Studies and Evolutionary Significance of the Hominin Skulls from Dmanisi, Republic of Georgia." *Journal of Human Evolution* 50:115–141.

Rightmire, P. 1990. *The Evolution of Homo erectus*. Cambridge: Cambridge University Press.

Robinson, J. T. 1953. "*Telanthropus* and Its Phylogenetic Significance." *American Journal of Physical Anthropology* 11:445–501.

Robinson, J. T. 1954. "The Genera and Species of the Australopithecinae." *American Journal of Physical Anthropology* 12:181–200.

Rosenblatt, K. P., Z. P. Sun, S. Heller, and A. J. Hudspeth. 1997. "Distribution of Ca^{2+} -activated K$^+$ -channel Isoforms along the Tonotopic Gradient of the Chicken's Cochlea." *Neuron* 19:1061–1075.

Russell, D. E., P. Louis, and D. E. Savage. 1967. "Primates of the French Early Eocene." *University of California Publications in Geological Science* 73:1–46.

Schindewolf, O. 1936. *Palaeontologie, Entwicklungslehre und Genetik*. Berlin: Bontrüager.

Schoetensack, O. 1908. *Der Unterkiefer den Homo heidelbergensis aus den Sanden von Mauer bei Heidelberg*. Leipzig: Wilhelm Engelmann.

Schwartz, J. H. 1999. *Sudden Origins: Fossils, Genes, and the Emergence of Species*. New York: John Wiley & Sons.

Schwartz, J. H. 2004. "Getting to Know *Homo Erectus*." *Science* 305:53–54.

Schwartz, J. H. 2006. " 'Race' and the Odd History of Paleoanthropology." *The Anatomical Record (Part B: The New Anatomist)* 289B:225–240.

Schwartz, J. H. 2008. "Cladistics." In *Icons of Evolution*, edited by B. Regal, 517–544. Westport, CT: Greenwood Press.

Schwartz, J. H. 2016. "Systematics and Evolution." *Reviews in Cell Biology and Molecular Medicine* 2: 1–43.

Schwartz, J. H., and I. Tattersall. 2000. "What Constitutes *Homo erectus*?" *Acta Anthropologica Sinica* Supplement to vol. 19:21–25.

Schwartz, J. H., and I. Tattersall. 2002a. "Bodo and the Concept of *Homo heidelbergensis*." In *25 Years of Bodo, Proceedings of the 4th Phillip V. Tobias Lecture on Human Evolution*, edited by H. Seidler and K. Begashaw, 107–127. Addis Ababa: National Museum.

Schwartz, J. H., and I. Tattersall. 2002b. *The Human Fossil Record, Volume One, Terminology and Craniodental Morphology of Genus Homo (Europe)*. Edited by J. H. Schwartz and I. Tattersall, *The Human Fossil Record*. New York: Wiley-Liss.

Schwartz, J. H., and I. Tattersall. 2003. *The Human Fossil Record, Volume Two, Craniodental Morphology of Genus Homo (Africa and Asia)*. Edited by J. H. Schwartz and I. Tattersall, *The Human Fossil Record*. New York: Wiley-Liss.

Schwartz, J. H., and I. Tattersall. 2015. "Defining the Genus *Homo*." *Science* 349:931–932.

Schwartz, J. H., I. Tattersall, and C. Zhang. 2014. "Comment on 'A Complete Skull from Dmanisi, Georgia, and the Evolutionary Biology of Early *Homo*.'" *Science* 344:360–361.

Schwartz, Jeffrey H., and Ian Tattersall. 1996. "Toward Distinguishing *Homo neanderthalensis* from *Homo sapiens*, and Vice Versa." *Anthropologie* 43:129–138.

Simpson, G. G. 1944. *Tempo and Mode in Evolution*. New York: Columbia University Press.

Simpson, G. G. 1945. "The Principles of Classification and a Classification of the Mammals." *Bulletin of the American Museum of Natural History* 85:1–350.

Simpson, G. G. 1952. "Review of O. Schindewolf, *Grundfragen der Paleontologie* and *Der Zeitfaktor in Geologie und Palaontologie*." *The Quarterly Review of Biology* 27:388–389.

Simpson, G. G. 1961. *Principles of Animal Taxonomy*. New York: Columbia University Press.

Spoor, F. 2013. "Small-Brained and Big-Mouthed." *Nature* 502:452–453.

Spoor, F., P. Gunz, S. Neubauer, Stelzer. S., N. Scott, A. Kwekason, and M. C. Dean. 2015. "Reconstructed *Homo habilis* Type OH 7 Suggest Deep-Rooted Species Diversity in Early *Homo*." *Nature* 519:83–86.

Stotz, K. 2006. "With 'Genes' Like That, Who Needs an Environment?: Postgenomics's Argument for the 'Ontogeny of Information.'" *Philosophy of Science* 73:905–917.

Strait, D. S., and F. E. Grine. 2004. "Inferring Hominoid and Early Hominid Phylogeny using Craniodental Characters: the Role of Fossil Taxa." *Journal of Human Evolution* 47:399–452.

Szalay, F. S. 1976. "Systematics of the Omomyidae (Tarsiiformes, Primates): Taxonomy, Phylogeny, and Adaptations." *Bulletin of the American Museum of Natural History* 156:157–450.

Tattersall, I., and Jeffrey H. Schwartz. 2000. *Extinct Humans*. Boulder, CO: Westview Press.

Trinkaus, E. 2006. "Modern Human Versus Neandertal Evolutionary Distinctiveness (with CA Comment)." *Current Anthropology* 47:597–620.

Villmoare, B, W. H. Kimbel, C. Seyoum, J. Campisano, E. DiMaggio, J. Rowan, D. R. Braun, J. R. Arrowsmith, and K. E. Reed. 2015. "Early *Homo* at 2.8 ma from Ledi-Geraru, Afar, Ethiopia." *Science* (347): 1352–1355.

Waddington, C. H. 1940. *Organisers and Genes*. Cambridge: Cambridge University Press.

Walker, A. C., and R. F. Leakey, eds. 1993. *The Nariokotome Homo erectus Skeleton*. Cambridge, MA: Harvard University Press.

Ward, S., and W. Kimbel. 1983. "Subnasal Alveolar Morphology and the Systematic Position of *Sivapithecus*." *American Journal of Physical Anthropology* 61:157–72.

Washburn, S. L. 1951. "The New Physical Anthropology." *Transactions of the New York Academy of Sciences* 13:298–304.

Watson, J. D., and F. H. C. Crick. 1953. "A Structure for Deoxyribose Nucleic Acid." *Nature* 171:737–738.

Weidenreich, F. 1937. "The Relationship of *Sinanthropus pekinensis* to *Pithecanthropus, Javanthropus* and Rhodesian Man." *Journal of the Royal Anthropological Institute* 67:51–65.

Weidenreich, F. 1943. "The Skull of *Sinanthropus pekinensis*: a Comparative Study of a Primitive Hominid Skull." *Palaeontologica Sinica* 10k:1–485.

Weidenreich, F. 1945. "Giant Early Man from Java and South China." *Anthropological Papers of the American Museum of Natural History* 40:1–134.

Weidenreich, F. 1946. *Apes, Giants, and Man*. Chicago: University of Chicago Press.

Wolpoff, M. 1989. "The Place of the Neandertals in Human Evolution." In *The Emergence of Modern Humans*, edited by E. Trinkaus, 97–141. Cambridge: Cambridge University Press.

Wolpoff, M., J. Hawks, and R. Caspari. 2000. "Multiregional, Not Multiple Origins." *American Journal of Physical Anthropology* 112:129–136.

Wolpoff, M., A. Thorne, A. J. Jelinek, and Y.-Q. Zhang. 1994. "The Case for Sinking *Homo erectus*: 100 Years of *Pithecanthropus* Is Enough!" *Courier Forschungs Senckenberg* 171:341–361.

Wolpoff, M. H., D. W. Frayer, and J. Jelínek. 2006. "Aurignacian Female Crania and Teeth from the Mladeč Caves, Moravia, Czech Republic." In *Early Modern Humans at the Moravian Gate*, edited by M. Teschler-Nicola, 273–340. Vienna: Springer.

Wolpoff, M. H., J. Hawks, D. W. Frayer, and K. Huntley. 2001. "Modern Human Ancestry at the Peripheries: a Test of the Replacement Theory." *Science* 291:293–297.

Wood, B. 1992. "Origin and Evolution of the Genus *Homo*." *Nature* 355:783–790.

Wood, B., and T. Harrison. 2011. "The Evolutionary Context of the First Hominins." *Nature* 470:347–352.

Wood, B. A., and M. Collard. 1999a. "The Changing Face of the Genus *Homo*." *Evolutionary Anthropology* 8:195–207.

Wood, B. A., and M. Collard. 1999b. "The Human Genus." *Science* 284:65–71.

Wright, S. 1929. "Evolution in a Mendelian Population." *Anatomical Record* 44:287.

Wright, S. 1931. "Evolution in Mendelian Populations." *Genetics* 16:97–159.

Wu, P., T.-X. Jiang, S. Suksaweang, R. B. Widelitz, and C. M. Chuong. 2004. "Molecular Shaping of the Beak." *Science* 305:1465–1466.

Zollikofer, C., M. Ponce de León, A. Margvelashvili, G. P. Rightmire, and D. Lordkipanidze. 2014. "Response to Comment on 'A Complete Skull from Dmanisi, Georgia, and the Evolutionary Biology of Early *Homo.*'" *Science* 344: 360. doi: 10.1126/science.1250081.

5 To Tree or Not to Tree *Homo sapiens*

**Rob DeSalle, Apurva Narechania, Martine Zilversmit,
Jeff Rosenfeld, and Michael Tessler**

Introduction

Understanding the origins of modern humans has recently become an exciting and innovating field in a new way due to the availability of whole-genome sequences for many living humans, as well as recently extinct ancestors and relatives. Tree-building methods to examine our history have been used to look at both our biological and linguistic evolution. The scholarly website "The Genealogical World of Phylogenetic Networks" (http://phylonetworks.blogspot.com/) goes into fascinating detail about the present and historical use of these methods. Here, David Morrison and colleagues point out that trees have been used since the early 1800s to illustrate relationships of humans to each other.

Keith (1915) constructed a tree of human races followed by Sparks (1930), Hooton (1946), Weidenreich (1946), and several others who addressed the same questions. Sparks shows clear branching of major "lineages" of *H. sapiens* in his colorful Histomap drawn in 1930. On the other hand, Weidenreich (1946) and Hooton (1946) depict the relationships humans as reticulating webs or trellises. Keith (1915) refrained from showing reticulation among his human racial terminals and left the relationships of races as an unresolved polychotomy.

The differences in the trees of these early genealogists are to a certain extent repeated in modern attempts to decipher human racial grouping and relationships of those groups. To represent *H. sapiens* population relationships, researchers still use bifurcating and fully resolved trees (Bodmer 2015; Krause et al. 2010; Nei and Roychoudhry 1993; Nievergelt et al. 2007), reticulating diagrams (Campbell and Tischkoff 2010; Zerjal et al. 2003), or a combination of both (Pickrell and Pritchard 2012; Pickrell et al. 2012). Rieppel (2010, p. 475) summarizes this problem succinctly in the following quote, "The history of biological systematics documents a continuing tension between classifications in terms of nested hierarchies congruent with branching diagrams (the 'Tree of Life') versus reticulated relations."

While trees are clearly an important tool for reconstructing the history of life on our planet, they are often misused and misinterpreted (see boxes for a primer on how phylogenetic trees are constructed). Until recently the most commonly used markers for comparing

Box 5.1

Molecular data

Unlike binary discrete morphological data (i.e., presence or absence of morphological traits) DNA sequence data have four character states corresponding to the four bases in DNA—guanine (G), adenine (A), cytosine (C), and thymidine (T). In some phylogenetic studies using DNA sequence data, a fifth character state is introduced to accommodate gaps in sequences. The first step in any analysis that infers relationships using DNA sequence information is to establish topological homology of the characters involved. For anatomical characters this is usually done on the basis of topological position in the anatomy of organisms or based on some ontogenetic or developmental criteria. For DNA sequences this is done by sequence alignment or in the case of human populations and closely related species based on ascertainment. Similar to morphological data, the endgame of sequencing is to obtain a matrix of character state data. In addition to DNA sequence data the translated protein sequences are often times used. This approach increases the number of character states to 20 because of the existence of 20 amino acids in the genetic code. For DNA any change from one individual to another is termed a single nucleotide polymorphism or SNP.

Box 5.2

Phylogenetic trees

Phylogenetic trees are hierarchical branching diagrams that represent the branching patterns of the divergence of the terminals used in a study. The terminals are simply the individuals, the species, the genera, or whatever taxonomic level one uses in a phylogenetic analysis. The figure in this box shows a generic tree with some of common terminology that is used when discussing trees. Trees can be rooted or unrooted implying that direction of evolution can be inferred if one assumes the root to be ancestral. Unrooted trees are more like networks than trees, and such networks lack the ability to infer ancestralness or derivedness of the organisms in the tree. Rooting of phylogenetic trees can be done a priori by assuming that the root should occur at the outgroup taxon/taxa. Outgroup taxa can be inferred based on their exclusion from the ingroup under analysis. So for instance, if we are generating a phylogenetic tree for taxa in the genus *Homo*, a potential outgroup would be the genus *Pan* or even the genus *Gorilla*. One should not impose an outgroup that is too distant from the ingroup taxa, as the rooting will occur randomly on the ingroup branches (Wheeler 1992). The goal of a phylogenetic analysis is to take the available data and derive a branching diagram that best represents the history of the organisms involved and at the same time best represents the evolution of the data on the tree. Hence optimality criteria are used to pick from the possible trees the best solutions to the problem. There are in general three optimality criteria used to assess the appropriateness of a particular branching order—minimum evolution (distance), maximum parsimony, and maximum likelihood. Minimum evolution attempts to optimize the least evolutionary distance or change on the branches of the potential tree. Maximum parsimony uses the optimality criterion of minimizing the number of steps on an evolutionary tree. Maximum likelihood uses the optimality criterion of maximizing the likelihood of character transitions on the tree.

Box 5.3

Robustness of inference

The incorporation of likelihood estimates and Bayesian approaches into phylogenetics in the last two decades lends a statistical framework for tests of robustness of tree inference to phylogenetic analysis. In the Bayesian approach each node is assigned a posterior probability that indicates the robustness of the node. Other methods of assessing robustness can be used in phylogenetic analysis and these include bootstrap and jackknife approaches that allow metrics to be applied to nodes in a phylogenetic tree. Bootstraps and jackknives vary from 0 to 100% and indicate the robustness (more toward 100%) or lack of such (usually under 50%). The Bayesian posteriors, bootstraps, or jackknifes all are placed onto trees to indicate the robustness of a node.

lineages of *H. sapiens* came from two clonally evolving sources of DNA—mitochondrial DNA (mtDNA) and Y chromosomal DNA. However, with the current technologies, genome sequencing can be accomplished on single individuals, adding a second general category of genetic markers—recombining nuclear genomic markers. This gives us two dimensions in what we will call the "analysis space of human genetics." In addition, because researchers can obtain genome-level information from subfossil remains, a third dimension of time needs to be added to the analysis space in human history studies (figure 5.1).

In this paper, two critical claims made about genealogy in modern human genomic studies are examined to explore the nuances of data treatment at the level of populations within our species that we see in figure 5.1. The first is that recent and, in some cases, living human populations from different geographic regions can be arranged into discrete biological units (Wade 2014). The second claim is that these geographical entities can be arranged in a hierarchical relationship (Bodmer 2015; Krause et al. 2010; Nei and Roychoudhry 1993; Nievergelt et al. 2007).

Discreteness and Hierarchy

The examination of the discreteness and potential hierarchy of human populations has a long history (Tattersall and DeSalle 2012; Yudell 2014). But perhaps the most notorious depiction of the state of the art is Wade's (2014) recent and highly visible claim that biological *H. sapiens* races are real. Wade's (2014) argument for discrete races is not based on trees, but it very well could be (DeSalle and Tattersall 2014). Instead, Wade (2014) relies on the clustering approaches in several papers to make the claim that biological races exist; specifically he cites a clustering algorithm called STRUCTURE (Falush et al. 2003; Falush et al. 2007; Hubisz et al. 2009; Pritchard et al. 2000) as applied to the data sets of Rosenberg et al. (2005) and Li et al. (2008). The Rosenberg et al. (2005) dataset

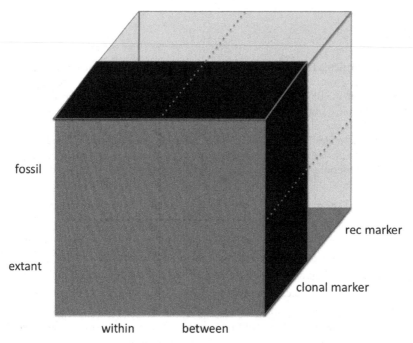

Figure 5.1
The analysis space of human genetics. There are three axes in the space and hence discrete octants. Most, if not all, human genetics studies go into a single point within the space. The Y axis represents whether a study includes a fossil sequence or not. The X axis represents whether the focus of the study is "within" a population or between discrete entities. The Z axis represents whether the analysis uses clonal markers (Y chromosome or mtDNA). The darkened part of the space is where phylogenetic trees are ontologically unsound.

consisted of 783 microsatellites and 210 insertion/deletion markers, and the Li et al. (2008) study used 650,000 common single nucleotide polymorphisms (SNPs).

STRUCTURE is based on a model in which the number of input populations (K) can vary under MCMC simulation with Bayesian statistical analysis. The model attempts to take advantage of the fact that the major population genetic effect of admixture is to create long-range correlation among linked markers. Such correlation, known as linkage, is seen in continuous sections of chromosomal DNA that show correlated patterns of ancestry. STRUC-TURE attempts to capitalize on this feature within a fairly simple model and to determine probabilities on ancestry patterns for each individual in a data set. The end product has two interesting applications. First, the optimal K (i.e., the number of populations the heuristic determines is the number of discrete populations) can be determined for specific data sets and second the probability that each individual in the data set belongs to each of the K populations is given. However, we point out that STRUCTURE also assumes random mating and

Hardy Weinberg equilibrium. Such assumptions are more than likely not met in many human populations (Fujimara et al. 2014).

Before testing any hypothesis, it is important to examine the behavior of the techniques that are used. One of the reasons why tree-based analyses do not agree with those methods that use a reticulate model is that different genes, chromosomes, and genomes have distinct evolutionary mechanisms and histories. A review of the degree of incongruence of trees for diploid eukaryotic organisms (Rosenfeld et al. 2012 and discussed in DeSalle 2016 reveals that from 22% (higher primate comparisons (Hobolth et al. 2011) to 35–50% (yeast and fruitflies, Rokas et al. 2003 and Pollard et al. 2006 respectively) of the single genes in a genome are incongruent with the taxonomically accepted topology for the same species. In one study of subspecies in the species *Mus musculus* (White et al. 2009), only 33% of the genes analyzed agreed in topology with the accepted genealogy for the subspecies (Rosenfeld et al. 2012).

The problem of incongruence among gene trees is largely related to coalescent theory. Coalescent theory suggests that even if there is a strong and irrefutable species tree, individual gene trees have a high likelihood of being incompatible. This observation from coalescent theory suggests two very important things about gene trees and species trees. First, there is a basic theoretical and functional difference between gene trees and species trees. Second, the problem of unraveling the history of species divergence becomes one of inferring species trees from gene trees. Several approaches have been postulated to implement such inference. These range from concatenation (Gatesy and Springer 2013, 2014; Springer and Gatesy 2014, 2016 to coalescent analysis of multiple gene trees (Liu et al. 2009a, 2009b, 2010).

The most obvious example of this concerns the clonally inherited genes of mtDNA and the Y chromosome, which evolve without recombination. Because they do not recombine with each generation, as almost all other nuclear loci do, the inferences made in the literature

Box 5.4

Incongruence of trees

> Two trees are incongruent when they do not agree with respect to the phylogenetic inference they express. There are differing degrees of incongruence that trees can have with each other. Hence several tests have been described that determine the statistical significance of the incongruence of two trees. In addition, it should be noted that not all genes or all stretches of DNA in a data set will give the same tree, or in other words they will be incongruent with each other. This phenomenon has resulted in systematists utilizing coalescent based methods to interpret large genome-sized data sets in a phylogenetic context. Coalescent methods (Edwards et al. 2007; Liu et al. 2009a, b; Liu et al. 2015), are suggested to be able to extract patterns that certain genes support to give an overall phylogenetic hypothesis for a data set, but these approaches have been criticized by some authors as inaccurate indicators of overall phylogenetic signal of organisms (Gatesy and Springer 2013, 2014; Springer and Gatesy 2014, 2016).

using these markers give well-resolved trees that are interpretable as histories of the mtDNA and the Y chromosome, as they should. On the other hand, trees can be generated for the 20,000 or so individual genes in the human genome, and some of the individual genes would also give resolved trees. But because of recombination, and the lack of clonality of the individual genes, such trees will represent the history of the gene, recombination and all. If enough genomes are included as terminals in separate phylogenetic analysis of all the genes in the human genome (the 1000 Genomes Project has sequenced over 2,000 human genomes), the probability that any two genes out of the 20,000 or so genes in the human genome would give the same tree is extremely small if not zero. It should be pointed out that this incongruence does not mean that gene trees or the information from discrete gene regions cannot be used to generate species trees.

For this contribution, SNPs called for the Phase 1 data set of the 1000 Genomes Project were used for 514 male individuals in the data set and called for mtDNA, Y chromosomal DNA, X chromosomal DNA and DNA from Chromosome 20 (matrices are available upon request from the authors). We constructed an ancestral sequence inferred by consensus for the purposes of rooting the tree. This method of rooting is arbitrary and will often times produce an inaccurate root. However, we prefer to use this method so that we can more easily cross-compare the topologies of the trees we generate for this study. Males were chosen because they will, by definition, have markers from all four chromosomal elements.

We generated phylogenetic trees separately for these four genomic elements (figure 5.2). Note the strongly discernable differences in topology of the trees, indicating the different evolutionary histories and phylogenetic resolution inherent in each genomic element. Nonrecombining genomic regions (mtDNA and Y chromosomal DNA) should show a high degree of congruence of phylogenetic signal within each region relative to those that recombine. When the phylogenetic patterns from the different genomic regions are compared, the "tanglegrams" in figure 5.3 are obtained. Not surprisingly, there is little overlap in the signal coming from the two clonal markers. This result should be obvious, and one need look no further than their very own genealogy to comprehend this pattern. Tattersall and DeSalle (2011) have made this same point using Darwin's genealogy by tracing his paternal lineage and his maternal lineage from his genealogy published in the 1930s. Darwin's paternal ("Y chromosomal") tree is quite different than his maternal ("mtDNA") tree. Furthermore, there is very little relationship between either clonal marker and geography, corresponding to the ability of so called races to experience gene flow and/or admixture. While chromosome 20 and the X chromosome show less "tangling" this is largely because of the almost complete lack of resolution of the trees derived from these two elements.

When all six pairwise comparisons are made among the X chromosome, Y chromosome, Chromosome 20, and mtDNA to examine for statistical significance of congruence (using the Incongruence Length Difference—ILD test), only one comparison (Chromosome 20 versus X chromosome) shows significant congruence. This latter result is almost surely

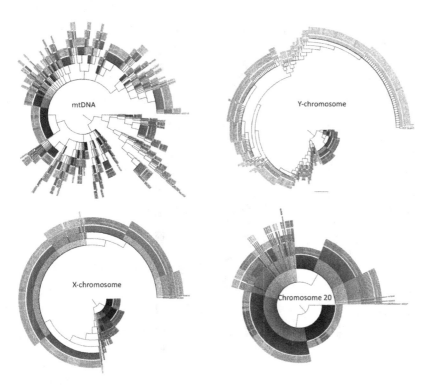

Figure 5.2
Phylogenetic trees for the four "chromosomal" entities discussed in the paper. Data from the 1000 Genomes Project were used to produce the trees, and individuals were colored according to the 1000 Genomes lists for continent of origin.
Purple = African American, Blue = Africa, Red = Asia, Green = Europe, and Orange = Central America.
The trees were generated in TNT (Goloboff et al. 2008) using a bootstrap (100 replicates), and only nodes with bootstraps greater than 50% are retained in each of the trees.

caused by the lack of resolution of both the overall X chromosome and Chromosome 20 trees (figure 5.3). As Tattersall and DeSalle (2011) point out, "The bottom line here, then, is that hierarchical structuring of humans using phylogenetic trees based on the entire genome gives an unrecognizable and unresolved bush."

The Bottom Line on Making Trees of Human Individuals

A very straightforward test of a hypothesis of the biological relevance of genetic clusters of people would be the following:

H_0 = There are n "races" of *H. sapiens* (A) that correspond to the n geographical divisions (often taken to be Africa, Asia, and Europe) that we see on the planet today.

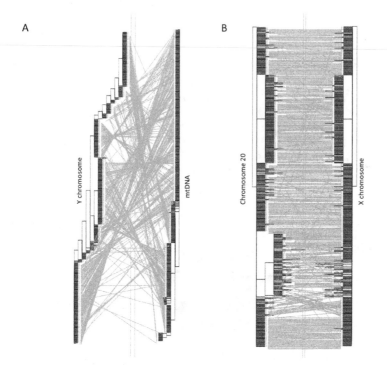

Figure 5.3
Tanglegrams of the individual chromosomal trees paired for clonal markers (on the left; mtDNA and Y chromosomal DNA) and recombining markers (on the right; Chromosome 20 and X chromosome).

To reject this hypotheses using a tree-based approach, one would expect a tree with rampant polyphyly intermixing individuals from groups corresponding to geography. Figure 5.4 shows a bootstrap phylogenetic tree of the 514 human males where data from the Y chromosome, X chromosome, mtDNA, and chromosome 20 have been combined into a single matrix. Indeed, this analysis gives an incredibly unresolved phylogenetic tree. In fact, of the 1,583 possible nodes in a fully bifurcating tree for the 514 ingroup terminals, only 15 show bootstraps greater than 85%, meaning that only 1% of the possible nodes in the tree are resolved at bootstrap values greater than 85%. In addition, only one node implying monophyly of a geographic region (Asia) exists in the tree (note that these are not monophyletic in clonal regions, and the sampling is geographically limited to Chinese and Japanese people). Clearly the hypothesis mentioned above can largely be rejected using this tree-building approach. The rejection of H_0 is furthered by reviewing the discordance between each of the trees produced for the four genomic region (see previous).

It should be noted though that the test performed here is agnostic with respect to ascertainment. A biased ascertainment approach will result in a different, more resolved result.

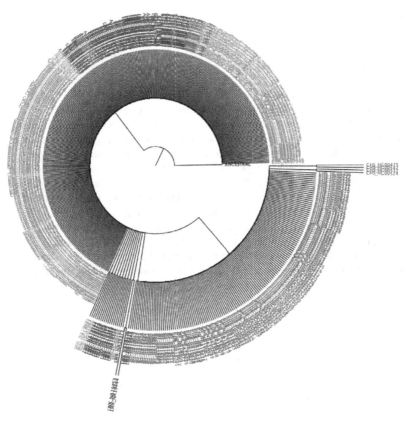

Figure 5.4
Bootstrap tree of combined Phase 1 1000 Genomes male dataset using TNT (Goloboff et al. 2008). Data from Phase 1 of the 1000 Genomes Project were used to produce the tree and individuals were colored according to the 1000 genomes lists for continent of origin. Purple = African American, Blue = Africa, Red = Asia, Green = Europe, and Orange = Central America. The trees were generated in TNT using a bootstrap (100 replicates) and only nodes with bootstraps greater than 85% are retained.

Ancestral Informative Markers (AIMs) are an extreme form of ascertainment bias, and indeed analysis using AIMs show highly biased results (Kittles and Weiss 2003). In order to obtain AIMs for *H. sapiens* corresponding to the three geographic regions mentioned in H_0 above, we constrained a tree to have these three groups as monophyletic and estimated the consistency of all of the SNPs in the data set. We then sorted the SNPs on this constrained tree by consistency. In this way we obtained lists of the top 1% consistent SNPs (AIMs) and the bottom 1% SNPs (AIMs). Figure 5.5A shows a tree constructed from the top 1% markers with respect to ancestry (top 1% AIMs) in the 1000 Genome data set, and figure 5.5B shows the tree generated from the bottom 1% AIMs. Again, in a systematic context there is no real reason to prefer the top 1% AIMs. In fact there are

Box 5.5

Ascertainment

Human genomics researchers up until recently have been constrained by the cost and intensity of whole genome sequencing. Hence they have often used an approach called ascertainment to obtain panels of genes that they postulate to be informative to questions about human ancestry and relatedness. Ascertainment bias or the "systematic deviation of population genetic statistics from theoretical expectations" due to choosing a subset of SNPs for analysis (Lachance and Tishkoff 2013) can be a problem with certain ascertainment approaches. This bias more than likely will be less of a problem as sequencing methods get faster and cheaper, and as whole genomes are used in human population studies. But because many of the more recent inferences about human population history have been made on ascertained genotyping arrays, it is important to examine the method and its impact on the inferences made. Lachance and Tishkoff (2013) state, "Unless the whole genome of every individual in a population is sequenced there will always be some form of SNP ascertainment bias." Most ascertainment strategies start out by examining only a few individuals. The small sample size used to do ascertainment to construct genotyping arrays produces several effects. First, SNPs that are older and have deeper coalescent properties are retained, and SNPs of intermediate frequency are also retained. Both of these effects can have a huge impact on inferences made from genotyping arrays based on ascertainment subsets of SNPs.

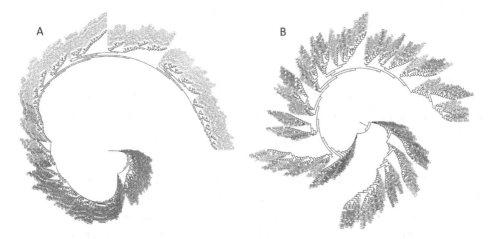

Figure 5.5
A. Phylogenetic tree constructed using TNT (Goloboff et al. 2008) using the top 1% SNPs that are Ancestral Informative Markers (AIMs) as defined in the text. Purple=African American, Blue=Africa, Red=Asia, Green=Europe, and Orange=Central America. B. Phylogenetic tree constructed using the TNT using the bottom 1% SNPs that are Ancestral Informative Markers (AIMs) as defined in the text.

strong philosophical and empirical arguments (Carpenter and Nixon 1992; Gatesy and Springer 2013, 2014; Kluge 1989; Springer and Gatesy 2014, 2016) that the combination of all of the data available is the most valid way to discover relationships in a phylogenetic analysis. In the present case consideration of all of the data produce a lack of hierarchy and discreteness. And this is exactly what we would expect for a reticulating group of genomes as we will discuss next. This failed exercise of using AIMs to infer relationships simply points to the fact that AIMs were not developed to generate phylogenetic or population level patterns. Doing so is really an abuse of their initial purpose, which was to provide a means for mapping of genomic regions in individuals using admixture as a tool.

Choosing a Sound, Empirically Based Philosophical Position

Since the Enlightenment, one of the guiding principles of systematic biology is to categorize organisms in a hierarchical context, and this remains a guiding principle for the field (Yoon 2010). Trees have been used to represent hierarchical relationships of entities for over 300 years. These bifurcating diagrams have indeed been an important aspect of the development and expansion of modern evolutionary biology, but sometimes have been used indiscriminately on all levels of biological organization. Early on in the development of phylogenetic systematics though, Hennig (1966) challenged the notion that trees could be used indiscriminately with his famous figure 6 (Brower et al. 1996; Goldstein and DeSalle 2000; Goldstein et al. 2000). This figure points to the problems with using tree-based approaches to describe biological systems that are reticulating (his so-called tokogenetic level).

Brower et al. (1996) articulated this difference in a Popperian context, suggesting that two competing scientific disciplines are relevant at this level of organismal examination—population genetics and systematics. They point out that scientific hypotheses are based on plausible prior theories that are background knowledge for a particular system. The acceptance of a particular background knowledge (an ontology) is justifiable according to Popper. According to Brower et al. (1996, p. 425) epistemology "is the questioning of the adequacy of these assumptions: How do we construct a rational framework to interpret and explain our observations? Epistemological questioning diminishes when a discipline reaches a consensus about an ontological paradigm; at that point the discipline becomes a 'normal science.'"

Systematics and population genetics both rely on distinct but perhaps equally entrenched ontologies and thus can be considered Kuhnian normal sciences. Population genetics can be viewed as a deterministic normal science that deduces evolutionary inferences from the established laws of heredity. Continuity of process is an inherently important aspect of the workings of population genetics and this discipline attempts to explain and detail the origin and subsequent maintenance of genetic diversity at the population level. Systematics, on the other hand, is a purely historical science that attempts to discover and describe the intricacies evident in the hierarchical pattern of nature. As Brower et al. (1996, p. 425) suggest, "Systematics seeks to document hierarchical patterns among disjunct entities and needs to postulate little except that a tree-like hierarchy exists and is recoverable by studying attributes of individual organisms." This view has led to the suggestion that a "line of death" exists at this level of biological analysis. And indeed the results of phylogenetic analysis of SNP data in this paper demonstrate this "line of death." There are cases, however. where a breach of the line of death can be made in showing divergence of populations into discrete lineages (Greenbaum et al. 2016).

Let's return to the human ancestry analysis space of figure 5.1. If we accept the argument above about tokogenetic populations, we should eliminate trees as a valid way of representing reticulating individuals within a population. This does not mean, though, that trees won't be useful at this level for other endeavors. In fact, clonally inherited parts of the genome (mtDNA and Y chromosomal DNA) are perfectly suited to tree-based analyses below the species level. There are several autosomal genomic regions too that experienced little or no recombination and these hotspots too would be reasonable regions for use in phylogenetic approaches. Hence trees should not be used in the indicated part of figure 5.1. But, as the test of H_0 above indicates, trees can be useful tools to test hypotheses of population discreteness when rampant polyphyly is used as the criterion for hypothesis testing. What is not valid in this ontological framework, though. is the representation of reticulating individual *H. sapiens* as part of any phylogenetic tree.

Unfortunately, even when researchers do recognize this important limit of trees, they continue to use them anyway. Why is this? Gower (1972, p. 10) has a good answer: ". . . The human mind distinguishes between different groups because there are correlated characters within the postulated groups." The underlying correlation of data that Gower (1972) mentions brings into play the so-called "Lewontin's fallacy" (Edwards 2005). In the 1970s, based on genetic information, Lewontin suggested that there is no hierarchy with respect to human populations. Edwards (2005) claimed that Lewontin's conclusion ignores the correlation structure of data and in doing so misses the fact that populations can be distinguished and delineated from each other using this underlying correlation structure. Modern phylogenetic systematic theory would not agree that a fallacy exists though. What Edwards (2005) is asking us to do is to ignore or "throw out" some of the information in the overall data set. Actually, the approach would be to throw out a great deal of the data. Modern phylogenetic systematics would search for signal in the data set not only for the information that has underlying correlation structure, but also in the rest of the data. Using just the information with underlying correlation structure is akin to clique analysis or compatibility analysis (Le Quesne 1969), in systematics. Clique analysis utilizes correlated characters to strengthen phylogenetic hypotheses, but it has been shown to be an unsound approach to phylogenetic reconstruction (Farris 1983; Farris and Kluge 1979).

Conclusion

The above discussion notwithstanding, someone may ask: If the ontological constraints of systematic analysis presented here imply that the underlying correlation structure cannot be used to extract hierarchy from a data set in a systematic context, what good is it? We suggest that while the underlying correlation structure is telling us something about ancestry, it is completely unlinked to delineating the existence of populations within our species or with establishing hierarchy of individuals in our species. Ancestry, in the age of genomics, has become the identification of parts of the genome in two individuals that can be traced back to a common point in the history of those genomes. Ancestry really should be the focus of studies using genome information of *H. sapiens* on a large scale. However, an exploration of ancestry does not need to use phylogenetic trees. In fact, to use trees would be illogical when the goal is to understand the ancestry of individual *H. sapiens*. There are more appropriate network-based methods for examining the genetics of individuals at this level (Greenbaum et al. 2016). Focusing on ancestry and away from hierarchy would also place the endeavor of human genome analysis on the sounder philosophical grounds of not making the assumption that population trees do exist.

Acknowledgments

The authors would like to acknowledge the continued support of the Sackler Institute for Comparative Genomics at the AMNH and the Korein Family Foundation. The authors also want to thank the Konrad Lorenz Institute and Dr. J. Schwartz for the invitation to participate in the conference

References

Bodmer, W. 2015. "Genetic Characterization of Human Populations: From ABO to a Genetic Map of the British People." *Genetics* 199(2): 267–279.

Brower, A. V. Z., R. DeSalle, and A. Vogler. 1996. "Gene Trees, Species Trees, and Systematics: A Cladistic Perspective." *Annual Review of Ecology and Systematics* 27: 423–450.

Campbell, M. C. and S. A. Tishkoff. 2010. "The Evolution of Human Genetic and Phenotypic Variation in Africa." *Current Biology* 20(4): R166–R173.

DeSalle, R. 2016. "What Do Our Genes Tell Us about Our Past?" *Journal of Anthropological Sciences* 94: 1–8.

DeSalle, R. and I. Tattersall. 2014. "Mr. Murray, You Lose the Bet." *GeneWatch* 27: 2–5.

Ding, Q., Y. Hu, S. Xu, J. Wang, and L. Jin. 2014. Neanderthal Introgression at Chromosome 3p21. 31 Was Under Positive Natural Selection in East Asians. *Molecular Biology and Evolution* 31(3): 683–695.

Edwards, A. W. F. 2003. "Human Genetic Diversity: Lewontin's Fallacy." *BioEssays* 25(8): 798–801.

Edwards, S. V., L. Liu, and D. K. Pearl. 2007. "High-Resolution Species Trees Without Concatenation." *Proceedings of the National Academy of Sciences* 104(14): 5936–5941.

Eriksson, A. and A. Manica. 2012. "Effect of Ancient Population Structure on the Degree of Polymorphism Shared between Modern Human Populations and Ancient Hominins." *Proceedings of the National Academy of Sciences* 109(35): 13956–13960.

Falush, D., M. Stephens, and J. K. Pritchard. 2003. "Inference of Population Structure Using Multilocus Genotype Data: Linked Loci and Correlated Allele Frequencies." *Genetics* 164(4): 1567–1587.

Falush, D., M. Stephens, and J. K. Pritchard. 2007. "Inference of Population Structure Using Multilocus Genotype Data: Dominant Markers and Null Alleles." *Molecular Ecology Notes* 7(4): 574–578.

Farris, J. S. 1983. "The Logical Basis of Phylogenetic Analysis." In *Advances in Cladistics: Proceedings of the Second Meeting of the Willi Hennig Society*, edited by N. I. Platnick and V. A. Funk, 7–36. New York, NY: Columbia University Press.

Farris, J. S. and A. G. Kluge. 1979. "A Botanical Clique." *Systematic Zoology* 28(3): 400–411.

Fujimura, Joan H., Deborah A. Bolnick, Ramya Rajagopalan, Jay S. Kaufman, Richard C. Lewontin, Troy Duster, Pilar Ossorio, and Jonathan Marks. 2014. "Clines Without Classes: How to Make Sense of Human Variation." *Sociological Theory* 32(3): 208–227.

Gatesy, J. and M. S. Springer. 2013. "Concatenation Versus Coalescence Versus "Concatalescence." *Proceedings of the National Academy of Sciences* 110(13): E1179–E1179.

Gatesy, J. and M. S. Springer. 2014. "Phylogenetic Analysis at Deep Timescales: Unreliable Gene Trees, Bypassed Hidden Support, and the Coalescence/Concatalescence Conundrum." *Molecular Phylogenetics and Evolution* 80: 231–266.

Goldstein, P. Z. and R. DeSalle. 2000. "Phylogenetic Species, Nested Hierarchies, and Character Fixation." *Cladistics* 16(4): 364–384.

Goldstein, P. Z., R. DeSalle, G. Amato, and A. P. Vogler. 2000. "Conservation Genetics at the Species Boundary." *Conservation Biology* 14(1): 120–131.

Goloboff, P. A., J. S. Farris, and K. C. Nixon. 2008. "TNT, a Free Program for Phylogenetic Analysis." *Cladistics* 24(5): 774–786.

Gower, J. C. 1972. "Measures of Taxonomic Distance and Their Analysis." In *The Assessment of Population Affinities in Man*, edited by J. S. Weiner and J. Huizinga, 1–24. Oxford: Clarendon Press.

Greenbaum, G., A. R. Templeton, and S. Bar-David. 2016. "Inference and Analysis of Population Structure Using Genetic Data and Network Theory." *Genetics* 202:1299–1312.

Hennig, W. 1966. *Phylogenetic Systematics*. Champaign: University of Illinois Press.

Hobolth, A., J. Y. Dutheil, J. Hawks, M. H. Schierup, and T. Mailund. 2011. "Incomplete Lineage Sorting Patterns among Human, Chimpanzee, and Orangutan Suggest Recent Orangutan Speciation and Widespread Selection." *Genome Research* 21(3): 349–356.

Hooton, E. A. 1946. *Up from the Ape*. New York, NY: Macmillan.

Hubisz, M. J., D. Falush, M. Stephens, and J. K. Pritchard. 2009. "Inferring Weak Population Structure with the Assistance of Sample Group Information." *Molecular Ecology Resources* 9(5): 1322–1332.

Keith, A. 1915. *The Antiquity of Man*. London, Williams and Norgate.

Kittles, R. A. and K. M. Weiss. 2003. "Race, Ancestry, and Genes: Implications for Defining Disease Risk." *Annual Review of Genomics and Human Genetics* 4(1): 33–67.

Krause, J., Q. Fu, J. Good, B. Viola, M. Shunkov, A. Derevianko, and S. Pääbo. 2010. "The Complete Mitochondrial DNA Genome of an Unknown Hominin from Southern Siberia." *Nature* 464(7290): 894–897.

Lachance, J., and S. A. Tishkoff. 2013. "SNP Ascertainment Bias in Population Genetic Analyses: Why It Is Important, and How to Correct It." *BioEssays* 35(9): 780–786.

Le Quesne, W. J. 1969. "A Method of Selection of Characters in Numerical Taxonomy." *Systematic Zoology* 18(2): 201–205.

Li, J. Z., D. M. Absher, H. Tang, A. M. Southwick, A. M. Casto, S. Ramachandran, H. M. Cann, G. S. Barsh, M. Feldman, L. L. Cavalli-Sforza, and R. M. Myers. 2008. "Worldwide Human Relationships Inferred from Genome-Wide Patterns of Variation." *Science* 319(5866): 1100–1104.

Liang, M. and R. Nielsen. 2011. "Q&A: Who Is *H. sapiens* Really, and How Do We Know?" *BMC Biology* 9(1): 20.

Liu, L., L. Yu, D. K. Pearl, and S. V. Edwards. 2009a. "Estimating Species Phylogenies Using Coalescence Times among Sequences." *Systematic Biology* 58: 468–477.

Liu, L., L. Yu, L. S. Kubatko, D. K. Pearl, and S. V. Edwards. 2009b. "Coalescent Methods for Estimating Phylogenetic Trees." *Molecular Phylogenetics and Evolution* 53: 320–328.

Liu, L., L. Yu, and S. V. Edwards. 2010. "A Maximum Pseudo-likelihood Approach for Estimating Species Trees under the Coalescent Model." *BMC Evolutionary Biology* 10: 302.

Liu, L., Z. Xi, S. Wu, C. C. Davis, and S. V. Edwards. 2015. "Estimating Phylogenetic Trees from Genome-Scale Data." *Annals of the New York Academy of Sciences* 1360(1): 36–53.

Mailund, T., K. Munch, and M. H. Schierup. 2014. "Lineage Sorting in Apes." *Annual Review of Genetics* 48: 519–535.

Martin, S. H., J. W. Davey, and C. D. Jiggins. 2015. "Evaluating the Use of ABBA–BABA Statistics to Locate Introgressed Loci." *Molecular Biology and Evolution* 32(1): 244–257.

Morrison, D. 2012. "Human Races, Networks and Fuzzy Clusters." Accessed April 1, 2016. http://phylonetworks.blogspot.com/2012/08/human-races-networks-and-fuzzy-clusters.html.

Nei, M., and A. K. Roychoudhury. 1993. "Evolutionary Relationships of Human Populations on a Global Scale." *Molecular Biology and Evolution* 10(5): 927–943.

Nievergelt, C. M., O. Libiger, and N. J. Schork. 2007. "Generalized Analysis of Molecular Variance." *PLoS Genetics* 3(4): e51.

Pickrell J. K., N. Patterson, C. Barbieri, F. Berthold, L. Gerlach, T. Güldemann, B. Kure, S. W. Mpoloka, H. Nakagawa, C. Naumann, and M. Lipson. 2012. "The Genetic Prehistory of Southern Africa." *Nature Communications* 3: 1143.

Pickrell, J. K., and J. K. Pritchard. 2012. "Inference of Population Splits and Mixtures from Genome-wide Allele Frequency Data." *PLoS Genetics* 8(11): e1002967

Pollard, D. A., V. N. Iyer, A. M. Moses, and M. B. Eisen. 2006. "Widespread Discordance of Gene Trees with Species Tree in Drosophila: Evidence for Incomplete Lineage Sorting." *PLoS Genetics* 2(10): e173.

Pritchard, J. K., M. Stephens, and P. Donnelly. 2000. "Inference of Population Structure Using Multilocus Genotype Data." *Genetics* 155(2): 945–959.

Reich, D., R. E. Green, M. Kircher, J. Krause, N. Patterson, E. Y. Durand, B. Viola, A. W. Briggs, U. Stenzel, P. L. Johnson, and T. Maricic. 2010. "Genetic History of an Archaic Hominin Group from Denisova Cave in Siberia." *Nature* 468(7327): 1053–1060.

Rieppel, O. 2010. "The Series, the Network, and the Tree: Changing Metaphors of Order in Nature." *Biology and Philosophy* 25(4): 475–496.

Rokas, A., B. L. Williams, N. King, and S. B. Carroll. 2003. "Genome-scale Approaches to Resolving Incongruence in Molecular Phylogenies." *Nature* 425(6960), 798–804.

Rosenberg, N. A., S. Mahajan, S. Ramachandran, C. Zhao, J. K. Pritchard, and M. W. Feldman. 2005. "Clines, Clusters, and the Effect of Study Design on the Inference of Human Population Structure." *PLoS Genetics* 1(6): e70.

Rosenfeld, J. A., A. Payne, and R. DeSalle. 2012. "Random Roots and Lineage Sorting." *Molecular Phylogenetics and Evolution* 64(1): 12–20.

Sankararaman, S., S. Mallick, M. Dannemann, K. Prüfer, J. Kelso, S. Pääbo, N. Patterson, and D. Reich. 2014. "The Genomic Landscape of Neanderthal Ancestry in Present-Day Humans." *Nature* 507(7492): 354–357.

Sparks, J. A, 1932. Histomap of Evolution. Rand McNally & co. HZ 011.

Springer, M. S. and J. Gatesy. 2014. "Land Plant Origins and Coalescence Confusion." *Trends in Plant Science* 19(5): 267–269.

Springer, M. S. and J. Gatesy. 2016. "The Gene Tree Delusion." *Molecular Phylogenetics and Evolution* 94: 1–33.

Tattersall, I. and R. DeSalle. 2011. *Race?: Debunking a Scientific Myth* no. 15. College Station: Texas A&M University Press.

Wade, N. 2014. *A Troublesome Inheritance: Genes, Race and Human History*. London: Penguin.

Weidenreich, F. 1946. *Apes, Giants, and Man*. Chicago, IL: University of Chicago Press.

White, M. A., C. Ané, C. N. Dewey, B. R. Larget, B. A. Payseur. 2009. "Fine Scale Phylogenetic Discordance across the House Mouse Genome." *PLoS Genetics* 5(11): e1000729.

Yoon, C. K. 2010. *Naming Nature: The Clash between Instinct and Science*. New York, NY: W.W. Norton and Company.

Yudell, M. 2014. *Race Unmasked: Biology and Race in the Twentieth Century*. New York, NY: Columbia University Press.

Zerjal, T., Y. Xue, G. Bertorelle, R. S. Wells, W. Bao, S. Zhu, R. Qamar, Q. Ayub, A. Mohyuddin, S. Fu, and P. Li. 2003. "The Genetic Legacy of the Mongols." *The American Journal of Human Genetics* 72(3): 717–721.

6 Hypothesis Compatibility Versus Hypothesis Testing of Models of Human Evolution

Alan R. Templeton

Introduction

Popper (1959) argued that testing hypotheses is a central activity of science and often distinguishes scientific knowledge from other forms of human knowledge. There are many forms of hypothesis testing, but two are particularly common in science. The first is logical hypothesis testing, in which an observation is made that is logically incompatible with a given hypothesis, thereby leading to a rejection of that hypothesis. When an observation is logically compatible with the hypothesis, it is often regarded as strengthening that hypothesis, but logical compatibility does not prove that the hypothesis is true unless every possible alternative is rejected. Rarely in science can we articulate all possible alternatives, so the "hard" inference (Platt 1970) is rejecting a model through logical incompatibility. The other major form of hypothesis testing is statistical, which in turn can be subdivided into two major types. The first is the formulation of a null hypothesis. Observations are made, and a statistic based on the observations is calculated. The probability of this statistic is then calculated under the assumption that the null hypothesis is true. If this probability is low, the null hypothesis is rejected. The probability of the statistic under the null model gives a quantitative assessment of how well the observations fit the predictions of the null hypothesis. Often a threshold probability level is used called a "p-value." In much science, if $p \leq 0.05$, the null hypothesis is rejected. As with logical hypothesis testing, the "hard" inference is the rejection of the null hypothesis. A p-value above the threshold is regarded as a failure to reject the null hypothesis and does not necessarily indicate that the null hypothesis is true. The second form of statistical hypothesis testing is the testing of alternative hypotheses. In this case, two or more alternative hypotheses are modeled, and the relative probabilities of the statistic under the alternatives are calculated. As with the other forms of hypothesis testing, the "hard" inference is rejecting one or more alternative hypotheses relative to other alternatives. The "favored" hypothesis with the highest *relative* probability is not necessarily true; rather, it is only the best of the alternatives considered.

In all the forms of scientific hypothesis testing described above, scientific knowledge advances by rejecting hypotheses. Compatibility with a hypothesis is a weak inference that merely shows that the hypothesis has not been rejected, not that it is true. Sometimes only compatibility of the data with a favored hypothesis is considered without any attempt at rejecting the favored hypothesis or comparing it logically or statistically to alternative hypotheses. In this case, hypothesis compatibility is highly subjective and suspect. Hence, there is a great effort among many scientists to formulate their models as falsifiable hypotheses and to consider alternatives. Despite the perceived importance of falsifiable hypotheses in science, in practice many distinguished scientists often ignore, dismiss, or rationalize away data that are incompatible with a favored model (Kuhn 1962). Kuhn (1962) further argued that falsification of a previously favored model was finally accepted in past paradigm shifts "once an alternative candidate is available to take its place" (Kuhn 1962, p. 77). The Out-of-Africa replacement model dominated human evolution over the past quarter of a century despite the fact that it was repeatedly and strongly falsified when placed in hypothesis-testing frameworks, and that alternative models of human evolution did exist that were not falsified. Hence, models of human evolution represent a persistent abandonment of the principle of falsification even more extreme than found in most past paradigm shifts.

The Out-of-Africa Replacement Hypothesis

The "classic" model of human evolution that was popular in the first half of the twentieth century had Old World hominins subdivided into many divergent evolutionary branches, with one branch later expanding out into the regions inhabited by the others, driving them to complete extinction (Dobzhansky 1944). Howells (1942) regarded these regional branches as so divergent as to constitute different species of humans. As the great population geneticist Theodosius Dobzhansky (1944) pointed out, this "classic" model had no role for any type of genetic interchange (either gene flow or admixture) between these hominin branches. Rather, the main factors besides natural selection that dominated human evolution were population dispersal and replacement. In the first half of the twentieth century, it was not clear which regional branch of humanity was the "winner" in this dispersal/replacement process, but subsequent fossil finds indicated that many anatomical features associated with living humans first arose in sub-Saharan Africa and only later spread out of Africa. Accordingly, the classic model became the Out-of-Africa replacement (OAR) model (Stringer and Andrews 1988). OAR posits that anatomically modern humans arose first in sub-Saharan Africa, and then around 60,000 years ago (although some argue for 100,000 to 130,000 years ago), expanded out of Africa, driving to complete extinction all the other human populations encountered in Eurasia. OAR predicts that all living humans trace their genetic ancestry exclusively to sub-Saharan Africa, with no genetic input from other regions of the Old World. As a result, modern humanity has a single, specific geographical location of origin—sub-Saharan Africa.

The OAR model quickly became the dominant model for recent human evolution after the publication of a mitochondrial DNA (mtDNA) haplotype tree (an evolutionary tree of the current haplotypes defined by the accumulated mutations over the evolutionary history of the human mtDNA molecule back to a common ancestral molecule; Cann et al. 1987). Their mtDNA haplotype tree (figure 6.1), as well as subsequent ones with better sampling and genetic resolution, indicated that the first split in the haplotype tree was between a clade (branch) of mitochondrial haplotypes found exclusively in sub-Saharan Africa versus a clade with a world-wide distribution. This pattern indicates that the root of the mtDNA haplotype tree was located in sub-Saharan Africa. Cann et al. (1987) inferred from this that all of present-day humanity was descended from the African population containing this mtDNA haplotype, thereby supporting OAR. This inference was reinforced when the bearer of this ancestral mtDNA haplotype was dubbed "mitochondrial Eve," thereby tapping into a strong cultural image of descent from a single woman (since mtDNA is maternally inherited).

The inference from figure 6.1 of all humanity coming from a sub-Saharan population is based on equating the mtDNA haplotype tree to an evolutionary tree of human populations. All figure 6.1 really shows is that human mtDNA is rooted in Africa, and it does not mean that all of modern humanity's gene pool came from a single African population. Coalescent theory in population genetics demonstrates that all the copies of any homologous DNA region in any species will coalesce to a common ancestral molecule at some time and location in the past regardless of population history (Templeton 2006). For example, Reece and colleagues (2010) studied moray eels, a species with extremely high dispersal capabilities due to a prolonged pelagic larval phase such that it is virtually panmictic throughout the entire Indo-Pacific Ocean—about two-thirds of the world. There is no population tree in such a species as there is only a single, world-wide panmictic population. Yet, this species still has well-defined mtDNA and nuclear-DNA haplotype trees, including its own mitochondrial Eve (i.e., the ancestral root of the mtDNA tree). Hence, the observation of a mitochondrial Eve is non-informative because it is universal. All models of human evolution are compatible with and even require a mitochondrial Eve. This illustrates why hypothesis testing is needed in addition to hypothesis compatibility to advance scientific knowledge.

Pseudo-Hypothesis Testing of OAR

Misrepresenting hypotheses

Cann et al. (1987) did attempt to do some logical hypothesis testing, but their attempt was flawed by two fundamental mistakes that were common at the time. First, they portrayed the state of human evolutionary models as having just two alternatives: the OAR and the "Multi-Regional Model." In fact, there were/are many alternatives, such as the candelabra model. This model, like the OAR, posited isolated Pleistocene lineages of humans, but unlike OAR had no uniregional expansion and replacement but rather independent evolution of all

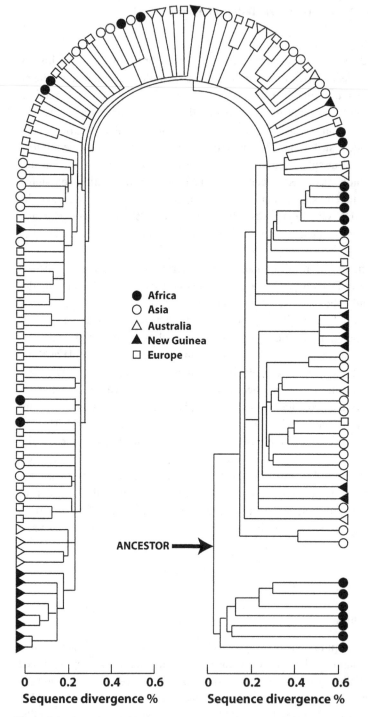

Figure 6.1
The mtDNA haplotype tree from Cann et al. (1987), with the ancestral root inferred by mid-point rooting.

these isolated lineages independently evolving into modern humans (Coon 1962). Weidenreich (1946) proposed the Multi-Regional Model as an alternative to the classic uniregional model (that later become OAR) and the candelabra model. His Multi-Regional Model portrayed human populations over time on a lattice or trellis, not a tree, because of recurrent genetic interchange among Pleistocene populations (Weidenreich 1946). Multiregional evolution was the opposite of the candelabra model, as it was based on humanity evolving into its current form as a single evolutionary lineage due to recurrent genetic interchange among local populations, whereas the candelabra model had complete isolation and parallel evolution in the total absence of genetic interchange. The second mistake of Cann and colleagues was to equate the candelabra model to the Multi-Regional Model (Templeton 2007). It was this mistake that allowed them to falsify the "Multi-Regional Model." Under the candelabra model, humans in Eurasia and Africa last shared a common ancestor when *Homo erectus* expanded out of Africa, now estimated at nearly two million years ago. However, the coalescence of all mtDNA molecules in humans from the entire globe occurred between 140,000 to 290,000 years ago, as estimated by Cann and colleagues (1987). This was logically incompatible with the candelabra model as there was no way an African mtDNA molecule could replicate and spread to Eurasia under the candelabra model's assumption of complete genetic isolation during this time period. This pattern does indeed falsify the candelabra model, but it does not falsify Weidenreich's Multi-Regional Model. Under the Multi-Regional Model, mtDNA (and other DNA as well) can spread at any time due to recurrent gene flow or during episodes of admixture. Consequently, the estimated coalescence time and place of mtDNA is completely compatible with both OAR and multiregional evolution.

Interestingly, Dobzhansky (1944) argued for the implausibility of the candelabra model and the classic uniregional model on the basis of their assumption of complete reproductive isolation of human populations for a million years or more. He further argued that the candelabra model was particularly implausible from an evolutionary perspective, as it required an extremely unlikely degree of parallel evolution during the evolution of human modernity from a *Homo erectus*-like ancestor. Dobzhansky supported the Multi-Regional Model because it did not require the unlikely persistence of reproductive isolation over long periods of time, and gene flow/admixture would eliminate the need for parallel evolution.

The claim that the mtDNA tree was incompatible with the "multiregional" model was false, but even repeated efforts by the proponents of multiregional evolution to correct this misrepresentation (Wolpoff 1996; Wolpoff and Thorne 1991; Wolpoff et al. 1994) did little to overcome the popular view that the mtDNA results were incompatible with the multiregional hypothesis. Some proponents of OAR eventually began to draw a few double-headed arrows between the main tree branches to represent weak gene flow between otherwise isolated lineages of humans in Africa and Eurasia, but the representation of complete isolation of Africans from Eurasians still persists even in papers that document some admixture from ancient DNA studies (e.g., figure 3 in Reich et al. 2010).

The ecological fallacy

A more subtle form of pseudo-hypothesis testing is based upon the ecological fallacy (Robinson 1950). The ecological fallacy is a type of illogical argument in which aggregate data (in this case, population level patterns of genetic variation, genetic distances, and so on) are regarded as proving an underlying individual process model (e.g., individual dispersal patterns, individual mating patterns such as admixture, and so forth). The problem is that many underlying process models can generate the same observable aggregate patterns, so model compatibility with the aggregate data does not prove that that model is true. The ecological fallacy is particularly common when dealing with the many computer simulations of human evolution that have been used to support OAR and other models of human evolution. For example, two simulation papers came out within one week of another that reached completely opposite conclusions about human evolution even though both explained the aggregate data on human genetic variation and genetic distances between populations. One was entitled "Genomics Refutes an Exclusively African Origin of Humans" (Eswaran et al. 2005) and claimed that their computer simulations showed that up to 80 percent of all nuclear loci came from non-African archaic humans. In contrast, Ray et al. (2005) claimed that their simulations "unambiguously distinguish between a unique origin and a multiregional model," and favored a complete African replacement with 0 percent of human variation coming from non-African archaic humans. Thus, two very different models fit "unambiguously" similar aggregate data sets, but the underlying process models were completely different. Despite the strong wording of both groups, neither group provided any statistical assessment of goodness of fit or any statistical hypothesis testing of the favored process model versus alternative models. Both groups simply committed the ecological fallacy, and thereby had no logical basis for their claims.

Incoherent inference

Fagundes et al. (2007) attempted to avoid the ecological fallacy by coupling computer simulations with a rigorous statistical framework known as Approximate Bayesian Computation (ABC). Process models of human evolution have many parameters whose values are unknown, and the investigators placed "prior probability distributions" upon these parameters. ABC is a method for sampling these prior distributions over many simulations and comparing the fit to various observed summary statistics (aggregate patterns) to approximate a "posterior distribution" of the parameters given the aggregate patterns. This allows the investigators to use the framework of Bayesian statistics to estimate parameters and test models, including direct statistical comparisons of different simulated models of human evolution. ABC is a legitimate statistical framework that allows hypothesis testing among a finite number of fully specified models.

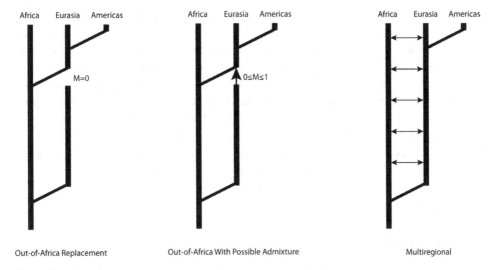

Figure 6.2
The three classes of models of human evolution simulated for an ABC analysis. M measures the amount of admixture of an expanding African population with archaic Eurasians. The double-head arrows in the Multi-Regional Model represent gene flow between Africans and Eurasians. Modified from Fagundes et al. (2007).

Figure 6.2 shows the three basic models of human evolution that were simulated in the ABC analysis. Fagundes et al. (2007) estimated the ABC posterior probabilities of these three models as 0.781 for the Out-of-Africa replacement model, 0.001 for the Out-of-Africa with possible admixture model, and 0.218 for the Multi-Regional Model. Although these results were portrayed as strong statistical support for OAR, note that there was no significant statistical discrimination between the OAR and Multi-Regional Models using any commonly accepted probability threshold in the scientific literature. However, the admixture model was strongly rejected in their analysis. Herein lies a major problem in measure theory. Note from figure 6.2 that both of the Out-of-Africa models are identical except for the range of the admixture parameter M. For OAR, M is set to zero; for the Out-of-Africa with possible admixture, they allowed M to take on any value between 0 and 1 inclusively with a uniform prior. This means that OAR is formally a special case (M=0) of the more general model with possible admixture ($0 \leq M \leq 1$). In terms of set theory, OAR is a proper logical subset of the general admixture model. Probability measures have the following property for nested models: let A be a nested subset of a more general model B, then the probability of A must be less than or equal to the probability of B. This is an absolute requirement of any probability measure. In this case, however, the probability of OAR is 0.781, whereas the probability of the general admixture model that contains OAR as a special case is 0.001—an impossible outcome for a probability measure. This is an example of incoherence—a technical term in statistics in which probability measures violate the constraints of formal logic and measure theory (Gabriel 1969). Hence, the rejection of admixture by Fagundes et al. (2007) is an incoherent inference (Templeton 2010). The reason for this incoherence was not the ABC method *per se*, but rather arose because Fagundes et al. (2007) incorrectly treated all three models as mutually exclusive and exhaustive events in defining their model priors and in their use of posteriors. Interestingly, a well-established coherent Bayesian test (Lindley 1965) was applied to their ABC posterior distributions with the result that the OAR model was rejected relative to the admixture model with a p≤0.025 (Templeton 2010). The relative probabilities of the two models thus changed by five orders of magnitude by going from incoherent to coherent inference. Thus, the ABC procedure actually rejected OAR, and the claim that it supported OAR was based solely on incoherent inference.

Hypothesis Testing

Logical hypothesis testing

Recall that the candelabra model of human evolution was logically falsified by the mtDNA haplotype tree because of the geographical impossibility of mtDNA spreading from Africa to Eurasia less than 300,000 years ago under the isolation assumed by that model. As DNA technology developed, it became increasingly feasible to perform surveys and construct haplotype trees on nuclear genomic regions displaying no or very little recombination in addition to nonrecombining mtDNA. One strong prediction of OAR was that *all* regions of the genome should have an African origin because of African replacement of all Eurasian populations. This prediction leads to another geographical/temporal logical test of the model: under OAR, no DNA regions should coalesce in Eurasia before the out-of-Africa expansion event (variously timed between 55,000 to 130,000 years ago).

Harding and colleagues (1997) reported haplotypes at the beta-hemoglobin locus that coalesced in Asia more than 200,000 years, thereby logically falsifying OAR. Soon many other nuclear DNA regions were found to coalesce in Eurasia in this early time range (Templeton 2007), thereby strengthening the falsification of OAR. However, a mixture of Eurasian and African coalescent events before 130,000 years ago was compatible with the multiregional or admixture models.

Although proponents of OAR accepted the falsification of the candelabra model with mtDNA, they did not accept the parallel logical falsification of OAR from nuclear DNA. For example, Takahata et al. (2001) inferred the geographical roots of 10 nuclear genes, one of which had an old Asian root. Instead of acknowledging that their results falsified OAR (which they called "uniregionality"), they concluded that "However, emphasizing the overwhelming genetic contribution of only one founding population is equivalent to uniregionality." However, replacement predicts only one founding population for *all* genes, not just a majority. Interestingly, Dobzhansky had addressed this issue back in 1944. As noted earlier, Dobzhansky doubted the "classic" uniregional model, but did add that "The 'classic' theory is probably justified to the extent that some of the races of the past have contributed more germ plasm than others to the formation of the present humanity" (Dobzhansky 1944, p. 263). Dobzhansky therefore had no problem with one geographical region dominating in its contribution to the modern gene pool (what is now called the mostly-out-of-Africa model), but he clearly regarded this as a special case of Weidenreich's Multi-Regional Model and not the "classic" uniregional model. In both the Multi-Regional Model and the mostly-out-of-Africa model, the main evolutionary forces are natural selection coupled with gene flow and/or admixture. In contrast, the main evolutionary forces of OAR are natural selection coupled with extinction and replacement with no role whatsoever for gene flow or admixture. Moreover, the main evolutionary consequence of both the Multi-Regional Model and mostly-out-of-Africa model is that the modern human gene pool has been pieced together from multiple regions due to gene flow or admixture, whereas in OAR the modern gene pool is exclusively of African origin. Thus, the mostly-out-of-Africa hypothesis is not logically "equivalent" to OAR, but rather is a special case of multiregional evolution. The equivalency statement made by Takahata et al. (2001) is false, but it allowed the proponents of OAR to keep advocating a clearly falsified model of human evolution.

Statistical hypothesis testing

As noted above, a coherent ABC Bayesian test rejected OAR relative to a model that incorporated small amounts of admixture (Templeton 2010). ABC is a parametric, model-based approach to statistical inference in which detailed models of human evolution must be specified beforehand, including the specification of all parameter values or prior distributions of parameter values. ABC then allows testing of the pre-specified models against one another. Like all Bayesian procedures, constructing informative and mathematically appropriate priors is a critical step because the resulting inferences can be very sensitive to the priors (Oaks et al. 2013). There are many articles in the primary statistical literature on prior construction, with one of the more commonly cited guidelines being Garthwaite et al. (2005). One of the primary recommendations from that paper is *not* to use uniform priors, and especially not when the range of the uniform prior is only a subset of the possible range of values. Doing so can create highly biased results that seemingly have great statistical confidence. Unfortunately, uniform priors with restricted ranges are the norm in much Bayesian phylogenetic inference. For example, Fagundes et al. (2007) defined 32 priors, all of which were either uniform priors or log-uniform priors. Only one of these 32 priors covered the entire possible range. Therefore, all 32 priors violated the standard recommendations for prior construction in a Bayesian analysis, and 31 of them egregiously so. It is therefore not surprising that none of the major conclusions of that paper have been vindicated by subsequent data and discoveries.

A very different approach to testing the events influencing recent human evolution is multi-locus, nested-clade phylogeographic analysis (MLNCPA) (Templeton 2002;

Templeton 2015). Unlike ABC and other model-based approaches, MLNCPA does not require a pre-specified model of human evolution. Rather, MLNCPA uses the data to predict specific events or processes that may have influenced human evolution and then subjects them to statistical testing through null hypotheses. A model of human evolution emerges directly from the results of hypothesis testing. Consequently, MLNCPA can discover events or processes not expected or predicted from any of the standard models of human evolution.

There are four main steps in inference with MLNCPA. First, MLNCPA only uses haplotype data from regions in human genomes that show little or no recombination. Evolutionary history is written most clearly in such regions, and this also allows some of the most powerful predictions from coalescence theory to be used to make inferences. Mitochondrial DNA and much of the Y chromosome do not recombine, and recombination in the human nuclear genome is concentrated into recombinational hotspots, with the areas in between the hotspots having little to no recombination (Templeton et al. 2000). Consequently, many nuclear genomic regions can also be analyzed with MLNCPA. Haplotype trees are estimated for each genomic region, with uncertainty in the haplotype being quantified through a Bayesian procedure (Templeton et al. 1992). The branching pattern of the tree defines a nested statistical design that can embrace the uncertainty in the tree (Templeton and Sing 1993). The second step is to test the null hypothesis of no association of haplotypes or clades of haplotypes with geography. If this null hypothesis is not rejected, there is no evidence that past events have left a detectable mark upon the haplotype trees with the available sample sizes and geographical coverage. This null hypothesis is tested with an exact, nested permutation test that simulates the null distribution of no geographical associations. Third, the pattern of association within each clade showing a significant geographical association is examined using coalescent theory to predict the type of event indicated by that pattern, such as past fragmentation events (population splits), range expansions, gene flow, and so on. These coalescent predictions were validated with an extensive data set of actual phylogeographic events known from prior evidence and were found to be accurate (Templeton 2008, Templeton 2009a). Moreover, all events or processes inferred from NCPA that were not known *a priori* in these data sets were regarded as a conservative indicator of the false positive rate (as some may have actually occurred even though not known from other evidence). The per-clade false positive rate was about 5 percent, the set nominal rate, and since nested clades are statistically independent, it is a simple matter to correct for multiple testing across the whole haplotype tree. The validation data set also showed that multiple historical inferences could be made from the same haplotype tree without any detectable interference with one another; so multiple events can be inferred from a single tree.

No other method of phylogeographic inference has been so extensively validated with real data sets. These low false-positive rates have also been validated by computer simulations executed by Knowles and Maddison (2002) and by Panchal and Beaumont (2010). Both of these simulation papers reported high false-positive error rates, but that was solely because they did not actually implement MLNCPA. Once that error was corrected, the simulation results of Knowles and Maddison (2002) actually yield a false positive error rate of 0 percent (Templeton 2009b) and those of Panchal and Beaumont (2010) an error rate less than the nominal rate of 5 percent (Templeton 2015).

The fourth and final step is to integrate the inferences across the haplotype trees through a cross-validation procedure. Only those inferences that are detected by more than one genomic region that match in inference type (e.g., a range expansion event) and geographical location (e.g., a range expansion event out of Africa into Eurasia) are retained. A likelihood ratio testing framework is then used to test the null hypothesis that all matching events occurred at the same time. If this null hypothesis is not rejected, this is regarded as a single event, not because it is necessarily true but simply because there is no significant evidence to indicate that more than one event occurred. If the null hypothesis is rejected, the individual inferences are subject to homogeneity tests to see if there is evidence for multiple events of the same type occurring in the same regions but at different times, with each inference of a separate time having to be supported by multiple loci as well. Because gene flow is a recurrent event, inferences of gene flow between the same two geographical regions are tested for concordance by testing the null hypothesis that the two regions were genetically isolated for a given time period. If this null hypothesis is rejected, the inference of gene flow between the regions in that time period is retained. The likelihood ratio tests of these null hypotheses concerning time have been extended to test other phylogeographic hypotheses. For example, as shown earlier, the logical testing of replacement (no admixture) was based on observing events in the area assumed to have been replaced that occurred before the assumed replacement event. Likelihood ratio tests were also developed to test the null hypothesis that no events occurred in the geographical area of interest before the assumed replacement, using the oldest tail of the 95 percent confidence region of the estimated time of the event as a conservative threshold for "before" (Templeton 2004a; Templeton 2004b; Templeton 2009a).

Figure 6.3 shows the MLNCPA inferences based on statistically significant, cross-validated inferences of human haplotype data from mtDNA, Y-DNA, and nuclear DNA. Fifteen out of 24 genomic regions have a strong signal of a population expansion out of Africa into Eurasia, but the null hypothesis that all of these range expansions occurred at the same time was strongly rejected ($p=3.89\times10^{-15}$). The 15 out-of-Africa expansions cluster into three distinct groupings, and in all cases there was strong concordance within a group ($p=0.95$ for the most recent expansion out-of-Africa, $p=0.51$ for the middle expansion, and $p=0.62$ for the oldest expansion), with genetically estimated times of 130,000 years ago, 650,000 years ago, and 1.9 mya (million years ago) respectively. The oldest range expansion out of Africa at 1.9 mya represents the initial colonization of Eurasia by the

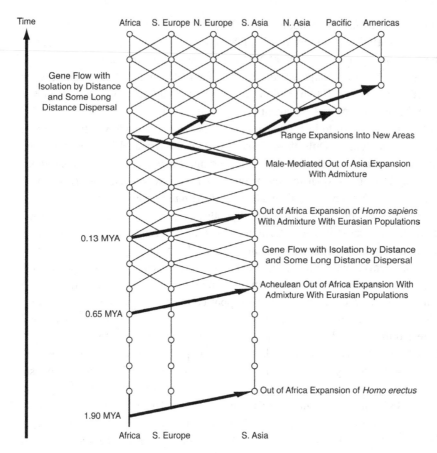

Figure 6.3
Inferences about recent human evolution based upon hypothesis testing with no prior model of human evolution using MLNCPA. The earliest event detected is an expansion out of Africa into Eurasia dated at 1.9 mya (million years ago), as shown by the thick arrow near the bottom, followed by two subsequent expansions of humans out of Africa. The thick arrows showing these subsequent expansions are overlaid upon the Eurasian human lineages to indicate significant admixture. After the mid-Pleistocene expansion, humans displayed significant gene flow among geographic populations, as indicated by the thin diagonal lines. There were additional expansions of human populations during the last 50,000 years followed by the establishment of gene flow among populations.

genus *Homo*. Although several genomic regions indicated limited gene flow between Africa and Eurasia between the times of this initial expansion and the mid-Pleistocene expansion at 0.65 mya, the null hypothesis of isolation was not rejected, so no trellis-like structure is indicated in this time period in figure 6.3.

The null hypothesis that the mid-Pleistocene expansion involved no admixture with Eurasian populations is rejected (p=0.0346). Hence, the mid-Pleistocene expansion of people out of Africa was marked by interbreeding with Eurasian populations. This mid-

Pleistocene expansion was a novel feature identified by MLNCPA that was not part of any of the other models of human evolution discussed earlier, including OAR.

Interbreeding continued through recurrent gene flow between Africa and Eurasia in the time interval between the mid-Pleistocene expansion and the last major out-of-Africa population expansion at 0.13 mya (the null hypothesis of no gene flow in the time interval between these expansion events is rejected with $p = 1.8 \times 10^{-8}$), so now a trellis-like structure is indicated in figure 6.3. Obviously, humans by 650,000 years ago had the capability of moving both in and out of Africa and did so on a recurrent basis, at least on a time scale of tens of thousands of years (the limit of the genetic resolution in this time interval). The inference of recurrent dispersal and gene flow is concordant with archaeological and paleoclimatic data (Groucutt et al. 2015; Jennings et al. 2015; Parton, Farrant, et al. 2015; Parton, White, et al. 2015).

The last of the major expansions out of Africa is genetically dated to 130,000 years ago, but was primarily limited to the southern tier of Eurasia, with a later expansion into Northern Eurasia, the Pacific, and finally the Americas (figure 6.3). The null hypothesis of no admixture (replacement) associated with this most recent out-of-Africa expansion is rejected with $p < 10^{-17}$. This is a definitive rejection of OAR as this most recent expansion is the one hypothesized to be the replacement event under OAR. MLNCPA yields a mostly out-of-Africa model in which there was a major expansion of modern humans out-of-Africa but with limited admixture with the Eurasian populations that they encountered rather than total replacement. The amount of admixture with all Eurasian populations was estimated to be about 10 percent. This inference has been supported by ancient DNA studies that indicate an admixture rate of 3.4 to 7.3 percent with Neanderthals (Lohse and Frantz 2014) and 4 to 6 percent from the Denisovan specimen in modern Melanesians and other modern Asian and Indonesian populations (Mendez et al. 2012; Povysil and Hochreiter 2014; Reich et al. 2010). Hence, as inferred originally through MLNCPA and coherent ABC hypothesis testing, OAR has been overwhelmingly statistically falsified in favor of limited admixture.

Integrating Genetic and Non-Genetic Data

The initial genetic papers supporting OAR through hypothesis compatibility ignored much of the relevant paleontological, archaeological, and paleoclimatic literature (Clark and Lindly 1989). Unfortunately, this feature continues to this day. Inferences about human evolution would be greatly strengthened by integrating genetic data with more traditional anthropological data.

A Bayesian framework provides a straightforward method for such integration. The relevant paleontological, archaeological, and paleoclimatic information, as well as previous genetic-based studies, can all be used to define the priors used in the analysis of new genetic or anthropological data. As mentioned earlier, Fagundes et al. (2007) presented a Bayesian analysis of models of human evolution, albeit flawed in its treatment of the logical relationships among these models and their construction of priors. The priors used by Fagundes and colleagues are given in their online supporting information. What is most noticeable is the complete absence of any literature review in defining these priors. For example, they placed a prior probability of 0 upon the mid-Pleistocene expansion out of Africa despite prior supporting evidence from genetic studies (Templeton 2002), archaeological studies (Bar-Yosef and Belfer-Cohen 2001, Hou et al. 2000), paleontological evidence (Aguirre and Carbonell 2001, Bar-Yosef and Belfer-Cohen 2001), and paleoclimatic studies (Trauth et al. 2007). All of these studies and more were available to suggest a non-zero prior for a mid-Pleistocene expansion, but Fagundes and colleagues (2007) not only ignored this literature, but constructed a prior that was incompatible with this literature and effectively claimed that all of this prior evidence was false with absolute certainty. As another example, Fagundes et al. (2007) assigned a uniform prior over the interval of 1,600 to 4,000 generations ago for the timing of the last out-of-Africa expansion event. Using a generation time of 20 years, this corresponds to a uniform prior over the time interval of 32,000 to 80,000 years ago. However, prior genetic studies estimated the time of expansion at 130,000 years ago from mtDNA, Y-DNA, and multiple nuclear loci (Templeton 2004b)—one of the few multi-locus estimates available in 2007. Paleontological (Grun et al. 2005) and archaeological (Vanhaeren et al. 2006) studies existed before 2007 that indicated that modern humans had already spread outside of sub-Saharan Africa more than 100,000 years ago. The uniform prior used by Fagundes and colleagues was incompatible with all this prior evidence, effectively assuming that all these prior studies were false with absolute certainty. Note also that the time of an event in the past has a natural range from 0 (very recent) to infinity (very old). Concerning the use of a uniform prior with a finite range on a variable that can exceed that range, Garthwaite et al. (2005, p. 688) wrote "... Unless the range [a, b] represents absolute physical limits to the possible values of the quantity, ... it is unreasonable to give zero probability to the event that the quantity lies outside the range." Hence, the prior used by Fagundes, et al. violates recommended Bayesian practice. In contrast, Templeton (2004a, 2004b) used gamma distributions for assigning probabilities to the timing of past events. The gamma probability distribution is defined on the range from 0 to infinity but with parameters that can concentrate the probabilities into a finite range without totally excluding the possibility that the event's time occurred outside the preferred range. Using a uniform distribution on a finite range when the variable can lie outside that range can cause serious statistical artifacts in a Bayesian analysis, so a gamma distribution is far more appropriate in any properly constructed analysis.

Although the Bayesian approach has much potential to integrate genetic and anthropological data, that potential has yet to be realized. It will require both geneticists and anthropologists to read the literature of the other area and to incorporate that information into priors of appropriate mathematical form. That has yet to be done.

MLNCPA provides another statistical framework for using genetic and anthropological data in making inference about human evolution. As noted above, MLNCPA inferred three major out-of-Africa expansions during the last 2 million years. The oldest expansion was genetically dated to 1.9 mya with a 95% confidence range of 0.99 to 3.10 mya. Fossil dating of the expansion of *Homo erectus* from Africa is by 1.85 mya into southern Eurasia (Ferring et al. 2011) and by 1.7 mya into far East Asia (Zhu et al. 2008). The null hypothesis that the genetic and fossil data are detecting the same event can be tested by examining the overlap of the paleontological times with the 95% confidence intervals of the genetic time estimator. There is great overlap, so the hypothesis that both types of data are detecting the same event cannot be rejected. Similarly, the 95% genetic time interval overlaps extensively with a wet cycle characterized by an expansion of grasslands and a faunal turnover favoring grassland species (Behrensmeyer 2006, Bobe and Behrensmeyer 2004, deMenocal 2011), so the null hypothesis is accepted that the genetic, paleoclimatic, and paleoecological data are all concordant.

The next out-of-Africa expansion is genetically dated at 0.65 mya with a 95% confidence interval of 0.3917–0.9745 mya—an interval that broadly overlaps with the expansion out of Africa into Eurasia of the Acheulean tool culture, whose sites are not widespread in Eurasia until about 0.6 to 0.8 mya, particularly eastern Asia (Hou et al., 2000). This 95% confidence interval also overlaps with the next major wet period in this region and expansion of grasslands (deMenocal 2011) and the expansion of several African savanna species into Eurasia (Martinez-Navarro and Rabinovich 2011). Once again, the null hypothesis of concordance of these diverse data sets is not rejected.

The last expansion event has a 95% genetic confidence range of 96,500 to 169,300 years ago. Many anatomically modern traits and archaeological features first appeared in sub-Saharan Africa around 200,000 years ago (McDougall et al. 2005), and then first appeared out of sub-Saharan Africa in northern Africa, the Arabian Peninsula, and Levant between 130,000 and 125,000 years ago (Armitage et al. 2011, Groucutt et al. 2015, Grun et al. 2005, Petraglia et al. 2010, Petraglia 2011, Vanhaeren et al. 2006). Anatomically modern human fossils have been found in far eastern Asia dating to perhaps as early as 127,000 years ago but no later than 110,000 years ago (Bae et al. 2014, Jin et al. 2009, Liu et al. 2010, Liu et al. 2015). Paleoclimatic and archaeological data also indicate dispersal events in this time interval (Jennings et al 2015, Parton, Farrant et al. 2015). Once again, the null hypothesis is accepted of concordance of the genetic date from MLNCPA with the anthropological and paleoclimatic evidence. In contrast, the date of around 55,000 to 60,000 usually given by proponents of OAR is clearly discrepant with the latest archaeological and fossil finds and is placed in the most severe dry period in the Sahara in the last 300,000 years (deMenocal 2011).

Discussion

Given that OAR had been definitively rejected both logically and statistically almost two decades ago by genetic data, why did geneticists and other scientists continue to embrace OAR until very recently (and some still do)? This instance of ignoring incompatible data does not fit into the mold of a Kuhnian paradigm shift because alternative paradigms were readily available throughout this entire time period that were compatible with the genetic, fossil, archaeological, and paleoclimatic data. Perhaps the answer resides outside the realm of science. The rapid and strong embrace of the phrase "mitochondrial Eve" reveals a strong cultural tie-in for both scientists and the general public in the Western world. The abandonment of the OAR paradigm would require not only abandonment of a scientific model, but of a cultural one as well. Moreover, this is a paradigm that is very close to us as humans—our very own roots writ large. Scientific objectivity often becomes more difficult to maintain when the subject of our science is ourselves. For example, human populations are commonly portrayed as distinct branches on a population tree (e.g., the first two panels in figure 6.2) despite the strong rejection of the null hypothesis of a tree-like structure in every human genetic data set that has been tested (Templeton 2013). These population trees that are published in prestigious scientific journals and textbooks represent not only bad scientific practice, but also actively mislead the general public about important social issues, such as the genetic basis of "race" in humans (Templeton 2013). All of this misleading science could end with the simple requirement that testable hypothesis should be tested, ideally with integrated genetic and anthropological data sets.

Acknowledgments

I am grateful for the support of the KLI Institute, and to Dr. Jeffrey Schwartz for organizing an excellent and intellectually exciting symposium, of which this paper was a part. Finally, I thank Dr. Rob DeSalle for his excellent suggestions on an earlier draft of this chapter.

References

Aguirre, E., and E. Carbonell. 2001. "Early Human Expansions into Eurasia: The Atapuerca Evidence." *Quaternary International* 75:11–18.

Armitage, S. J., S. A. Jasim, A. E. Marks, A. G. Parker, V. I. Usik, and H. Uerpmann. 2011. "The Southern Route out of Africa: Evidence for an Early Expansion of Modern Humans into Arabia." *Science* 331(6016):453–456.

Bae, C. J., W. Wang, J. Zhao, S. Huang, F. Tian, and G. Shen. 2014. "Modern Human Teeth from Late Pleistocene Luna Cave (Guangxi, China)." *Quaternary International* 354:169–183.

Bar-Yosef, O., and A. Belfer-Cohen. 2001. "From Africa to Eurasia—Early Dispersals." *Quaternary International* 75:19–28.

Behrensmeyer, A. K. 2006. "Climate Change and Human Evolution." *Science* 311(5760):476–478.

Bobe, R., and A. K. Behrensmeyer. 2004. "The Expansion of Grassland Ecosystems in Africa in relation to Mammalian Evolution and the Origin of the Genus *Homo.*" *Palaeogeography Palaeoclimatology Palaeoecology* 207(3–4):399–420.

Cann, R. L., M. Stoneking, and A. C. Wilson. 1987. "Mitochondrial DNA and Human Evolution." *Nature* 325:31–36.

Clark, G. A., and J. M. Lindly. 1989. "Modern Human Origins in the Levant and Western Asia: the Fossil and Archeological Evidence." *American Anthropologist.* 91:962–985.

Coon, C. S. 1962. *The Origin of Races.* New York: Knopf.

deMenocal, Peter B. 2011. "Climate and Human Evolution." *Science* 331(6017):540–542.

Dobzhansky, T. 1944. "On Species and Races of Living and Fossil Man." *American Journal of Physical Anthropology* 2(3):251–265.

Eswaran, V., H. Harpending, and A. R. Rogers. 2005. "Genomics Refutes an Exclusively African Origin of Humans." *Journal of Human Evolution* 49(1):1–18.

Fagundes, N. J. R., N. Ray, M. Beaumont, S. Neuenschwander, F. M. Salzano, S. L. Bonatto, and L. Excoffier. 2007. "Statistical Evaluation of Alternative Models of Human Evolution." *Proceedings of the National Academy of Sciences* 104(45):17614–17619.

Ferring, R., O. Oms, J. Agustí, F. Berna, M. Nioradze, T. Shelia, M. Tappen, A. Vekua, D. Zhvania, and D. Lordkipanidze. 2011. "Earliest Human Occupations at Dmanisi (Georgian Caucasus) Dated to 1.85–1.78 Ma." *Proceedings of the National Academy of Sciences* 108(26):10432–10436.

Gabriel, K. R. 1969. "Simultaneous Test Procedures—Some Theory of Multiple Comparisons." *Annals of Mathematical Statistics* 40:224–250.

Garthwaite, P. H., J. B. Kadane, and A. O'Hagan. 2005. "Statistical Methods for Eliciting Probability Distributions." *Journal of the American Statistical Association* 100(470):680–700.

Groucutt, H. S., M. D. Petraglia, G. Bailey, E. M. L. Scerri, A. Parton, L. Clark-Balzan, R. P. Jennings, L. Lewis, J. Blinkhorn, N. A. Drake, P. S. Breeze, R. H. Inglis, M. H. Devès, M. Meredith-Williams, N. Boivin, M. G. Thomas, and A. Scally. 2015. "Rethinking the Dispersal of *Homo sapiens* out of Africa." *Evolutionary Anthropology: Issues, News, and Reviews* 24(4):149–164.

Groucutt, H. S., E. M. L. Scerri, L. Lewis, L. Clark-Balzan, J. Blinkhorn, R. P. Jennings, A. Parton, and M. D. Petraglia. 2015. "Stone Tool Assemblages and Models for the Dispersal of *Homo sapiens* out of Africa." *Quaternary International* 382:8–30.

Grun, R., C. Stringer, F. McDermott, R. Nathan, N. Porat, S. Robertson, L. Taylor, G. Mortimer, S. Eggins, and M. McCulloch. 2005. "U-series and ESR Analyses of Bones and Teeth Relating to the Human Burials from Skhul." *Journal of Human Evolution* 49(3):316–334.

Harding, R. M., S. M. Fullerton, R. C. Griffiths, J. Bond, M. J. Cox, J. A. Schneider, D. S. Moulin, and J. B. Clegg. 1997. "Archaic African and Asian Lineages in the Genetic Ancestry of Modern Humans." *American Journal of Human Genetics* 60:772–789.

Hou, Y. M., R. Potts, B. Y. Yuan, Z. T. Guo, A. Deino, W. Wang, J. Clark, G. M. Xie, and W. W. Huang. 2000. "Mid-Pleistocene Acheulean-like Stone Technology of the Bose Basin, South China." *Science* 287(5458):1622–1626.

Howells, W. W. 1942. "Fossil Man and the Origin of Races." *American Anthropologist* 44(2):182–193.

Jennings, R. P., J. Singarayer, E. J. Stone, U. Krebs-Kanzow, V. Khon, K. H. Nisancioglu, M. Pfeiffer, X. Zhang, A. Parker, A. Parton, H. S. Groucutt, T. S. White, N. A. Drake, and M. D. Petraglia. 2015. "The Greening of Arabia: Multiple Opportunities for Human Occupation of the Arabian Peninsula during the Late Pleistocene Inferred from an Ensemble of Climate Model Simulations." *Quaternary International* 382:181–199.

Jin, C., W. Pan, Y. Zhang, Y. Cai, Q. Xu, Z. Tang, W. Wang, Y. Wang, J. Liu, D. Qin, R. L. Edwards, and H. Cheng. 2009. "The *Homo sapiens* Cave Hominin Site of Mulan Mountain, Jiangzhou District, Chongzuo, Guangxi with Emphasis on Its Age." *Chinese Science Bulletin* 54(21):3848–3856.

Knowles, L. L., and W. P. Maddison. 2002. "Statistical Phylogeography." *Molecular Ecology* 11:2623–2635.

Kuhn, Thomas S. 1962. *The Structure of Scientific Revolutions*. Chicago: University of Chicago Press.

Lindley, D. V. 1965. *Introduction to Probability and Statistics from a Bayesian Viewpoint*. Cambridge: Cambridge University Press.

Liu, W., C. Jin, Y. Zhang, Y. Cai, S. Xing, X. Wu, H. Cheng, R. L. Edwards, W. Pan, D. Qin, Z. An, E. Trinkaus, and X. Wu. 2010. "Human Remains from Zhirendong, South China, and Modern Human Emergence in East Asia." *Proceedings of the National Academy of Sciences* 107(45):19201–19206.

Liu, W., M. Martinon-Torres, Y. Cai, S. Xing, H. Tong, S. Pei, M. J. Sier, X. Wu, R. L. Edwards, H. Cheng, Y. Li, X. Yang, J. M. Bermudez de Castro, and X. Wu. 2015. "The Earliest Unequivocally Modern Humans in Southern China." *Nature* 526(7575):696–699.

Lohse, K., and L. A. F. Frantz. 2014. "Neandertal Admixture in Eurasia Confirmed by Maximum-Likelihood Analysis of Three Genomes." *Genetics* 196(4):1241–1251.

Martínez-Navarro, B., and R. Rabinovich. 2011. "The Fossil Bovidae (Artiodactyla, Mammalia) from Gesher Benot Ya'aqov, Israel: Out of Africa during the Early-Middle Pleistocene Transition." *Journal of Human Evolution* 60(4):375–386.

McDougall, I., F. H. Brown, and J. G. Fleagle. 2005. "Stratigraphic Placement and Age of Modern Humans from Kibish, Ethiopia." *Nature* 433(7027):733–736.

Mendez, F. L., J. C. Watkins, and M. F. Hammer. 2012. "Global Genetic Variation at OAS1 Provides Evidence of Archaic Admixture in Melanesian Populations." *Molecular Biology and Evolution* 29(6):1513–1520.

Oaks, J. R., J. Sukumaran, J. A. Esselstyn, C. W. Linkem, C. D. Siler, M. T. Holder, and R. M. Brown. 2013. "Evidence for Climate-Driven Diversification? A Caution for Interpreting ABC Inferences of Simultaneous Historical Events." *Evolution* 67(4):991–1010.

Panchal, M., and M. A. Beaumont. 2010. "Evaluating Nested Clade Phylogeographic Analysis under Models of Restricted Gene Flow." *Systematic Biology* 59(4):415–432.

Parton, A., A. R. Farrant, M. J. Leng, M. W. Telfer, H. S. Groucutt, M. D. Petraglia, and A. G. Parker. 2015. "Alluvial Fan Records from Southeast Arabia Reveal Multiple Windows for Human Dispersal." *Geology* 43(4):295–298.

Parton, A., T. S. White, A. G. Parker, P. S. Breeze, R. Jennings, H. S. Groucutt, and M. D. Petraglia. 2015. "Orbital-Scale Climate Variability in Arabia as a Potential Motor for Human Dispersals." *Quaternary International* 382:82–97.

Petraglia, M. D. 2011. "Archaeology: Trailblazers across Arabia." *Nature* 470(7332):50–51.

Petraglia, M. D., M. Haslam, D. Q. Fuller, N. Boivin, and C. Clarkson. 2010. "Out of Africa: New Hypotheses and Evidence for the Dispersal of *Homo sapiens* along the Indian Ocean Rim." *Annals of Human Biology* 37(3):288–311.

Platt, J. R. 1970. *Perception and Change: Projections for Survival*. Ann Arbor: The University of Michigan Press.

Popper, K. R. 1959. *The Logic of Scientific Discovery*. London: Hutchinson.

Povysil, G., and S. Hochreiter. 2014. "Sharing of Very Short IBD Segments between Humans, Neandertals, and Denisovans." *bioRxiv*. http://dx.doi.org/10.1101/003988.

Ray, N., M. Currat, P. Berthier, and L. Excoffier. 2005. "Recovering the Geographic Origin of early Modern Humans by Realistic and Spatially Explicit Simulations." *Genome Res.* 15(8):1161–1167.

Reece, J. S., B. W. Bowen, K. Joshi, V. Goz, and A. Larson. 2010. "Phylogeography of Two Moray Eels Indicates High Dispersal throughout the Indo-Pacific." *Journal of Heredity* 101(4):391–402.

Reich, D., R. E. Green, M. Kircher, J. Krause, N. Patterson, E. Y. Durand, B. Viola, A. W. Briggs, U. Stenzel, P. L. F. Johnson, T. Maricic, J. M. Good, T. Marques-Bonet, C. Alkan, Q. Fu, S. Mallick, H. Li, M. Meyer, E. E. Eichler, M. Stoneking, M. Richards, S. Talamo, M. V. Shunkov, A. P. Derevianko, J. Hublin, J. Kelso, M. Slatkin, and S. Paabo. 2010. "Genetic History of an Archaic Hominin Group from Denisova Cave in Siberia." *Nature* 468(7327):1053–1060.

Robinson, W. S. 1950. "Ecological Correlations and the Behavior of Individuals." *American Sociological Review* 15(3):351–357. doi: 10.2307/2087176.

Stringer, C. B., and P. Andrews. 1988. "The Origin of Modern Humans." *Science* 239:1263–1268.

Takahata, N., S. Lee, and Y. Satta. 2001. "Testing Multiregionality of Modern Human Origins." *Mol Biol Evol* 18(2):172–183.

Templeton, A. R. 2002. "Out of Africa Again and Again." *Nature* 416(6876):45–51.

Templeton, A. R. 2004a. "A Maximum Likelihood Framework for Cross Validation of Phylogeographic Hypotheses." In *Evolutionary Theory and Processes: Modern Horizons*, edited by S. P. Wasser, 209–230. Dordrecht, The Netherlands: Kluwer Academic Publishers.

Templeton, A. R. 2004b. "Statistical Phylogeography: Methods of Evaluating and Minimizing Inference Errors." *Molecular Ecology* 13(4):789–809.

Templeton, A. R. 2007. "Perspective: Genetics and Recent Human Evolution." *Evolution* 61(7):1507–1519.

Templeton, A. R. 2008. "Nested Clade Analysis: an Extensively Validated Method for Strong Phylogeographic Inference." *Molecular Ecology* 17(8):1877–1880.

Templeton, A. R. 2009a. "Statistical Hypothesis Testing in Intraspecific Phylogeography: Nested Clade Phylogeographical Analysis vs. Approximate Bayesian Computation." *Molecular Ecology* 18(2):319–331.

Templeton, A. R. 2009b. "Why Does a Method that Fails Continue to Be Used: the Answer." *Evolution* 63(4):807–812.

Templeton, A. R. 2010. "Coherent and Incoherent Inference in Phylogeography and Human Evolution." *Proceedings of the National Academy of Sciences* 107(14):6376–6381.

Templeton, A. R. 2013. "Biological Races in Humans." *Studies in History and Philosophy of Science Part C: Studies in History and Philosophy of Biological and Biomedical Sciences* 44(3):262–271.

Templeton, A. R. 2015. "Population Biology and Population Genetics of Pleistocene Hominins." In *Handbook of Paleoanthropology*, edited by W. Henke and I. Tattersall, 2331–2370. Heidelberg, New York, Dordrecht, London: Springer.

Templeton, A. R., A. G. Clark, K. M. Weiss, D. A. Nickerson, E. Boerwinkle, and C. F. Sing. 2000. "Recombinational and Mutational Hotspots within the Human *Lipoprotein Lipase* Gene." *American Journal of Human Genetics* 66:69–83.

Templeton, A. R., K. A. Crandall, and C. F. Sing. 1992. "A Cladistic Analysis of Phenotypic Associations with Haplotypes Inferred from Restriction Endonuclease Mapping and DNA Sequence Data. III. Cladogram Estimation." *Genetics* 132:619–633.

Templeton, A. R., and C. F. Sing. 1993. "A Cladistic Analysis of Phenotypic Associations with Haplotypes Inferred from Restriction Endonuclease Mapping. IV. Nested Analyses with Cladogram Uncertainty and Recombination." *Genetics* 134:659–669.

Trauth, M. H., M. A. Maslin, A. L. Deino, M. R. Strecker, A. G. N. Bergner, and M. Duhnforth. 2007. "High- and Low-Latitude Forcing of Plio-Pleistocene East African Climate and Human Evolution." *Journal of Human Evolution* 53(5):475–486.

Vanhaeren, M., F. d'Errico, C. Stringer, S. L. James, J. A. Todd, and H. K. Mienis. 2006. "Middle Paleolithic Shell Beads in Israel and Algeria." *Science* 312(5781):1785–1788.

Weidenreich, F. 1946. *Apes, Giants, and Man*. Chicago: University of Chicago Press.

Wolpoff, M., and A. Thorne. 1991. "The Case against Eve." *New Scientist* 22:37–41.

Wolpoff, M., A. G. Thorne, F. H. Smith, D. W. Frayer, and G. G. Pope. 1994. "Multiregional Evolution: a World-Wide Source for Modern Human Populations." In *Origins of Anatomically Modern Humans*, edited by M. H. Nitecki and D. V. Nitecki, 175–200. New York: Plenum Press.

Wolpoff, M. H. 1996. "Interpretations of Multiregional Evolution." *Science* 274(5288):704–706.

Zhu, R. X., R. Potts, Y. X. Pan, H. T. Yao, L. Q. Lü, X. Zhao, X. Gao, L. W. Chen, F. Gao, and C. L. Deng. 2008. "Early Evidence of the Genus *Homo* in East Asia." *Journal of Human Evolution* 55(6):1075–1085.

7 Out of Africa: The Evolution and History of Human Populations in the Southern Dispersal Zone

Michael D. Petraglia and Huw S. Groucutt

Introduction

The evolution and history of our species, *Homo sapiens*, is one of the hottest topics in paleoanthropology. Scholars, students, and the public at large have been fascinated with our origins and how our species spread around the world, to inhabit virtually every corner of the Earth. Many of the scholarly debates on our origins and dispersal from Africa have centered on historical questions, especially on the timing and place of population movements, which are commonly implied to resemble "migrations" of recent history. Somewhat less frequently, researchers have posed questions about the relations between macro- and microevolutionary processes and the selective forces in shaping human populations.

The aim of this chapter is to briefly review two of the most currently prominent Out-of-Africa dispersal models, examining in particular how geneticists and archaeologists have characterized the movements of human populations out of Africa and through southern Asia. The "single, rapid coastal dispersal model" and the "early multiple dispersal model" will then be examined relative to the latest findings to emerge from Asia. Though understanding the history of our species is of great interest, we will emphasize the need to reorient research in an evolutionary framework, examining the influence of climate change on structuring Late Pleistocene demography and cultural innovations.

Out-of-Africa Models: The Southern Dispersal

The geographic areas bordering the Indian Ocean have been the focus for studying the movement of modern humans towards Sahul (e.g., Field and Lahr 2005, Oppenheimer 2009, Petraglia et al. 2010, Boivin et al. 2013, Erlandson et al. 2015, Groucutt et al. 2015). The Horn of Africa and Northeast Africa have been viewed as the launching pads for human dispersals eastwards across Eurasia (e.g., Pagani et al. 2015, Breeze et al. 2016). While it has been recognized that *Homo sapiens* moved out of Africa by 100,000 years ago (ka), as clearly shown by fossil evidence in the Levant, this dispersal has been considered to be a failed event (e.g., Mellars 2006). Though various scenarios have been put forward to account for human dispersals, two that have emerged in recent years, and the subject of contemporary debate, are the "early" pre-Toba and the "late" coastal dispersal models (for background, see Groucutt et al. 2015). The models are united by focus on the importance of southern Asia as a key geographic zone, though they differ in important historical details of "when" and "how" populations moved across Asia.

The single, rapid coastal dispersal model at 60,000 years ago

Subsequent to the alleged failed dispersal of *Homo sapiens* to the Levant, the movement of modern humans towards Sahul has often been depicted as following a coastal pathway (e.g., Lahr and Foley 1994, 2000, Field and Lahr 2005). Highly influential genetic studies took this characterization to new heights, as mitochondrial DNA sampling of contemporary populations in Africa and across Eurasia indicated that haplogroup M was a marker of early population divergence, allowing mapping of the only "successful" movement of populations across southern Asia. Inferred coalescence ages suggested that this occurred around somewhere between 65–60 ka (Quintana-Murci et al. 1999). At the same time, the virtual absence of haplogroup M in the Levant, and its high frequency in India, was taken as indication that populations used coastlines to migrate to Southeast Asia and beyond. Echoing some of the same key points, Macaulay and colleagues (2005) sampled "relict populations" and proposed that at about 65 ka, there was a single dispersal from East Africa through to India and towards Sahul. Given the similar estimated ages of haplogroups M, N, and R, it was suggested that several hundred individuals exited Africa, and given the similarity of the haplogroups across southern Asia, a rapid movement, employing coasts, was surmised.

Claims for archaeological support for the genetic models came from Mellars (2006) who argued that there was support for a single successful dispersal that saw the rapid expansion of modern humans along the Indian Ocean rim at ca. 60 ka. In laying the foundation of this argument, Mellars contended that archaeological sites in India and Sri Lanka had small crescentic stone tools that were similar to those found in Howiesons Poort and analogous assemblages in southern and eastern Africa, thus indicating the original source for these stone tool industries. Moreover, similarities were observed in beads and in geometric designs. Expanding and reinforcing this model, Mellars and colleagues (2013) argued that genetic and archaeological data were in broad agreement, converging on a dispersal between 60–50 ka, by culturally and symbolically advanced populations of modern humans. In constructing this model, a forceful argument was made that *Homo sapiens* in the Levant represented a failed dispersal event between ca. 100 and 70 ka and that it was inconceivable that they could have expanded at this time and contributed to the gene pool of modern Eurasian populations (see also Mellars 2006).

The early multiple dispersal model

An alternative model concerning the expansion of *Homo sapiens* outside Africa was emphasized by our own interdisciplinary team, arguing that modern humans were present in India before and after the Toba volcanic super-eruption of ca. 75 ka (Petraglia et al.

2007). Lithic evidence was put forward that pre- and post-Toba Middle Paleolithic cores were similar to those in the Middle Stone Age of sub-Saharan Africa, thus suggesting the early presence of *Homo sapiens* in India. Additionally, it was argued that *Homo sapiens* was present based on resilience to a major volcanic event and the overlap with the earlier end of some genetic coalescence ages. Such arguments were reinforced with additional paleoenvironmental and archaeological field work across India (Petraglia et al. 2012), though others contended that the Toba super-eruption had catastrophic and decimating effects on ecological settings and hominins (Williams et al. 2009).

Several articles have since expanded on the early dispersal of *Homo sapiens* along the southern route, pulling together fossil, genetic, and archaeological information (Petraglia et al. 2010, Dennell and Petraglia 2012, Boivin et al. 2013). The gist of the argument was that multiple dispersals of *Homo sapiens* were probable, including both early and late phases, owing to ameliorated environmental conditions in the Saharo-Arabian zone, as favorable ecologies, including numerous rivers and lakes formed during pluvials (e.g., Scerri et al. 2014, Breeze et al. 2015). It was surmised, however, that substantial geographic differences in environments, and long-term climatic fluctuations, would have had a significant effect on Late Pleistocene populations, at times leading to population expansions and continuity and in other times, particularly in Arabia, population contractions and lineage extinctions. Though others argued that young mitochondrial DNA coalescence ages were due to sources of error in the data and the failure to examine multiple loci in a hypothesis-testing framework (e.g., Templeton 2002, 2005), others have explained this discrepancy by suggesting that early groups were demographically swamped by later population expansions and increases. The patchy geographic appearance of cultural innovations after ca. 45 ka, including symbolic items, bone tools, and microlithic industries, has been viewed as an adaptive response of hunter-gatherers to specific environmental challenges and increasing population sizes (Petraglia et al. 2009).

New Realizations: Fossils and Ancient Genomes

Though a number of arguments have been put forward to sustain the idea that human populations exited Africa after 60 ka, from both genetics (Fernandes et al. 2015) and archaeology (Kaifu et al. 2015), alternative genetic models and evidence have been advanced to support a more complex demographic scenario. Taking a hypothesis testing approach, Templeton (2002, 2005) examined multiple loci, arguing that the young mitochondrial DNA ages were due, in part, to sources of error in the genetic data, and secondly, that out of Africa expansions were much earlier, ranging as early as ca. 130 ka. More recent whole genome studies support this interpretation. Scally and Durbin (2012) have argued that mutation rates have been traditionally modeled as too fast, thereby calculating that the origin of our own species, and the dispersal of human populations across Asia, are older than realized. Genetic simulations of the L3 node have indicated that alternative models can be constructed, involving early and late dispersals (Groucutt et al. 2015).

Fossil evidence to demonstrate early Out-of-Africa movements has been geographically patchy, and early human fossils have often been difficult to assign to *Homo sapiens* and often lack precise chronological control (Dennell and Petraglia 2012, Dennell 2014). Though not without controversy, the recovery of fossils of *Homo sapiens* in Fuyan Cave, China, dating to at least 80 ka (Liu et al. 2015), has been taken as convincing evidence (Dennell 2015). This finding has now been shored up by the recovery of ancient DNA from a Neanderthal from the Altai, which showed interbreeding by ca. 100 ka, though an inferred lack of continuity with contemporary populations meant that this was also interpreted as a failed dispersal (Kuhlwilm et al. 2016). The recovery of ancient DNA was nevertheless striking, and there is now the distinct possibility that "relict" groups of contemporary populations in Asia will carry evidence for human expansion as early as 100 ka.

An Evolutionary Approach to Out-of-Africa Dispersals

Orbital and millennial-scale changes in climate across the Late Pleistocene, and the subsequent effects on terrestrial ecosystems, provide the context to understand some of the forces that shaped the evolution of human populations (Potts 2013). In order to understand the differentiation of human populations, emphasis has been placed on the need to understand small-scale demographic processes as well as spatial factors in influencing dispersals, contractions, and extinctions (Lahr and Foley 1998). The processes that underlay the evolution and differentiation of Pleistocene hominins, including *Homo sapiens*, have a long and contentious history in paleoanthropology. The purest forms of the multiregional and single origin models have been modified by their original proponents (Thorne and Wolpoff 2003, Stringer 2014) in favor of hybridization or assimilation models that argue for some degree of interbreeding and gene flow between differentiated populations (e.g., Templeton 2002, Trinkaus 2005).

Some of the foregoing theoretical propositions can be examined relative to the Late Pleistocene records of the Arabian Peninsula and Indian subcontinent. Though both landmasses form key and significant terrestrial routes as part of the southern dispersal, interdisciplinary archaeological fieldwork in these regions is still in its infancy (Petraglia et al. 2015). Nevertheless, as is immediately clear, the geographic and environmental characteristics and variability of both of these subcontinents differ substantially, which must have had a significant influence on the survival of human populations in the Late Pleistocene. Modeling of terrestrial environments clearly shows the difference of these landmasses on account of variations in monsoon weather systems across the last 125,000 years.

As illustrated by Boivin and colleagues (2013), for instance, during pluvials Arabia experienced increased rainfall, leading to the formation of sub-desert ecozones and Sahel vegetation (grassland and savanna, with some areas of woodland and shrubland). Though Arabia would have experienced increased levels of precipitation, resulting in the formation of rivers and lakes in most areas, there was substantial geographic and ecological structure. The central and southern zone would have been mostly characterized by plateaus, sand seas, savannahs and woodlands whereas the northern regions, especially in the Nefud Desert, would have been large sand seas with some grasslands. It is important to recognize that while Arabia experienced ameliorated conditions and wet environments at times in the Late Pleistocene, the region remained a relatively marginal setting, as arid landscapes and sand seas appear to have been present even during pluvials. On the other hand, during pluvials, the Indian subcontinent received increased levels of plentiful monsoonal rainfall, which would have resulted in a wide range of ecological conditions, including a tropical savannah/woodland-grassland mosaic, dry tropical woodland, moist tropical woodland and grassland mosaic, moist tropical woodlands, tropical montaine vegetation, and warm temperate hill and sub-montaine vegetation zones. During these wet periods, there would have been a reduction in less favorable zones, such as the Thar Desert (Blinkhorn 2013), though these sand seas would have been fully bordered by sub-desert ecozones and Sahel vegetation, and fringed by riverine corridors, marshes, and gallery forests especially along the nearby Indus River system.

With respect to the broader picture of hominin expansions in the southern zone, a compelling case can be mounted that humans would have moved across the Saharo-Arabian zone during wet periods. The combination of riverine routes and plentiful potable freshwater lakes in this region, coupled with a range of ecological settings that would have had a plentiful supply and range of fauna (e.g., small and large mammals, aquatic resources, fish, birds) and flora (e.g., seeds, tubers, fruits), would have facilitated movements of foraging populations. The lack of clarity over the spatial ranges covered by different hominin species at different times makes a variety of demographic scenarios possible. For instance, it's possible that Neanderthals extended further south than often implied, and that areas such as Arabia may have seen interactions between them and *Homo sapiens*. The ecological differences and the relative stability of environments in Arabia and the Indian subcontinent, however, likely influenced human settlement and mobility systems, the impression being that in Arabia, many Middle Paleolithic sites represent relatively short-term occupations (Scerri et al. 2015).

In glacial periods, the habitability of Arabia and India for hominins was radically different, which suggests a major divergence in the demographic trajectories of human populations. The Arabian Peninsula would have witnessed the drying-up of surface water and the expansion of arid deserts with minimal plant and animal resources. Limited areas, in southern Arabia, would have trained spatially reduced sub-desert/Sahel vegetation zones and limited woodlands in the mountainous zones. On the whole, populations facing increased aridity and deterioration of ecosystems would have contracted, and possibly competed for resources in the spatially limited, but more favorable, habitats. Groups in such geographically limited areas would have been isolated from other populations, thereby leading to fewer genetic interchanges and cultural isolation. While patches of somewhat more productive environments may have allowed populations to survive for longer, it is perhaps unlikely that they acted as long-term refugia. For the great majority of the Arabian Peninsula, environments would have been hostile to sustained human presence, and given that hunter-gatherers could not all contract to favorable zones, regional population extirpations and lineage extinctions are entirely possible, and in fact probable.

In contrast, and most importantly from an evolutionary and demographic perspective, even in glacial periods, the Indian subcontinent would have had a significant range of terrestrial environments, with only geographic reconfiguration and a spatial remodeling of ecological settings and faunal populations (Roberts et al. 2014). In times of reduced rainfall in South Asia, there would have been a substantial increase in tropical/savannah woodlands at the expense of the moist tropical woodland and grassland mosaic. Though the Thar Desert would have increased in size, perennial river systems such as the Indus would have persisted with the region still being flanked by sub-desert/Sahel vegetation zones. Hunter-gatherer populations would have had to meet the challenges of environmental downturns but also contingent events, such as the Toba volcanic super-eruption.

In general, after 50 ka, populations in South Asia appear to have been growing, with increasing levels of population structure (Karmin et al. 2015). As populations increasingly packed into the geographic areas of India, downturns in environments would have placed increasing subsistence stress on populations. A response to this may have been the emergence and development of advanced toolkits and weaponry between ca. 45–35 ka and increased levels of symbolism after 35 ka (Petraglia et al. 2009, Clarkson et al. 2009). Increased demographic trends in Terminal Pleistocene and expansion of Holocene farmers may have had a profound demographic influence on original hunter-gatherer populations, perhaps leading to genetic swamping of original hunter-gatherers and even replacement of populations.

Discussion and Conclusion

The story of Out-of-Africa has largely been historical in orientation, with a fascination on questions revolving around "who," "what," "when," and "where." Journals such as *Nature* and *Science* tend to encourage this historical approach, as articles on paleoanthropology often report the "oldest" new discovery or the most significant new fossil or archaeological find from a particular place and time. While there is certainly a place for sustained fieldwork and new discoveries, an evolutionary approach should be emphasized and encouraged.

Here we have emphasized examining population level changes through time, and examining how variability in environments through time and across geographic space influences the demography of hunter-gatherers. A variety of recent discoveries are radically changing knowledge on the origin and dispersal of *Homo sapiens*. These include the increasing evidence for ancient population structure (e.g. Gunz et al. 2009, Scerri et al. 2014), which has important implications for how genetic data is modeled. Likewise, phylogeographic interpretations of single locus genetic systems are increasingly problematic, as recently shown by the discovery of mitochondrial M haplogroups in early Europeans (Posth et al. 2016). In terms of hominin paleontology, we can expect to see the diversity of human populations as fossil discoveries are made (e.g., Crevecoeur et al. 2016).

New developments in the record of human evolution are rapidly rendering the simplistic certainties of the single rapid coastal 60 ka dispersal model obsolete. Interdisciplinary approaches, integrating genetics, biological anthropology, archaeology, and environmental sciences certainly help to contextualize new field findings, allowing us to address population expansions, contractions, and extirpations. An evolutionary approach, incorporating well-integrated interdisciplinary approaches, is essential to understand the selective forces that shaped *Homo sapiens*. Developments such as the increasing application of ancient DNA promise radical changes in our understanding of human evolution, but these are strongest when situated in a genuinely multidisciplinary framework. As we have explored in this chapter, archaeological and environmental records are crucial to explain the nature of changes in Southwest and South Asia in the Late Pleistocene. The notion that our species appeared one day in East Africa and subsequently spread in a manner akin to a historical period migration must be firmly rejected. The emerging story is much more complicated, and much more interesting.

Acknowledgments

Funding for field research in Arabia and South Asia was provided by the European Research Council (No. 295719), the British Academy, the Leverhulme Trust, the National Geographic Foundation, and the Leakey Foundation. Permission to conduct archaeological fieldwork was granted by the Saudi Commission for Tourism and National Heritage and the Archaeological Survey of India. Petraglia thanks Jeff Schwartz and the KLI for the invitation to participate in the stimulating workshop in Vienna. Robin Dennell and Alan Templeton provided helpful comments on a draft of this chapter.

References

Blinkhorn, J. 2013. "A New Synthesis of Evidence for the Upper Pleistocene Occupation of 16R Dune and Its Southern Asian Context." *Quaternary International* 300: 282–291.

Boivin, N., D. Q. Fuller, R. Dennell, R. Allaby, and M. D. Petraglia. 2013. "Human Dispersal across Diverse Environments of Asia during the Upper Pleistocene." *Quaternary International* 300: 32–47.

Breeze, P. S., H. S. Groucutt, N. A. Drake, T. S. White, R. P. Jennings, and M. D. Petraglia. 2016. "Palaeo-hydrological Corridors for Hominin Dispersals in the Middle East~ 250–70,000 Years Ago." *Quaternary Science Reviews* 144: 155–185.

Clarkson, C., M. Petraglia, R. Korisettar, M. Haslam, N. Boivin, A. Crowther, P. Ditchfield, D. Fuller, P. Miracle, C. Harris, K. Connell, H. James, and J. Koshy. 2009. "The Oldest and Longest Enduring Microlithic Sequence in India: 35 000 Years of Modern Human Occupation and Change at the Jwalapuram Locality 9 Rockshelter." *Antiquity* 83(320): 326–348.

Crevecoeur, I., A. Brooks, I. Ribot, E. Cornelissen, and P. Semal. 2016. "Late Stone Age Human Remains from Ishango (Democratic Republic of Congo): New Insights on Late Pleistocene Modern Human Diversity in Africa." *Journal of Human Evolution* 96: 35–57.

Dennell, R. 2014. "Smoke and Mirrors: the Fossil Record for *Homo sapiens* between Arabia and Australia." In *Southern Asia, Australia and the Search for Human Origins,* edited by R. Dennell and M. Porr, 33–50. Cambridge: Cambridge University Press.

Dennell, R. 2015. "Palaeoanthropology: *Homo sapiens* in China 80,000 years Ago." *Nature* 526: 647–648.

Dennell, R., and M. D. Petraglia. 2012. "The Dispersal of *Homo sapiens* across Southern Asia: How Early, How Often, How Complex?" *Quaternary Science Reviews* 47: 15–22.

Erlandson, J. M., and T. J. Braje. 2015. "Coasting out of Africa: The Potential of Mangrove Forests and Marine Habitats to Facilitate Human Coastal Expansion via the Southern Dispersal Route." *Quaternary International* 382: 31–41.

Fernandes, V., P. Triska, J. B. Pereira, F. Alshamali, T. Rito, A. Machado, Z. Fajkošová, B. Cavadas, V. Černý, P. Soares, and M. B. Richards. 2015. "Genetic Stratigraphy of Key Demographic Events in Arabia." *PLoS ONE* 10: e0118625.

Field, J. S., and M. M. Lahr. 2005. "Assessment of the Southern Dispersal: GIS-Based Analyses of Potential Routes at Oxygen Isotopic Stage 4." *Journal of World Prehistory* 19: 1–45.

Groucutt, H. S., M. D. Petraglia, G. Bailey, E. M. L. Scerri, A. Parton, L. Clark-Balzan, R. P. Jennings, L. Lewis, J. Blinkhorn, N. A. Drake, P. S. Breeze, R. H. Inglis, M. H. Devès, M. Meredith-Williams, N. Boivin, M. G. Thomas, and A. Scally. 2015. "Rethinking the Dispersal of *Homo sapiens* out of Africa." *Evolutionary Anthropology* 24: 149–164.

Gunz, P., F. L. Bookstein, P. Mitteroecker, A. Stadlmayr, H. Seidler, and G. W. Weber. 2009. "Early Modern Human Diversity Suggests Subdivided Population Structure and a Complex out-of-Africa Scenario." *Proceedings of the National Academy of Sciences* 106: 6094–6098.

Kaifu, Y., M. Izuho, and T. Goebel. 2015. "Modern Human Dispersal and Behavior in Paleolithic Asia." In *Emergence and Diversity of Modern Human Behavior in Paleolithic Asia,* edited by Y. Kaifu, M. Izuho, T. Goebel, H. Sato, and A. Ono, 535–566. College Station: Texas A&M University Press.

Karmin, M., L. Saag, M. Vicente, M. A. W. Sayres, M. Järve, U. G. Talas, S. Rootsi, A.-M.Ilumäe, R. Mägi, M. Mitt, L. Pagani, T. Puurand, Z. Faltyskova, F. Clemente, A. Cardona, E. Metspalu, H. Sahakyan, B. Yunusbayev, G. Hudjashov, M. DeGiorgio, E.-L. Loogväli, C. Eichstaedt, M. Eelmets, G. Chaubey, K. Tambets, S. Litvinov, M. Mormina, Y. Xue, Q. Ayub, G. Zoraqi, T. S. Korneliussen, F. Akhatova, J. Lachance, S. Tishkoff, K. Momynaliev, F.-X. Ricaut, P. Kusuma, H. Razafindrazaka, D. Pierron, M. P. Cox, G. N. N. Sultana, R. Willerslev, C. Muller, M. Westaway, D. Lambert, V. Skaro, L. Kovačević, S. Kurdikulova, D. Dalimova, R. Khusainova, N. Trofimova, V. Akhmetova, I. Khidiyatova, D. V. Lichman, J. Isakova, E. Pocheshkhova, Z. Sabitov, N. A. Barashkov, P. Nymadawa, E. Mihailov, J. W. T. Seng, I. Evseeva, A. B. Migliano, S. Abdullah, G. Andriadze, D. Primorac, L. Atramentova, O. Utevska, L. Yepiskoposyan, D. Marjanovic', A. Kushniarevich, D. M. Behar, C. Gilissen, L. Vissers, J. A. Veltman, E. Balanovska, M. Derenko, B. Malyarchuk, A. Metspalu, S. Fedorova, A. Eriksson, A. Manica, F. L. Mendez, T. M. Karafet, K. R. Veeramah, N. Bradman, M. F. Hammer, L. P. Osipova, O. Balanovsky, E. K. Khusnutdinova, K. Johnsen, M. Remm, M. G. Thomas, C. Tyler-Smith, P. A. Underhill, E. Willerslev, R. Nielsen, M. Metspalu, R. Villems, and T. Kivisild. 2015. "A Recent Bottleneck of Y Chromosome Diversity Coincides with a Global Change in Culture." *Genome Research* 25: 459–466.

Kuhlwilm, M., I. Gronau, M. J. Hubisz, C. de Filippo, J. Prado-Martinez, M. Kircher, Q. Fu, H. A. Burbano, C. Lalueza-Fox, M. de La Rasilla, A. Rosas, P. Rudan, D. Brajkovic, Z. Kucan, I. Gusic, T. Marques-Bonet, A. M. Andres, B. Viola, S. Paabo, M. Meyer, A. Siepel, and S. Castellano. 2016. "Ancient Gene Flow from Early Modern Humans into Eastern Neanderthals." *Nature* 530: 429–433.

Lahr, M. M., and R. Foley. 1994. "Multiple Dispersals and Modern Human Origins." *Evolutionary Anthropology* 3: 48–60.

Lahr, M. M., and R. A. Foley. 1998. "Towards a Theory of Modern Human Origins: Geography, Demography, and Diversity in Recent Human Evolution." *Yearbook of Physical Anthropology* 41: 137–176.

Liu, W., M. Martinón-Torres, Y.-J. Cai, S. Xing, H. W. Tong, S. W. Pei, M. J. Sier, X.-H. Wu, R. L. Edwards, H. Cheng, Y.-Y. Li, X. X. Yang, J. M. Bermudez de Castro, and X. J. Wu. 2015. "The Earliest Unequivocally Modern Humans in Southern China." *Nature* 526: 696–699.

Macaulay, V., C. Hill, A. Achilli, C. Rengo, D. Clarke, W. Meehan, J. Blackburn, O. Semino, R. Scozzari, F. Cruciani, A. Taha, N. K. Shaari, J. M. Raja, P. Ismail, X. Zainuddin, W. Goodwin, D. Bulbeck, H. J. Bandelt, S. Oppenheimer, A. Torroni, and M. Richards. 2005. "Single, Rapid Coastal Settlement of Asia Revealed by Analysis of Complete Mitochondrial Genomes." *Science* 308: 1034–1036.

Mellars, P. 2006. "Why Did Modern Human Populations Disperse from Africa ca. 60,000 Years Ago? A New Model." *Proceedings of the National Academy of Sciences* 103: 9381–9386.

Mellars, P., K. C. Gori, M. Carr, P. A. Soares, and M. B. Richards. 2013. "Genetic and Archaeological Perspectives on the Initial Modern Human Colonization of Southern Asia." *Proceedings of the National Academy of Sciences* 110: 10699–10704.

Oppenheimer, S. 2009. "The Great Arc of Dispersal of Modern Humans: Africa to Australia." *Quaternary International* 202: 2–13.

Pagani, L., S. Schiffels, D. Gurdasani, P. Danecek, A. Scally, Y. Chen, Y. Xue, M. Haber, R. Ekong, T. Olijira, E. Mekonnen, D. Luiselli, N. Bradman, E. Bekele, P. Zalloua, R. Durbin, T. Kivisild, and C. Tyler-Smith. 2015. "Tracing the Route of Modern Humans out of Africa by Using 225 Human Genome Sequences from Ethiopians and Egyptians." *The American Journal of Human Genetics* 96: 986–991.

Petraglia, M., R. Korisettar, N. Boivin, C. Clarkson, P. Ditchfield, S. Jones, J. Koshy, Lahr, M. M., C. Oppenheimer, D. Pyle, R. Roberts, J.-L. Schwenninger, L. Arnold, and K. White. 2007. "Middle Paleolithic Assemblages from the Indian Subcontinent before and after the Toba Super-eruption." *Science* 317: 114–116.

Petraglia, M., C. Clarkson, N. Boivin, M. Haslam, R. Korisettar, G. Chaubey, P. Ditchfield, D. Fuller, H. James, S. Jones, T. Kivisild, J. Koshy, M. M. Lahr, M. Metspalu, R. Roberts, and L. Arnold. 2009. "Population Increase and Environmental Deterioration Correspond with Microlithic Innovations in South Asia ca. 35,000 Years Ago." *Proceedings of the National Academy of Sciences* 106: 12261–12266.

Petraglia, M. D., P. Ditchfield, S. Jones, R. Korisettar, R., and J. N. Pal. 2012. "The Toba Volcanic Super-eruption, Environmental Change, and Hominin Occupation History in India over the Last 140,000 Years." *Quaternary International* 258: 119–134.

Petraglia, M. D., M. Haslam, D. Q. Fuller, N. Boivin, and C. Clarkson. 2010. "Out of Africa: New Hypotheses and Evidence for the Dispersal of *Homo sapiens* along the Indian Ocean Rim." *Annals of Human Biology* 37: 288–311.

Petraglia, M. D., A. Parton, H. S. Groucutt, and A. Alsharekh. 2015. "Green Arabia: Human Prehistory at the Crossroads of Continents." *Quaternary International* 382: 1–7.

Posth, C., G. Renaud, A. Mittnik, D. G. Drucker, H. Rougier, C. Cupillard, F. Valentin, C. Thevenet, A. Furtwangler, C. Wissing, M. Francken, M. Malina, M. Bolus, M. Lari, E. Gigli, G. Capecchi, I. Crevecoeur, C. Beauval, D. Flas, M. Germonpre, J. van der Plicht, R. Cottiaux, B. Gely, A. Ronchitelli, K. Wehrberger, D. Grigorescu, J. Svoboda, P. Semanl, D. Caramelli, H. Bocherens, K. Harvati, N.J. Conard, W. Haak, A. Powell, and J. Krause. 2016. Pleistocene Mitochondrial Genomes Suggest a Single Major Dispersal of Non-Africans and a Late Glacial Population Turnover in Europe. *Current Biology* 26: 827–833.

Potts, R. 2013. "Hominin Evolution in Settings of Strong Environmental Variability." *Quaternary Science Reviews* 73: 1–13.

Quintana-Murci, L., O. Semino, H.-J. Bandelt, G. Passarino, K. McElreavey, and A. S. Santachiara-Benerecetti. 1999. "Genetic Evidence of an Early Exit of *Homo sapiens sapiens* from Africa through Eastern Africa." *Nature Genetics* 23: 437–441.

Roberts, P., E. Delson, P. Miracle, P. Ditchfield, R. G. Roberts, Z. Jacobs, J. Blinkhorn, R. L. Ciochon, J. G. Fleagle, S. R. Frost, C. C. Gilbert, G. F. Gunnell, T. Harrison, R. Korisettar, and M. D. Petraglia. 2014. "Continuity of Mammalian Fauna over the Last 200,000 y in the Indian Subcontinent." *Proceedings of the National Academy of Sciences* 111: 5848–5853.

Scally, A., and R. Durbin. 2012. "Revising the Human Mutation Rate: Implications for Understanding Human Evolution." *Nature Reviews Genetics* 13: 745–753.

Scerri, E. M. L., P. S. Breeze, A. Parton, H. S. Groucutt, T. S. White, C. Stimpson, L. Clark-Balzan, R. Jennings, A. Alsharekh, and M. D. Petraglia. 2015. "Middle to Late Pleistocene Human Habitation in the Western Nefud Desert, Saudi Arabia." *Quaternary International* 382: 200–214.

Scerri, E. M. L., N. A. Drake, R. Jennings, and H. S. Groucutt. 2014. "Earliest Evidence for the Structure of *Homo sapiens* Populations in Africa." *Quaternary Science Reviews* 101: 207–216.

Stringer, C. 2000. "Palaeoanthropology: Coasting out of Africa." *Nature* 405: 24–27.

Stringer, C. 2014. "Why We Are Not All Multiregionalists Now." *Trends in Ecology & Evolution* 29: 248–251.

Templeton, A. 2002. "Out of Africa Again and Again." *Nature* 416: 45–51.

Templeton, A. 2005. "Haplotype Trees and Modern Human Origins." *Yearbook of Physical Anthropology* 48: 33–59.

Thorne, A. G., and M. H. Wolpoff. 2003. "The Multiregional Evolution of Humans (revised paper)." *Scientific American* 13: 46–53.

Trinkaus, E. 2005. "Early Modern Humans." *Annual Reviews Anthropology* 34: 207–230.

Williams, M. A. J., S. H. Ambrose, S. van der Kaars, C. Ruehlemann, U. Chattopadhyaya, J. Pal, and P. R. Chauhan. 2009. "Environmental Impact of the 73ka Toba Super-eruption in South Asia." *Palaeogeography, Palaeoclimatology, Palaeoecology* 284: 295–314.

8 The Phylogenomic Origins and Definition of *Homo sapiens*

Peter J. Waddell

Introduction

In this chapter I examine *Homo sapiens* from a phylogenetic perspective and, using a variety of phylogenetic methods (Swofford et al. 1996, Felsenstein 2004), look at the definition and origins of this species. Phylogenetics is sometimes viewed as a subset of genetics, but the reverse also has validity. Every single nucleotide in every single organism follows a tree, and that is the basis of phylogenetics/phylogenomics. Genetic methods, in practice, often aim to avoid the difficulty of knowing each of these trees in order to yield tractable methods and interpretable results, which may or may not be trees or networks of descent (Felsenstein 2004).

Taking a phylogenetic view of *Homo sapiens* requires a definition of that species (Wilkins 2011). Darwin (1871) even once pondered the question of whether there was one or more species of living human. Although before World War II numerous species as well as genera of fossil hominids had been proposed, post-World War II there was a radical shift towards uniting nearly all these into the genus *Homo*, with many falling into one super-species, or a species swarm, designated with the suffix *sensu lato*. On the belief that one species graded imperceptibly into the other, Wolpoff et al. (1994) even went so far as to suggest lumping all specimens subsumed in *H. erectus* into *H. sapiens sensu lato*, and getting rid of the former altogether. Although extreme, this perspective derives from and characterizes the Multi-Regional (MR) hypothesis of human evolution (e.g., Wolpoff 1989, Wolpoff et al. 2001).

In contrast, the "Out of Africa" (OA) set of hypotheses (Cann et al. 1987, Vigilant et al. 1991), for the origin of living human population differences—which, according to the model, emerged as the descendants of an African ancestor migrated from that continent and spread throughout the Old World only ~100-60kya (~means approximately) is neatly tailored to a precise phylogenetic species definition. Nevertheless, the boundaries of this taxically constrained species are often vaguely delineated, even by those who embrace some elements of the OA model. One example is the desire to call fossil forms such as LH-18 anatomically modern *Homo sapiens* (e.g., Rightmire 1998), while others will call it *Homo heidelbergensis* (e.g., Profico et al. 2016). Thus, the first task of this chapter is to define *Homo sapiens* using the same phylogenetic principals (Hennig 1966, Jefferies 1979) used to define anthropological groups all the way up to the mammalian super-order containing Primates, Euarchonta.

In this regard, I begin by reviewing how morphological data can objectively help us diagnose what may and may not be *Homo sapiens*, phylogenetically defined. Recent reevaluations of fossil skulls make it clear that the term "anatomically modern *H. sapiens*" must be refined (Schwartz 2016). Independently, quantitative phylogenetic analysis of skull shape is yielding statistically robust results supporting the OA model (Waddell 2015) and, perhaps, instances of hybridization (Waddell 2014). Importantly, these are broadly consistent with recent subjective views (e.g., Stringer 2016). Surprisingly, there are few "traditional" qualitative morphological character-based phylogenetic analyses (Swofford et al. 1996, Felsenstein 2004) of the origins of *H. sapiens*.

The ongoing efforts to sequence modern and ancient, or aDNA, genomes are making major impacts on the field (Der Sarkissian et al. 2015). These studies have a huge advantage over single nucleotide polymorphism (SNP) panels in not having ascertainment bias, which is due to only incorporating SNP's already seen to be variable in previously sequenced genomes. Large amounts of genomic DNA have the advantage over mtDNA of having considerably more informative sites, at many unlinked loci, of which many are effectively near neutral and thus amenable to mathematical models of evolution. aDNA sequencing at the full genome level has progressed rapidly (Pääbo 2014), and now includes extensive sampling of individuals (e.g., Fu et al. 2016). Sequenced ancient *Homo* genomes now extend back to about 100+ kya (kya = one thousand years ago) for exceptionally preserved specimens (Prufer et al. 2014), and partial genomes have been analyzed from specimens as old as ~420 kya (Meyer et al. 2016).

The analysis of these genomes requires methods compatible with population genetics. They are able to detect introgression/hybridization, even when the DNA comes from an unknown form, a "ghost lineage," as is the case in suggestions of an early diverging (*Homo erectus*-like?) lineage leaving genes in the Denisova genome (Waddell 2013, Prufer et al. 2014 Kuhlwilm 2016). A wide range of methods, including Buneman, P and D-statistics, plus well-established phylogenetic methods, are considered. A full spectrum of a site-patterns approach allows a detailed look at residuals (Waddell 1995, Waddell et al. 2011b). Computationally expensive Bayesian and Hidden Markov models are also providing a richer view of phylogenies (Gronau et al. 2011, Schiffels and Durbin 2014). An example of how unexpected splits in the autosomal genome data are still being discovered is the suggested introgression of a non-*sapiens* lineage into West Africans analyzed below. This contrasts with Chromosome X, which shows clear introgression between distinct populations of *Homo sapiens*, but much less evidence of non-*sapiens* introgression.

It should also be emphasized that the individuals/populations selected for these studies are often somewhat atypical of most recent modern populations, which often show high levels of mixing between lineages, regions, or continents due to major migratory events of the past ~10 kya. The hope of building trees and networks of populations that were relatively isolated 10 kya is to recover the major deep patterns of where *H. sapiens* came from and how it spread around the world. At the beginning and end of this chapter, I will examine how the OA model was conceived in the early 1990s, versus how it is supported and/or contradicted by the latest results.

Out-of-Africa Predictions: Early 1990s

For the last three decades the OA hypothesis (Cann et al. 1987, Vigilant et al. 1991) has been the favored explanation for the origin of *Homo sapiens*. Waddell and Penny (1996) reassessed and reanalyzed the key data available at that time using novel statistical analyses, to examine the When, Where, Who, and How set of hypotheses. To recap, these are:

When: Origin of *Homo sapiens* almost certainly less than 200 kya.

Where: Most consistently Africa, almost certainly sub-Saharan, with the very deepest lineages dominated by pygmies of Central Southern Africa.

Who: The actual people that left Africa seem to have been closer to the ancestors of East and West African, rather than Southern African populations.

How: OA involved a population bottleneck, followed by a relatively rapid expansion across Eurasia and then down into Sahul (Australia and New Guinea), replacing earlier inhabitants with no apparent evidence of interbreeding.

In terms of when, the occupation of Sahul at about 50–60 kya (Bowler et al. 2003, Roberts et al. 1990) remains a key calibration point for any genomic OA study, and is a valid test of the rather wide range of current rate estimates.

The dividing line between an OA-like scenario that emphasizes the basic genetic origins of a species or lineage in one region (with perhaps a co-adapted genome) versus a MR-like model that emphasizes essential genetic components/modules inherited from widely scattered independent lineages, is not easy to draw. Here, I propose that a rough line can be drawn if descendant lineages average more than 80% from a single parental population, and if genetic regions coming in from independent lineages also show clear depletion in most functional genes and few or no inherited genetic modules. In this case, evolution is likely reflected in a model of independent lineages with limited introgression (not necessarily adaptive overall). Templeton 2017 (this volume) argues that there is evidence suggesting over 10% of the genomes of non-Africans came from earlier Eurasian lineages, such as Neanderthals. However, for MR supporters, it is not clear how little gene flow or essential surviving genes there must be before they would reject the MR model. Thus, some researchers may never be able to accept that the OA model represented here was ever different from an MR model.

To contrast these two models another way, one may ask: If all hominids outside of Africa were exterminated immediately after the first lineage split among living humans, would

H. sapiens still exist (the essence of OA)? If not, the answer might include the historic MR model. In this regard, my conclusion diverges from Templeton's (2017, this volume).

A Phylogenetic Definition of *Homo sapiens*

Thanks to Hennig (1966), a rigorous approach to taxonomic classification—phylogenetic systematics—has increasingly dominated biology, and will be followed here. The desire to have a formal objective classification founded on monophyletic groups (a group or clade = ancestor and all its descendants) has, in turn, driven and been driven by a major effort to estimate the phylogeny (a tree or more complicated network of descent) of a set of homologous (same origin) and reliably aligned DNA sequences. Because a hypothesis of monophyly is precise and testable, it brings with it a level of rigor and objectivity to the understanding of "species" and "higher taxa" lacking in other approaches.

The most powerful phylogenetic definition of a species is that of an "extant crown group." For living organisms, and in order to be fully testable, such a hypothesis should include a list of all members of that group. If it shows precise congruence with at least some of the properties of other popular concepts of "species" (Wilkins 2011), all the better, but equivalence is not essential. For *H. sapiens*, the most inclusive living crown group to be considered is the last ancestral population of all living humans and all descendants of that population.

It is increasingly being demonstrated that one ancestral population is the source of >90% of all living human DNA. This is somewhat surprising since the genus *Homo*, being relatively recent (~2.5 million years old) and encompassing mobile, adaptive individuals, has been expected to show elements of a species complex, including a significant amount of introgression/hybridization. The MR hypothesis is based on such an expectation. In contrast, the OA model consists of a set of hypotheses that are consistent with *H. sapiens* being a phylogenetically recent and easily defined species. However, a consequence of *H. sapiens* being nested deep within the genus *Homo*, is that other proposed species, such as the purportedly widely distributed *Homo erectus*, are likely to be para- or polyphyletic, and themselves in need of taxonomic revision. Indeed, the disparate morphologies of specimens attributed to "*erectus*" suggest revision that goes well beyond allocating the East African specimens attributed to *H. erectus* to *H. ergaster* (Schwartz and Tattersall, 2002, 2003, 2005).

Why an Extant Crown-Based Definition of *Homo sapiens*?

In phylogenetic systematics, the basic principals should be the same across all taxonomic levels. For example, the clade and superorder Euarchonta (Waddell et al. 1999, 2001), to which humans belong, is defined as the common ancestor of the extant mammalian orders

Primates, Dermoptera, and Scandentia, plus all descendants. As there are living species on both sides of the deepest split within Euarchonta, it is considered a "living" or "extant crown group," which has attained special meaning in phylogenetics (stem and crown group terminology was developed by Jefferies [1979], based on the work of Hennig [1966]). By extension, it also has special meaning in computational biology, genomics, evolution, and other areas of biology that have been transformed or are emergent because of the methods of phylogenetics. For the last common ancestor of an extant crown group is, on balance, the ancestral population on which the best inferences of its biological properties can be made, which in turn, best predicts the properties of its descendants. Using extant crown groups, these ancestor-based hypotheses can be generated with a maximum amount of interpolation and a minimum amount of extrapolation. They are, therefore, expected to maximize the predictive power of the full range of biological hypotheses. In the case of the species *H. sapiens*, its crown encompasses the ancestral population of all living humans. As we shall see below, research over the past 30 years on sub-Saharan African populations and, with increasing focus on the Khoisan of Southern Africa, suggests they provide half of the deepest split amongst living humans.

Potential complication of this definition arises when a "species" is expected to present "special" properties: for example, there are at least two-dozen proposals along this line, many of which are contradictory (Wilkins 2011). Here I follow the view that species do not have any essential quality or typological essence, but that, given a sufficient amount of "evolutionary time," a separate lineage will not readily mix parts of their genomes with other species and tends to express emergent properties. With regard to the Biological Species Concept, there are various forms, each with different emphases on pre- or post-mating mechanisms, different ecologies, prolonged geographical isolation and adaptation, and distinct mate recognition systems. Another process may, however, also be at work with distinct co-adapted genomes: There may be introgression, but over time, purifying selection will remove most of the functionally distinct introgressed elements, leaving mostly near neutral genomic regions.

Unfortunately, any emergent property can come into conflict with phylogenetic classification. For example, when a single descendant population does not mix with its relatives, the latter would constitute a non-monophyletic group. Ultimately, what phylogenetics offers biology is the encouragement to map as accurately as possible the pattern of evolution, as well as guidelines for how this "map" might most informatively be divided up. Such a phylogeny can certainly have properties related to "failure to introgress" mapped upon it. Phylogenetic classification is the natural extension of taking phylogenies as fundamental in understanding biology.

Finally, phylogenetic systematics permits some genetic introgression. If it is great enough, a hybrid species may need to be diagnosed when its properties are sufficiently distinct from either parent lineage. On the other hand, the sharing of relatively small genomic blocks, particularly if they are not fixed and account for less than 10 percent of the genome, constitutes introgression, which is expected to be a common pattern among closely related species that do not always have strongly developed pre- and post-zygotic barriers to gene flow.

Semantic Corollaries of an Extant Crown-based Definition of *Homo sapiens*

This phylogenetic definition of *H. sapiens* leads to certain semantic corollaries. Firstly, if it is accepted that human equals *H. sapiens* then, in the absence of a better equivalence, Neanderthals and other non-*Homo sapiens* lineages should not be referred to as human. Another is that "modern human" and "human" effectively become synonymous, with living humans remaining a distinct category. Such precision is particularly important in reference to specimens, such as those from Qafzeh (about 90–100 kya Israel). Specifically, Qafzeh 9 is the oldest fossil that appears to be consistent with the current understanding of morphological variability in living humans. It might even be shown to be morphologically indistinguishable from *Homo sapiens* and even nestled well inside that envelope of variability. Nevertheless, as argued here, Qafzeh 9 would not be identified as *H. sapiens* if it derived most of its genetic material from a population that predates the last common (ancestral) population of living humans. While such a strict use of the term "human" may seem pedantic, even discriminatory, I suggest it is appropriate if the goal is to come to solid scientific conclusions. Where there is the potential for ambiguity, *H. sapiens* with a phylogenetic definition should be used.

As mentioned earlier, the term "anatomically modern humans" (AMH) should be used with more caution. According to Schwartz (2016), only extant humans and a very small set of fossils can be accurately called anatomically modern, that is, morphologically indistinguishable from living humans. Much of the literature is full of imprecise uses of the term AMH, for specimens that, are likely "pre-*Homo sapiens*." More precisely, and, for instance, with regard to Neanderthals, "pre-*H. sapiens*" would specify all lineages that originated along the lineage between the common ancestor of *H. sapiens* and the common ancestor of *H. sapiens* and *H. neanderthalensis*, but not including *H. sapiens*. Further, in recognition of the the need to refer to a specimen that is either *H. sapiens* or pre-*H. sapiens*, one might consider using the phrase "pro-*H. sapiens* vis-à-vis *H. neanderthalensis*," to mean "of the clade that is closer to *H. sapiens* than to *H. neanderthalensis*."

The currently popular terms "archaic human" or "archaic *Homo*" (AH) basically means any member of the genus *Homo* that is not *H. sapiens* according to some definition of that species (e.g., Schwartz 2016). The term "archaic" is nearly synonymous with "primitive" in general biology (which in turn typically equals "ancestral"). The term "primitive" in phylogenetics and evolutionary biology needs to be used very carefully, as it often means nonderived. As discussed below, there may have been many lineages (species in their own right with their own sets of derived features) during the last ~500 kya that were closely related to *Homo sapiens*, of which none were ancestral for *H. sapiens*. In this case, a more balanced term is simply "non-*H. sapiens*."

At present, we do not have a name for the often-referred-to crown-group within *Homo* that comprises *H. sapiens*, *H. neanderthalensis*, and all other descendants of their common ancestor, such as the Denisovan individual. I propose that this clade of three genetically characterized hominids should be identified as "HND," for human, Neanderthal, and Denisovan. This clade has the unusual property of being a "DNA-extant" crown group that includes nearly fully sequenced genomes on each descendant lineage, while lacking living individuals on all but one descendant lineage.

Finally, proposed subspecific names (for example, *H. sapiens sapiens*, *H. s. idaltu*) have no validity in this phylogenetic classification. *H. s. sapiens* is synonymous with *H. sapiens*, while *H. s. idaltu* is almost certainly neither *H. sapiens* (Schwartz 2016) nor a subspecies of *H. sapiens*, in which case *H. idaltu* is available.

Morphological Phylogenetic Information

Details of morphology can help diagnose, but not phylogenetically define (see previous sections) those specimens that may be *H. sapiens* and those that are not. These details should be analyzed in ways that are consistent with the major methodologies of phylogenetics. In this regard, a broad collaboration, representing a wide range of knowledge and disagreement, would seem to be desirable in order to arrive at a viable data matrix (e.g., O'Leary et al. 2013).

Derived features (evolutionary novelties) present in certain lineages (e.g., *H. sapiens*) that can be described succinctly and without need for measurement are qualitative or discrete characters. These can be encoded into a character data matrix for analysis by numerous valid phylogenetic methods, including parsimony (Swofford et al. 1996, Felsenstein 2004).

Properties of specimens that can be represented as measures or some other numerically continuous variable, such as shape or size, are quantitative. Various ways of divvying these up into discrete characters have been proposed, but they are problematic for reasons that include loss of information and bias (Swofford et al. 1996, Felsenstein 2004).

While there has been stagnation in the development of phylogenetic methods/models for quantitative geometric morphometric data (Adams et al. 2013), multiple new methods are now forthcoming. For example, extended distance-based phylogenetic techniques, which constitute a form of compound likelihood, have been shown to have considerable promise (Waddell 2014, 2015). Maximum likelihood and Bayesian analyses are also feasible, but methods for analyzing this specific type of data are still in need of refinement (e.g., Theobald and Wuttke 2008, Felsenstein 2015). Unlike gene frequencies, geometric morphometric data points are not independent characters, which makes problematic their analysis using methods designed for allele frequencies; however, analyses can rest on a simple additive genetic model wherein neutral drift gives rise to Brownian motion (Felsenstein 2002).

Qualitative Morphological Diagnosis of *Homo sapiens*

Schwartz and Tattersal (2000, 2002, 2003, 2005, 2010) have been particularly vigorous in trying to delineate derived qualitative features (apomorphies; Hennig 1966) found at high frequencies in living humans, that, because of their rarity or absence in fossils, should help in identifying *H. sapiens*. Among these, an inverted-T-shaped chin and a particular form of a bipartite brow seemed to be most promising. The major problem with the bipartite brow as a diagnostic character is that it has much less than 100 percent penetrance (e.g., young males and any female may lack it); it does not appear fully until later in life, and the *H. sapiens* unexpressed and the ancestral state appear to be the same (Schwartz 2016). Other potentially powerful diagnostic characters listed in (Schwartz 2016) need further investigation. A lacuna of all these studies, however, is that Khoisan, Pygmy, and Papuan/ Australian populations have not been assessed.

The inverted T-shaped chin appears an excellent diagnostic character as its characteristic morphologies emerge early in fetal development and remain crisply defined in young children, after which, growth, differential bone deposition and resorption has the effect of accentuating or softening, but not obliterating, the basic features; it therefore seems to show 100 percent penetrance and persistence. It is interesting to note that Oase 1, with~10% Neanderthal SNPs (Fu et al. 2015), presents a well-defined inverted-T-shaped chin (Schwartz 2016). Since chin shape emerges early in development, it might also be linked to fundamental developmental programs that do not easily admit to disruption with non-*sapiens* genes. Absence of this mandibular character, and/or supraorbital adornment that is not non-bipartite, exclude nearly all skulls and mandibles older than 45 kya from *Homo sapiens* (Schwartz 2016).

The oldest specimens clearly presenting a T-shaped chin are from Border Cave and Klasies River Mouth (~100 kya); penecontemporaneous Qafzeh 8, 9, and 10 may also show it. This seemingly removes many specimens widely proclaimed as *H. sapiens* from this species, including Chinese specimens older than Tianyuan (~45 kya) (e.g., Zhiren-dong, Liu et al. 2010). Further, and unexpectedly, the absence of this feature in various recent fossils from South Africa (Boskop perhaps 12–14 kya, Fish Hoek ~9 kya) and China (Curnoe et al. 2012, ~11–14.5 kya) suggests that multiple non-*sapiens Homo* survived until relatively recently (Schwartz 2016). The South African specimens lacking chins are particularly interesting because their basic skull shapes closely overlap those of living *H. sapiens*, which suggests their particularly close relationship. Their DNA would provide an excellent test of the utility of the inverted-T-shaped chin to diagnose *H. sapiens*.

Qualitative Morphological Phylogenetic Analysis

While it is useful to look for the best discrete characters to diagnose *Homo sapiens*, this endeavor does not replace a carefully constructed, and hopefully widely discussed, critiqued, and annotated character data matrix (Swofford et al. 1996, Felsenstein 2004). Matrices such as that of Mounier et al. (2011) have been analyzed using objective phylogenetic methods (e.g., Waddell 2013), but the characters are in serious need of scrutiny and repeatable description. For example, a key "discrete" character in Mounier et al.'s (2016) matrix is skull volume. However, the cut-off of 1200 cc, which is used to produce a binary character, has no objective justification. Indeed, this decision may derive from preconceived notions of which are the earlier- or later-diverging specimens. If so, this and similar "characters" are analogous to the old joke about Dwayne, the undisputed best shot in the county, whose bullet always hits the target dead center. The problem, as it turns out, is that Dwayne draws the target around his bullet hole. Thus, while the trees obtained to date are interesting, they contain relatively few specimens and are not highly resolved by unweighted parsimony (Mounier et al. 2016). The challenges of developing well-substantiated characters are discussed in Wagner (2001).

Quantitative Phylogenetic Morphological Analysis

Over relatively short periods of time, myriad genes can influence aspects of morphology, such as shape, without leaving easily agreed-upon discrete characters (Felsenstein 2002, Adams et al. 2013). The shape of the upper skull/calvaria is one example. Phylogenetic shape analysis (= geometric morphometric phylogenetics) aims to capture and quantify such change according to an explicit statistical model, one that often assumes Brownian motion.

In terms of the actual processing of 3D measurements, currently points in space are generally scaled, centered, and rotated to coincide with each other with a minimum sum of squared coordinante differences; this yields a Procrustes, or a minimum-scaled Euclidean, distance (Adams et al. 2013). A matrix of all such distances between specimens can serve as the starting point for distance-based phylogenetic analysis, for example, Caumul and Polly (2005).

Recent analyses employing improved methods indicate that there is a phylogenetic signal, with some apparently well-supported branching patterns, in the skullcap shape of specimens allocated to *Homo* (Waddell 2014, 2015). A reanalysis of that data (Harvati et al. 2011) with a refined set of constraints is shown in figure 8.1. The node where the two recent Khoisan skulls meet encapsulates approximately two-thirds of 58 Khoisan skulls and 62 percent of all 242 recent (last 10 kya) human skulls sampled (OLS+ tree not shown). Other Khoisan skulls fall around most of the sub-tree containing all recent *H. sapiens* skulls, and also near the particularly deep European female 012 (EurF012H).

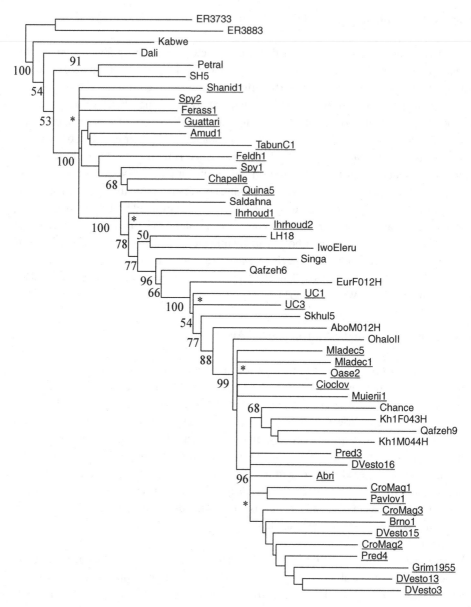

Figure 8.1

The Ordinary Least Squares (OLS+) tree (Swofford 2000) of fossil *Homo* skulls based on a pairwise Procrustes distance (q=0.5). Residual resampling support values of >50% are shown (+2% increase in sampling variance to counter bias). Underlined skulls constrained *a priori* to fall together on the tree based upon the expectations expressed in Schwartz and Tattersal (2002, 2003, 2005), that is, no suggestion of more than one morph within that set (from top to bottom, Neanderthals, Jebel Irhoud site, Zhoukoudian Upper Cave) or timing and/or cultural artifacts (pre-Gravettian Europe and Gravettian Europe).

This data is consistent with a strong phylogenetic signal via multiple tests. These include an application of the statistics of Hillis and Huelsenbeck (1992), yielding a g1 statistic of −2.80 and a g2 of −15.94 (both larger than in the reduced dataset of Waddell [2014]), and the substantial levels of residual resampling support (Waddell et al. 2011a) on many internal edges. The tree is also in good agreement with the interpretations of some morphologists (e.g., Klein 2009, Stringer 2016), but suggests more specific detail. African specimens of the last ~300 kya fall closer to *H. sapiens* than to *H. neanderthalensis* in a phylogenetic, not simply a phenetic, sense. The results also suggest that *H. heidelbergensis* is not monophyletic. Even African specimens such as Kabwe and Saldanha appear well separated in the tree (Waddell 2015). Figure 8.1 also suggests that there may have been an accelerated evolutionary rate towards *H. sapiens*, as well as a branching pattern of multiple pre-*sapiens* lineages that may have emerged with some regularity in Africa after a split with Neanderthals ~500–800 kya.

A question of considerable debate is whether the specimens at ~100 kya Middle Eastern sites, such as Qafzeh and Skhul, are the result of hybridization (Trinkaus 2007) or if they represent different, even non-*sapiens*, forms (morphs) (Schwartz and Tattersal 2010, Schwartz 2016). In figure 8.1, Qafzeh 9 consistently falls in that part of the tree populated by *H. sapiens* skulls. If the positions of Chancelade and Ohalo II are ignored, Qafzeh 9 forms an exclusive group with the Khoisan skulls 85 percent of the time, while the pre-Gravettian skulls diverge more deeply than the Gravettian + Qafzeh 9 + Khoi skulls 99 percent of the time. Since Qafzeh 9 may be the oldest specimen of *H. sapiens* (Schwartz and Tattersal 2003), it might merit the accolade "the skull of Eve" (Waddell 2015) in similitude with mitochondrial Eve (Cann et al. 1987). Another specimen, Skhul 5, falls deeper in the tree than most *H. sapiens*, which is consistent with Schwartz and Tattersall's (2005; also Schwartz 2016) suggestion that it represents a distinct morph.

In contrast, Qafzeh 6 diverges deeply (figure 8.1) and yields a result consistent with it being a hybrid when analyzed with Neighbor Net (figure 7 in Waddell 2014). Its ancestry seems to be approximately two-thirds close to *H. sapiens*, and approximately one-third close to *H. neanderthalensis*. These results would appear to make probable the existence of distinct morphs and of hybrids at Skhul and Qafzeh, respectively.

In figure 8.1, European pre-Gravettians (>32 kya) diverge significantly deeper than Gravettians (~22–32 kya), which are deeper than most living humans (the Khoisan pair), which is consistent with suggestions of a decrease in Neanderthal features (Trinkaus 2007) and DNA in Europe over time (Fu et al. 2016). Curiously, the pre-Gravettian skulls that show the most evidence of discrete features characteristic of *H. neanderthalensis* (see Trinkaus 2007) are not necessarily those that diverge deeper (Waddell 2015), thereby suggesting that overall shape and specific features are not tightly linked genetically. Another example of this effect might be seen amongst Neanderthals. Of the notably late Spy Neanderthals, in contrast to Spy 1, which has been suggested to show *Homo sapiens*-like features, the divergence of Spy 2 is closer to *H. sapiens* (Waddell 2015). Another example seems to be Irhoud 1, which in Figure 8.1 diverges more deeply than Irhoud 2 (Waddell 2015); the Mounier et al. (2016) analysis of discrete characters yields the reverse.

Specimens from Upper Paleolithic/Upper Cave Zhoukoudian, China (UC) (~13–30 kya) show a position in figure 8.1 that is consistent with even more inflow of non-*sapiens* genes than the European specimens. Indeed, UC1 is the most divergent skull yet (Waddell 2015), showing an inverted-T-shaped chin (Schwartz 2016). Inasmuch as the scattering of developmental outcomes is a common feature following hybridization (e.g., Ackermanna et al. 2006), early European, Qafzeh, and UC localities all yield skulls with very different positions on the tree, each including some skulls that are much closer than others to those typical of living humans. These results tend to contradict the common assertion that Upper Paleolithic skulls look different (more robust, for example) than living *H. sapiens* simply because of their lifestyle and, instead, suggest a genetic component that was introduced by introgression from non-*sapiens* individuals.

Similar to the classical study of morphology (Schwartz 2016), phylogenetic analysis of skull shape argues that non-*Homo sapiens* lineages persisted until relatively recently. The ~12 kya Iwo Eleru skull from Nigeria, West Africa, is located in the tree as a deep pre-*sapiens* lineage (Waddell 2014), which is consistent with expectations of Harvati et al. (2011) based on non-evolutionary models.

The edge lengths in figure 8.1 suggest that the most marked acceleration of change in skull shape, and thus also of brain shape and organization, occurred in the lineage leading to *H. sapiens* (Waddell 2014, 2015). Further, the long edges leading to Iwo Eleru or to Singa, also suggests an acceleration of change in these lineages as well. These results are consistent with the view that the brain of Neanderthals is essentially a scaled-up version of the brain of its ancestor, while in the lineage leading to *H. sapiens* its shape and developmental program were altered (Lieberman 2011). Figure 8.1 suggests that this renewed phase of evolutionary experimentation began soon after the divergence of Neanderthals and *H. sapiens*.

Overall, figure 8.1 shows a "caterpillar-like" pattern consisting of many fairly long terminal lineages, particularly amongst pre-*H. sapiens* specimens, coming off the main trunk. However, this picture changes somewhat when using the BME criterion (Gascuel and Steel 2006). While the likelihood of this model is not as high as OLS+, quite a few groups of two, three, or four specimens appear in this part of the tree. For example, a wide range of analyses yield the sub-tree ((Iwo Eleru, (LH18, Saldahna)); [Waddell and Swofford, unpublished]). This offers the tantalizing possibility that these three enigmatic skulls may belong to an ancient African clade that was widespread in both space (West, East, and Southern Africa) and time (from ~12 to >300 kya) and, therefore, a potential long-term competitor with the lineage leading to *H. sapiens*.

This analysis, and others (e.g., Harvati et al. 2011, Stojanowski 2014, Waddell 2014), along with Iwo Eleru having "no pronounced chin" (Brothwell and Shaw, 1971), and an accessible type specimen, arguably present a stronger case for this specimen being recognized a distinct non-*sapiens* lineage than for *H. idaltu*, consistent with it being dubbed *H. iwoelerueensis* (Waddell 2014). As yet there are few clues to the capabilities of this lineage, other than its Middle Stone age context and an apparent, but crude, burial.

Although a tree is presented here, a major challenge of phylogenetics is to accurately reconstruct a more general network when there has been reticulation. While this may have been achieved in the diagnosis of Qafzeh 6 as a hybrid (Waddell 2014), an ongoing methodological challenge is to achieve this reliably for larger data sets such as this.

Molecular Phylogenetic Information

Mitochondrial (mt) DNA

As mentioned above, mtDNA data was key to developing the OA set of hypotheses. The initial data (Cann et al. 1987) were from restriction sites, soon to be followed by sequences of approximately 1,000 bases across the control region (Vigilant et al. 1991). By the late 1990s, sequencing the mtDNA in its entirety led to finer resolution of trees and a better estimation of edge lengths.

Although overall tree structure is similar to that of earlier studies, extensive sampling continues to find the deepest branches of this tree in southern and central southern African populations (e.g., Tishkoff et al. 2007, Gonder et al. 2007, Tishkoff and Gonder 2006). Using models of sequence evolution and calibration points (like those introduced by Waddell and Penny [1996]), the divergence time of the root of the mtDNA tree for living humans also remains similar. The confidence intervals for the age of the ancestral population, however, have not tightened dramatically because other sources of error—calibration error and coalescent dynamics—that are incorporated in the estimates of Waddell and Penny (1996) have not improved much. Further, questions—for example, the impact of slightly deleterious alleles—remain concerning the adequacy of the models of substitution in estimating relative edge lengths (e.g., Peterson and Masel 2009). Uncertainties about the rate of mutation in the genomes of *H. sapiens* apply equally to mtDNA. Trio sequencing, for example, of the genomes of parents and offspring suggests that the mutation rate over the last million years or so might have been about twice as slow as was assumed based on the overall rate after the human-chimpanzee divergence (Scally 2016).

The ultimate problem with mtDNA is that it is a single genetic locus and, simply because of the random fluctuation of allele frequencies between populations, it may give the wrong impression of the most closely related populations. An example might be the frequency of haplogroup M in living populations leading to the hypothesis of an early coastal *H. sapiens* migration route to Australia. This hypothesis suggests a later domination of Eurasia by inland routes whereby the order of splitting of lineages would be (Australia (Europe, Asia)), as has been proposed in some full-genome studies (e.g., Rasmussen et al. 2011), but not in others (e.g., Reich et al. 2010, Prufer et al. 2014).

Single nucleotide polymorphisms (SNPs)

Counting allele frequencies, first with allozymic protein techniques and eventually with DNA alterations, particularly SNPs, has lent considerable support for the OA model (e.g., Li et al. 2008) as well as for a (Europe [Asia, Australia]) migration pattern (Li et al. 2008).

Nevertheless, counting SNP frequencies suffers from the weakness that the variants must be pre-selected on the basis of known populations, such that whichever populations are chosen creates an ascertainment bias (that is, one cannot discover what one doesn't already know). This bias even exists when large data sets of ≥100,000 SNPs derived from across the genome are used. Other issues involve methodology, as is the case in interpreting the results of non-phylogenetic methods such as Principal Component Analysis. However, SNP-based studies continue to play an essential role, for example, to screen for the individuals that show the least evidence of admixture, but minimal inbreeding, as candidates for full genome sequencing (e.g., Kim et al. 2014).

Whole genome sequences

The advent of affordable, massively parallel sequencing, in conjunction with advances in sequencing "library" preparation, has made possible whole-genome sequencing of many individuals, whether recent or ancient (Der Sarkissian et al. 2015). These data are amenable to a diversity of phylogenetic methods, including detecting signals that contradict a single tree. Whole-genome sequences of a small number of specimens of Neanderthal, and a phalanx from Denisova Cave, Russia, present an increasingly detailed picture of how distinct lineages within *Homo* interacted (Green et al. 2010, Reich et al. 2010, Waddell et al. 2011b, Meyer et al. 2012, Prufer et al. 2014, Vernot and Akey 2015, Fu et al. 2016).

Each whole-genome sequence is ~3 billion bases long each; filtering and alignment from short reads often yield >1 billion well-aligned sites. The total number of discrete loci is unclear because recombination hot spots shift over time and recombination can occur anywhere in a sequence. Consequently, although analyses may produce estimates of the predominant tree for a given region of the genome, techniques that assume discrete loci can introduce bias and/ or cause most of the data to be overlooked. Alternatively, one can use marginal statistics, for example, pairwise distances (Swofford et al. 1996), rooted triples, or "unrooted" quartets (Bandelt and Dress 1992, Waddell et al. 2001, Reich et al. 2010), or a full spectrum of site pattern frequencies (basically, all binary numbers, Penny and Hendy 1993). These approaches may provide estimates of the splits in the data, both tree and non-tree.

Trees

Simple edit, mismatch, or Hamming distances between genomes of closely related organisms are essentially unrooted two-taxon trees that converge at the average coalescence time (Felsenstein 2004). As such, they integrate over all gene trees to produce a set of estimates, which is the set of all pairwise distances that, in spite of the randomness of the coalescent process in a purely splitting or tree model, are expected to be additive on the "species" tree. In this tree, edge lengths represent the expected number of substitutions, which are usually normalized by dividing by the total number of analyzed sites. This in turn, and assuming a neutral model of evolution, is dependent on mutation rate as well as changes in population size, which determine the rate at which mutations are "fixed" to become substitutions or are lost. By ignoring possible multiple changes at a nucleotide, these requirements permit application of consistent distance-based phylogenetic methods (Swofford et al. 1996, Felsenstein 2004).

Another potentially fruitful aspect of working with whole genomes is that, as long as branching events occurred many generations before, a single human genome can effectively sample an entire ancestral population. Assuming about one recombination event per generation, and per chromosome arm, each genome will undergo about 50 "cuts" per generation, wherein each "chunk" of genome might have been inherited from a different ancestor. That is, because random mating and recombination have divvied the ancestral genomes into many smaller pieces, an individual descendant autosomal genome samples two pieces at each autosomal position. Consequently, and multiplying that by 100 generations (about 3,000 years for humans), one might be sampling up to about 5,000 ancestors, which would be close to many effective population sizes.

The full genome sequences used in Reich et al. (2010) yielded a highly resolved tree: ((Neanderthal, Denisovan), (Khoisan, (West African, (European, (Asian, Papuan))))). Reanalyzing their data with a different method (figure 8.2) depicts the same clear structure in Africa, with the Khoisan diverging markedly deeper than the Yoruba following an even longer internal edge leading to non-African genomes, in which Asians and the Papuan group together with 85 percent residual resampling support (Waddell 2011b).

A re-sequencing of these genomes, in conjunction with additional genomes, yielded a tree among African populations that presents the San diverging first, then Mbuti Pygmy, followed by Yoruba, Mandenka, and Dinka branches, and then non-African groups (Meyer et al. 2012, Prufer et al. 2014). Nevertheless, there are differences between these two analyses. For example, Meyer et al. (2012) provide strong support for an edge separating Papuan and Han (Chinese) from French of length 4.5 ($\times 10^{-6}$). This edge shrinks to less than half this length with the data and analyses in Prufer et al. (2014). When reanalyzed with the methods in Waddell et al. (2011b), the Papuan moves outside other non-Africans, separated by an edge of length 1.5 ($\times 10^{-6}$). Further, the scale used is $\times 10^{-6}$, which means that, for every million well-aligned/sequenced bases, one should expect only ≤ 5 substitutions underlying this split, that is, a coalescence on that edge of the tree. This is also the approximate size of many reported signals of non-*sapiens*-*H. sapiens* interbreeding (e.g., Waddell 2013). Clearly, good-quality data and robust analytical methods are essential.

Networks from pairwise distances

A powerful aspect of these simple distances is that they jointly allow the inference of a system of non-tree splits, including reticulation or gene swapping (hybridization, introgression, etc.) between lineages of the main tree. A popular method for doing this is Neighbor Net, which is a generalization of the popular Neighbor-Joining method of tree reconstruction (Huson and Bryant 2006). Neighbor Net can infer a circular split system via a heuristic process. A circular split system is one in which all genomes can be placed, in any order, around a circle, with any or all splits being depicted as chords across this circle. If, however, a pattern of interbreeding becomes more complicated, a circular split system cannot represent it. Thus, problems of not being able to capture all splits in the data will increase the degree of misfit: for example, the residual sum of squared errors will increase. This, in turn, will cause a method such as residual resampling to downgrade its confidence in the splits in the graph (Waddell et al. 2011b). Nevertheless, splits that are missed in such a network might be revealed by scrutiny of the residuals (Waddell 2013, 2014).

Application of circular-split systems to Reich and colleagues' (2010) data yielded unexpected results, such as the largest non-tree split presented the Denisovan as splitting with the outgroup (*Pan*) rather than with other *Homo* genomes (Waddell et al. 2011b; figure 8.2). This split would be generated by the Denisovan interbreeding with a lineage that diverged somewhere along the edge of the chimpanzee outgroup (an unknown "ghost lineage," such as a pre-*H. sapiens*/*H. neanderthalensis* lineage). An *a priori* possibility is that individuals of a Denisovan lineage interbred with a lineage that might loosely be called *H. erectus*-like, as possibly represented by specimens from Zhoukoudian, which fall roughly in the correct time interval of the last ~800 kya. (Neurocranial shape is often used to lump these with other specimens into a spatially and temporally very broad group identified as *H. erectus sensu lato* [e.g., Lordkipanidze et al. 2013]).

Another difference generated by applying circular-split systems to Reich and colleagues' (2010) data is that the Papuan split not with the Denisovan, but with a Neanderthal-Denisovan-*Pan* grouping. In theory, removal of the Denisovan should group the Papuan with only the Neanderthal lineage if genetic introgression came from that direction only. However, doing this, the Papuan still splits off with Neanderthals and the outgroup (Waddell et al. 2013); this split is also much larger than expected due to the Denisovan lineage interacting with an earlier ghost lineage. Although Prufer et al. (2014) re-sequenced most of Reich et al.'s (2010) genomes, with the distances presented, the Papuan still presents a non-tree split with Neanderthals, the Denisovan, and *Pan* (result not shown). Even if the Denisovan lineage that introgressed into the Papuan was a distant relative of the sequenced Denisovan and not particularly coalesced with the patterns of that individual, this split is difficult to understand (Waddell et al. 2011b). Indeed, although Prufer et al. (2014) prefer this explanation, Kuhlwilm et al. (2016) depict the source of introgression into the Papuan as close to the sequenced Denisovan.

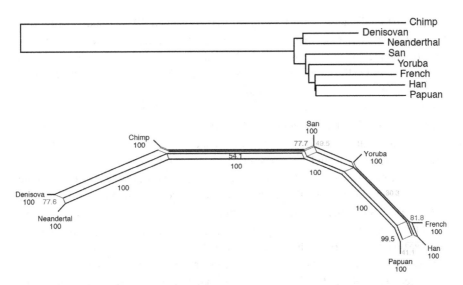

Figure 8.2
(a) The Weighted Least Squares (WLS+), power (*P*)=1, tree estimated in Waddell et al. (2011) from the data presented in Reich et al. (2010). Shown on the edges are the degree of confidence in a split using the technique of residual resampling for trees or networks (Waddell et al. 2011), unless 100%.
(b) The Neighbor Net with the external edges trimmed. Edge weights estimated with weighted least squares (WLS+) power (*P*)=1 (Felsenstein 2004). Note, the total residual for (b) (as measured by the ig%SD statistic, Waddell et al. 2011a), is less than 10 percent, suggesting that any or all "missing" signals will not be large enough to overwhelm the generally tree-like structure observed here.

A distance-based calculation that estimated the fraction of the Denisovan genome that came from the earlier "ghost" lineage gave a point estimate of about 2 percent introgression from an early "*erectus*-like" ancestor (Waddell et al. 2013). This point estimate, which resulted from different methods and dataset is in agreement with Kuhlwilm et al. (2016; ~0.2–1.2%); Prufer et al.'s (2014) estimate of ~2.7–5.8% seems rather high. This notwithstanding, Kuhlwilm (2016) suggested that the genes that introgressed into the Denisovan lineage came approximately 300 kya from a lineage that diverged over 2.5 mya (perhaps biased upwards by over a half million years, due to assuming a premature slowdown of the mutation rate). If over 2.5 mya is correct, since specimens identified as *H. erectus* significantly postdate that, a *H. erectus*-like Denisovan donor seems to be excluded. Thus, while there appears to be broad agreement on the reality of this introgression, there is disagreement over timing and percentages of contribution.

Buneman condition, P-statistics, D-statistics

Another way to learn about splits in the data that do not fit a single tree is via the analysis of all quartets, which in some cases are the same as rooted triplets. Under models with binary data, the differences between these quartets are equivalent to the Buneman four-point conditions (BFPC) of distances on a four-tipped tree (Buneman 1971, Bandelt and Dress 1992). This BFPC is used to build networks that are more complicated than trees, for example, Split Decomposition (Bandelt and Dress 1992), by adding a split only if that split is always one of the two strongest splits implied by distances for all possible quartets; when the network is drawn, the weight of that split is the minimum over all quartets.

Thus, if there is more disagreement, Split Decomposition collapses that split and, in this sense, is more conservative than Neighbor Net, which averages split differences (Huson and Bryant 2004). (Indeed, unless the data has a low incidence of rate of error or of conflict, and includes relatively few taxa, Split Decomposition often seems too conservative, while Neighbor Net, with residual resampling, seeks an appropriate balance [Waddell et al. 2011a.]) One should note that, with binary data, the BFPC is calculated from the frequencies of the site patterns 0011, 0101, and 0101, where 0 is ancestral and 1 the derived state. Here, pattern 0011 means that taxon 1 and taxon 2 have state 0, and taxon 3 and 4 the derived state 1. This is because no other patterns with four taxa change the basic relationships of the distances from which quartet structure is inferred, when multiple changes at a site are discounted

Rooted triplets test whether the data clearly favor one tree over others and are also consistent with a single tree (Waddell 1995, Waddell et al. 2001, Waddell et al. 2011b). They work with site patterns that take the form $f(0011)$, $f(0101)$, and $f(0110)$, where $f =$ frequency and the first taxon from the left is the outgroup, which is always assigned state 0. Rooted triplets can be applied to a variety of data, e.g. SINEs (short interspersed nucleotide sequences) and SNPs. Assuming that site patterns are unlinked, and $f(011)$ is the greatest, one can use likelihood ratio tests (Waddell et al. 2001, 2005), with the P2 probability assessing $f(101) = f(110)$.

When these tests are applied to Reich and colleagues' (2010) genome data for order *Pan* (outgroup), the San, the Yoruba, and the Han, f(San+Yoruba) and f(San+Han) are not significantly different (Waddell et al. 2011b). If there are no parallelisms, convergences, or mistakes in the data, this result is consistent with the interpretation that the San population was long separated from the other populations with little introgression (Waddell et al. 2011b). Indeed, analysis of a larger sample of genomes has demonstrated that the long run ancestral Khoisan population was considerably larger than the others and, until recently (~5 kya), isolated from them for a vast amount of time (Kim et al. 2014). Based on trio-based mutation rates (Scally 2016), this period of isolation may have been ~140–200 kya, which predates considerably the fossil record of *H. sapiens*.

In their genomic analyses, Green et al. (2010) use what they call "D-statistics" (i.e., $[f(0101) - f(0110)]/[f(0101) + f(0110)]$). Reich et al. (2011) also employ "S-statistics" of the form $f(0101) - f(0110)$ (a Buneman condition on binary data). If informative sites are assumed to be independent, the latter is the P2 statistic (above; Waddell et al. 2001). Green et al. (2010) used these and other statistics to infer a network of genetic interchange. Reich et al. (2010) then added to the data and analyses to estimate a Neanderthal introgression into non-Africans of 1–4%, and of Denisovan into the Papuan of 4–6%.

Challenges for all methods of building networks of descent include realizing that (1) finding the optimal network is NP Hard, which is a difficult search problem; (2) many different phylogenies may all look the same based on "distance" quartets; (3) systematic as well as sampling error must be accounted for when assessing robustness; and (4) detecting and dealing with convergence, parallelism, and even errors in the data are difficult indeed. With regard to the latter point, extension to true quartets, such as those obtained via a Hadamard conjugation on four sequences (Waddell et al. 2011b), may help. Further, true quartets have the potential of untangling more complex networks of descent than Buneman/D-statistics alone (Yang et al. 2013).

Full-Site Pattern Spectrum

In phylogenetic analysis, character-based methods tend to work with a full-site pattern spectrum (that is, the frequency of all possible assignments of 0 and 1 across all genomes [Hendy and Penny 1993]). However, these methods tend to discard sequence order, which is both an advantage and a disadvantage. If the method of analysis can integrate out the gene tree at each location in the genome, this can be a huge advantage since misidentification of a gene tree at each location can lead to major biases (Waddell 1995). Another advantage is that this approach does not need haplotype or phasing information.

A conceptually simple way to do this is to calculate, for a given species tree, what spectrum is predicted when all weighted gene trees and their frequencies under exact coalescent models of population change are taken into consideration. This can be greatly simplified using mathematical coalescent theory (Felsenstein 2004).

Waddell et al. (2011b) fitted such a model to the data of Reich et al. (2010), including a single flow of genes from Neanderthal to all non-Africans, and another from a Denisovan relative to the Papuan lineage. This significantly improved the fit of the model that allowed for these introgressions. A further advantage of these methods is that they identify the residual of every single site pattern with X^2-like statistics tuned to either high sensitivity (which helps spot the worst fitting part of the data), or high robustness (which gives more robust parameter estimates in spite of misfits) (Waddell et al. 2011b, 2012). Upon doing so, it turned out that one of the residuals using sensitive fit measures showed that while the Neanderthal shared with present-day humans a significant number of fixed, derived alleles, the Denisovan did not. This would be expected if a fraction of the Denisovan genome came from a more deeply diverging "ghost" lineage.

In addition to the above, the site pattern spectrum unequivocally rejects a strong form of the MR hypothesis. The site pattern frequency spectrum in such a fully mixed population is given by Tajima (1983). It predicts the same frequency for each site pattern of two, three, or more derived SNPs. Testing this series of linear invariants leads to rejecting fully mixing genomes, a result expected to become stronger as errors in the data are removed.

For example, testing that the frequency of the derived pattern FHP being equal to DFH yields a X^2 statistic of~4000. This alone is far larger than the misfit of all site patterns to an OA model that assumed limited introgression of genes from Neanderthals and the Denisovan (~2000 or less).

Sequencing-error patterns in this type of data can lead to biases in parameter estimates. When a complete likelihood-mixture model, produced by modeling and adding a spectrum of error, was added to the previous site pattern spectral model, optimal fitting indicated that sequence-error rates affecting more than one individual were much higher than expected based on singleton error (Waddell et al. 2011b). The predicted order of genome quality remained in agreement with Reich et al. (2010). However, with the best fitting model, the predicted proportion of "surviving" Neanderthal introgression into non-Africans drops from ~4.4% to ~1.9% (Waddell et al. 2011b). The latter figure is very close to a ~1.8% estimate based on effectively re-sequencing all these genomes (Prufer et al. 2014).

The X Chromosome

The human X chromosome (Reich et al. 2010, Meyer et al. 2012) is particularly deficient in alleles associated with non-*sapiens* interbreeding events (Waddell et al. 2011b, Meyer et al. 2012, Waddell 2013, Prufer et al. 2014, Sankararaman 2014). This could be due to mating with non-*sapiens* males (Waddell 2011b). However, the re-sequencing and remapping data of Meyer et al. (2012) reveals a near complete absence of the signal of non-*sapiens* alleles in the Papuan X chromosome (Waddell 2013). This suggests purifying-selection, because male-biased gene flow should depress non-*sapiens* allelic transfer on the X chromosome by one-third at most. Re-sequencing Neanderthal genomes further suggests negative (or purifying) selection against Neanderthal gene flow (Prufer et al. 2014). Indeed, when Sankararaman et al. (2014) studied the genomes of more than 1,000 Eurasian individuals, they discovered that Neanderthal alleles are less common in areas that are gene enriched: for example, genes showing high testes expression levels. On the X chromosome, which comprises numerous male hybrid sterility genes, overall Neanderthal ancestry is about one-fifth that of autosomal genes (Sankararaman et al. 2014). In contrast, non-tree splits found on the X chromosome involving only H, sapiens population (see Waddell 2013) far better match those of autosomal genes (figure 8.3 below).

Beyond Sequence Spectrums

The ongoing development of methods that analyze complete haplotypes and their phasing offer promise of inferring local genome-sequence history, including recombination, the timing of coalescence and population splitting, rates of migrants, and population size along each line of descent, for example, Multiple Sequentially Markovian Coalescent (MSMC; Schiffels and

Durbin 2014). These methods tend to embody many "hidden" parameters, such as unknown recombination points, that are dealt with using methods such as a Hidden Markov Model or a Bayesian framework (Felsenstein 2004). Another method does not look for recombination points, but uses spaced blocks of sequences that are hopefully unlinked, "non-functional," and individually single tree (i.e., they have not recombined; Gronau et al. 2011). Although these methods are presenting interesting insights, it remains unclear how they are related to methods reviewed above in terms of being unbiased, efficient, and robust. For example, while Kuhlwilm et al. (2016) corroborated the earlier-discussed ghost-lineage introgression hypotheses, there were unexpected twists. Using the methods of Gronau et al. and Schiffels and Durbin, Kuhlwilm et al. (2016) were able to place estimates of effective population size onto an expanded phylogeny, that included allele flow from another ghost population (*sapiens* or pre-*sapiens*) near the root of *H. sapiens* into the Altai Neanderthal genome.

Another Non-*sapiens* Introgression?

Figure 8.3 presents the Neighbor Network results for the autosomal data presented in Meyer et al. (2012) using analyses in Waddell (2013) for Chromosome X. The Denisovan is a particularly useful outgroup for examining what may have occurred with pre-*sapiens* in Africa. A large non-tree split between the Denisovan and Papuan is consistent with the earlier analyses of Reich et al. (2010),Waddell et al. (2011b) and Meyer et al. (2012). There is also evidence of a complex set of splits between Europeans, Native Americans, and East Asians that is consistent with, for example, the suggestion that Native American genomes consist of about one-third West Eurasian and about two-thirds East Asian genes (Fu et al. 2015).

Figure 8.4 shows a Q-Q plot of residuals; deviation away from the linear trend indicates significant outliers. A large negative residual between Mbuti and Dinka, and another between Mandenka and Denisova, suggests that these pairs are far too close; also the residual between Mbuti and Mandenka is too large. In the 3D plot of the residuals (figure 8.4b), the residuals of Denisova to Africans are negative leading up to Mandenka, where they become largest in absolute value and then abruptly switch sign and drop toward the Denisova to Papuan residual. The pattern is similar near the Mbuti-Dinka pair. These patterns may be reflective of a least squares fitted network (NeighborNet) attempting to accommodate unexpectedly small distance pairs; indeed, only a few hundred kilometers separate Northeastern Congolese Mbuti and South Sudanese Dinka. Interactions between these groups tend to be male Dinka with female Mbuti and may be the basis for their having the largest residual (figure 8.4). The large negative residual of the Mandenka/Denisova pair seems most consistent with genetic input from pre-*sapiens* individuals.

By reducing the expected distance from Denisova to Mandenka, the pre-*sapiens*-Mandingo hypothesis is implicitly modeled modeled with an "unseen" split of Denisova and Mandenka versus all other individuals. That is, minus twice the raw residual of Denisova

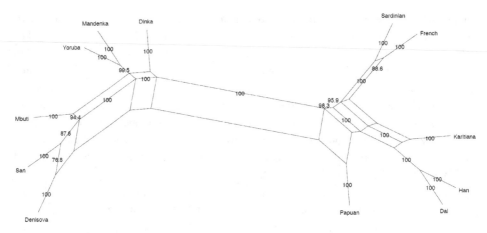

Figure 8.3
Neighbor Net with residual resampling estimated as in figure 1, but applied to the distances reported in Meyer et al. (2012). Only splits with greater than 70% support are shown (c.f. Hillis and Bull 1993). Compare with the same technique applied to the X-chromosome, figure 3c, Waddell (2013).

to Mandenka (corresponding to a split of 8.63×10^{-6}) was added to the observed Denisova to Mandenka distance to account for the fact that, in least-squares fitting, a misfit induced by pushing one residual down results in other residuals going up. The same adjustment was made for the Mbuti-Dinka pair (split size 12.49×10^{-6}). In comparison, the split modeling the Denisova + Papuan-versus-all-others splits on the same data is 15.53×10^{-6}.

Modeling these additional splits results in a considerable improvement in fit, with the residual weighted sum of squares going from ~0.12 to ~0.05. The lnL improves by over 29.5 units, with criteria such as AIC, BIC, and AICc favoring the revised model. Figure 8.4c shows there are now no obvious outliers in the Q-Q plot and figure 8.4d shows that not only the directly modeled residuals, but near neighbors (such as the two previously largest positive residuals) have approached zero. Further, similar to previously proposed interactions with pre-*sapiens* (Reich et al. 2010, Waddell et al. 2011b, Meyer et al. 2012, Waddell 2013, Prufer et al. 2014), there is much less evidence of gene transfer on the X chromosome for the Denisova/Mandenka signal (Waddell 2013). This contrasts with visible signals of introgression on the X chromosome amongst *sapiens* lineages, including Mbuti/Dinka (Waddell 2013).

The Pre-*sapiens* Mandingo Hypothesis

There is increasing evidence of pre-*sapiens* gene flow into the ancestors of different African groups (Hammer et al. 2011). Based on deeply diverging haplotypes, Mbuti pygmies present the greatest evidence for pre-*sapiens* gene flow (Hammer et al. 2011). Evidence in West Africa of a specific event or set of events includes: (1) deeply divergent autosomal haplotypes in the

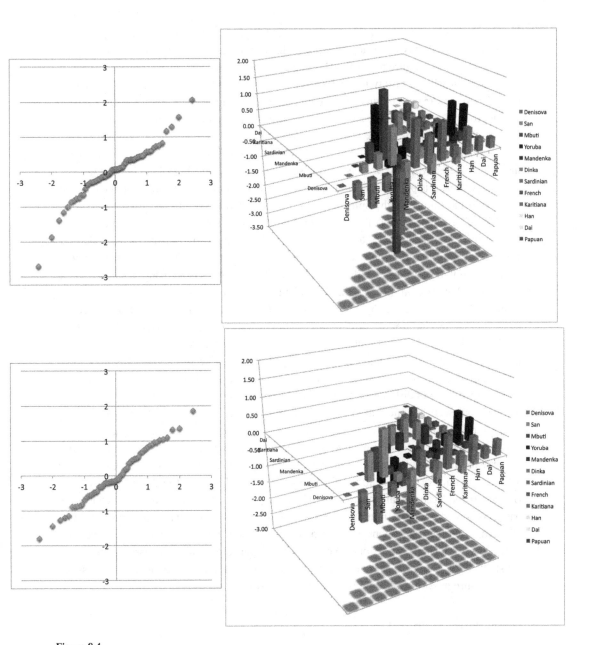

Figure 8.4
(a) A Q-Q plot of the standardized residuals from figure 8.3.
(b) A 3D plot of the standardized residuals. The large negative outliers are Denisova with Mandenka (–2.17) and Mbuti with Dinka (–2.41).
(c) As for (a), but the Q-Q plot after allowing for a Denisova + Mandenka and a Mbuti + Dinka split.

Yoruba (Hammer et al. 2011); (2) a late transition from a Middle Stone Age culture across large areas; (3) a deeply diverging pre-*sapiens* lineage represented by the the Iwo Eleru skull (Harvati et al. 2011; figure 8.1); (4) a few copies of the most divergent *H. sapiens* Y chromosome found in Mbo (almost twice as deep as that of the Khoisan; Cruciani et al. 2011, Mendez et al. 2016); and (5) a particularly large and unexplained residual in the Mandenka toward the outgroup (Denisova) in the autosomal genome but not in the X chromosome (figure 8.4). Further, the locations of Yoruba, Iwo Eleru, Mandenka, and Mbo fall in a rough line in West Africa trending toward the southeast for ~1000 km. This constitutes the "pre-*sapiens* Mandingo" hypothesis, whereby Mandenka refers to the language and Mandingo the people.

Using crania, it is possible to estimate when the Iwo Eleru lineage split from the Khoisan relative to the timing of the split between the latter and Neanderthals. First, is a simple "backbone" estimate (Waddell et al. 1999), which is the ratio of the path length from the Iwo Eleru/*H. sapiens* ancestor to Khoisan divided by the path length from the Neanderthal/*sapiens* ancestor to the same Khoisan (figure 8.1). The result is 79%; if the path lengths are traced to Iwo Eleru instead of Khoisan, it is 75%. In comparison, using the most filtered data from Mendez et al. (2016), the equivalent numbers of the edge length of Y chromosome A00 to a Neanderthal Y chromosome yield a ratio of 11/(11+12) (~48%). Since the tree suggests acceleration of skull-shape change toward *H. sapiens* and Iwo Eleru, these divergence-time estimates are likely to be too large. The evolutionary rate-change penalty of Kitazoe et al. (2007), which is specifically designed to accommodate this bias when making divergence-time estimates, should in future be applied.

The proportions of differences back to the *H. sapiens*/Neanderthal ancestor do not account for the "coalescence" time within populations, that is, here, the average difference in shape between two skulls. Using the difference between two Khoisan skulls as such an estimate, which, one should note, is similar to the difference between a pair of Neanderthal skulls, this fraction of the path length is removed, resulting in an estimate of 69% when traced to a Khoisan skull, or 57% if applied to the path to Iwo Eleru. If the divergence time of *H. sapiens* and Neanderthal is taken as 600 kya (e.g., Meyer et al. 2016, Kuhlwilm et al. 2016 based on slower *H. sapiens* mutation rates), it seems that the Iwo Eleru lineage diverged ~300–450 kya. A similar adjustment for Y chromosomes should produce less change, since their expected coalescence time is ~one-fourth as large as the autosomal genes assumed to generate skull shape. Based on these divergence times, and their uncertainty, it remains possible a Y A00 introgression was via a member of the Iwo Eleru lineage.

Out-of-Africa at Present

In this section I summarize how the latest data and analyses confirm, challenge, or enrich this hypothesis-set (Waddell and Penny 1996).

When?

The age of the split between Khoisan and other living humans seems to be ~100–200 kya, this time range being largely due to conflicting methods of rate calibration. As mentioned above, a recent estimate is ~250 kya, which is surprisingly old and will certainly need more validation. There is some suggestion that values may be trending toward ~150–200 kya (Scally 2016). However, values toward the lower end of this range tend to be favored by the archeological record, which does not show clear signs of a "package of modernity," found abundantly among the Khoisan, for example, until < 100 kya (Klein 2009). This raises the issue of why it took so long for *H. sapiens* to populate Africa. Perhaps part of this puzzle lies in the co-existence until the Holocene of *H. sapiens* with closely related, but competing hominids (e.g., Iwo Eleru, Fish Hoek, and Boskop specimens). Earlier, at 100 to 200kya, Africa and the Middle East may have been full of competing lineages, suggested by multiple morphs/lineages likely present in that period, including, LH18, Jebel Irhoud, Singa, Omo 2, Herto, and Skhul 5.

Where?

Autosomal diversity gradient genetic maps among African populations suggest Southern Africa (south of Kenya; e.g., Henn et al. 2011). mtDNA analyses are consistent with the geographic range of the Khoisan (Gonder et al. 2007), as are those of the Y chromosome, if the very deep rare A00 type is attributed to introgression (see above). A Southern African origin is also suggested by tree and split diagrams of full genomes (figures 8.2 and 8.3), which depict the divergence first of the Khoisan, and then Mbuti, West Africans, Dinka, and, finally, all non-Africans. The deep coalescence time, greatest genetic diversity, and largest long-term effective population size of the Khoisan are consistent with this conclusion (Kim et al. 2014).

Nevertheless, three lines of evidence suggest East or Southeast (e.g., Kenya, Uganda and Tanzania) Africa might have been the birthplace of *H. sapiens*. First, archeologists see in that region more evidence of a continuous trend towards technological modernity (e.g., stone tool forms) and of larger, long-term population sizes (Blome, et al. 2012). Second, the Hadza and Sandawe of Tanzania are similar to the Khoisan in having a "click" language, and harboring some of the same types of deep mtDNA and Y lineages (Tishkoff et al. 2007). After removing apparent recent (< 5 kya) West African (Bantu) introgression, a Hadza genome branches deeply as the sister to the Khoisan, with the suggestion that the former may have had the larger ancestral population (Pickrell 2012). Third, late-occurring, apparently pre-*sapiens* fossils from South Africa (Boskop ~12–14 kya, and Fishhoek (~6.7 kya) might also be consistent with Khoisan moving into the region, and taking some time to fully populate it. If the Khoisan harbor a significant component of a very late pre-*sapiens* population's genes, this could boost their genetic diversity relative to Hadza. It might also accentuate the "long fuse" issue of missing archeological evidence for *H. sapiens*, but would be difficult to detect with Khoisan at the root by themselves (accentuating the importance reconstructing the Hadza lineage genotypes pre-10 kya).

Who?

The tree/graph of figure 8.3 suggests the expansion, over a long period of time, of within-Africa *H. sapiens* populations that preceded the OA event, beginning with Mbuti, followed by West Africa, and then Dinka (Sudan). The timing of OA is linked to the age of the split with Khoisan, so OA was likely not older than ~100 kya, and not younger than 60 kya. The introgresssion of possibly *H. sapiens* genes into the Altai Neanderthal, the individual represented by the morphologically *H. sapiens* Qafzeh 9 skull, and the archeological record together suggest a population that existed well before 60 kya, albeit one apparently leaving no genetic legacy. Specimens in China (~100–150 kya) may indicate an earlier pre-*sapiens* wave. Although no genetic split is compatible with such earlier migrations (figure 8.3), this may be due to inadequacies of statistical power and sampling. In terms of who left Africa, and if heavily admixed genomes (e.g., from Egypt) are not considered, the South Sudanese Dinka currently appear to be the most representative of sampled living populations. A more detailed estimate will require reconstruction of ancient genomes from populations in areas such as Egypt, Ethiopia, Eritrea, and Somalia.

How?

It seems that human adaptive genetic/cultural advantage eventually overcame all pre-*sapiens* lineages. There is evidence of considerable interbreeding upon initial contact (Trinkaus 2007, Waddell 2014, figure 8.1) that may have been rapidly diluted by subsequent bands of *H. sapiens*, perhaps in conjunction with negative selection. In addition, male sterility may have been a factor, with purifying selection removing most, but not the adaptively advantageous (e.g., immune system, hair color [Sankararaman et al. 2014]), non-*sapiens* genes. Further, although Chromosome X is heavily depleted of non-*sapiens*-introgressed genes, there is evidence of abundant mixing of *H. sapiens* genes on the X chromosome between lineages, even when cultures were swamped. Surviving non-*sapiens* blocks in the genome probably globally average<4% of the genomes of living people (perhaps less than 2.5%). The larger portions of these blocks have little or no function, and/or differ minimally from the common ancestral form in function.

It is now also becoming clear that the effective population size of *H. sapiens* and its ancestors was probably always of a healthy size (e.g.,>20,000, and comparable to *Pan* subspecies [Prufer et al. 2014, Kuhlwilm et al. 2016]), which probably reflects the presence in Africa of more than one late *Homo* lineage, perhaps each with a geographically defined home territory (as seen with the 3 or 4 chimp subspecies). Further, there is no reliable evidence to support a "saltational," genetic drift origin of *H. sapiens*, which, in turn, reinforces the role of natural selection. These points are elaborated later in this chapter.

Morphological Evidence for *Homo sapiens* Origins

In terms of a fossil-based geographic origin of *H. sapiens*, the tree in figure 8.1 is generally consistent with morphology (e.g., Klein 2009, Schwartz and Tattersall 2005, Stringer 2016). Estimating trees and networks based on morphological data should help to move paleoanthropology away from a still-entrenched "Great Chain of Being" or "Chronospecies" view of human evolution, into which virtually every newly discovered fossil is inserted as a direct ancestor of something else. Improved phylogenetic analyses may even support hypotheses of distinct clades (e.g., Iwo Eleru, LH-18, Saldanha), whose individuals would have competed with *H. sapiens* as well as with its ancestors.

Based on analysis of fossils, the archeological record, and DNA, the Middle East, from the Sinai well into Eurasia, and between ~120 and ~60 kya may have been a region of hybridization. For example, qualitative phylogenetic analysis suggests that Qafzeh 9 might have been *H. sapiens*, while Qafzeh 6 may be ~one-third hybrid with a lineage close to Neanderthals (Waddell 2014). Skhul 5 may represent a distinct pre-*sapiens* lineage (or at least not *H. sapiens* [Schwartz 2016]). Evidence of *H. sapiens*/pre-*H. sapiens* genes in the very roughly dated 100 kya Altai Neanderthal might reflect this phase (Kuhlwilm et al. 2016). Since these remnant genes cannot be attributed to potential ancestors of living OA descendants, this might be indicative of at least one "failed" major movement of pro-*H. sapiens* into Eurasia at ~100kya or older. Interestingly, the shape of the 35–55 kya Shanidar 1 Neanderthal skull that draws it towards *sapiens* (Waddell 2015) also suggests the existence of a large and dynamic zone of hybridization.

To the east, China has yielded ~80–150 kya specimens that appear to be close to *H. sapiens*, although at least one from this period, from Zhirendong, appears to be pre-*sapiens* (or at least not *H. sapiens* [Schwartz 2016]). China also seems to have harbored very unusual and potentially early diverging forms until the very late Pleistocene-early Holocene, for example, from Maludong and Longlin (Curnoe 2014, Schwartz 2016). Although having inverted-T-shaped chins (Schwartz 2016), the Zhoukoudian Upper Cave skulls (~20 kya) are placed unusually deep in figure 8.1, which suggests possible hybridization of *H. sapiens* with earlier forms. In spite of some enthusiasm for China in the scenario of *H. sapiens* origins, the large, long-term effective population of this species' ancestors (e.g., Prufer et al. 2014, Kuhlwilm et al. 2016) suggests otherwise. That is, the hypothesized relatively large, ancestral population sizes, and their general continuity, do not seem consistent with a scenario of *H. sapiens*' ancestors enduring an earlier OA bottleneck, migrating to China, and then enduring another bottleneck when returning to Africa.

Figure 8.1 suggests that quantitative morphological phylogenetics can detect a shift in the shape of early European *H. sapiens* skulls over the period ~40–25 kya, with the earlier being more deeply divergent. Although this appears to be consistent with Trinkaus' (2007) proposal, specimens are highlighted differently in one versus the other analysis. The recent work of Fu et al. (2016) shows an ongoing decrease due to selection in the quantity of

Neanderthal-derived genes during this period. A tree in Waddell (2015) also shows a late European Neanderthal skull, Spy 2, drawn in the direction of *H. sapiens*, which is consistent with suggestions of *H. sapiens* gene flow into Neanderthals in Europe just prior to the latter species' extinction.

Given the frequency of potential hybrids in contact zones in Europe, the Middle East, and China, it seems possible that there was quite "leaky" pre-mating isolation of *H. sapiens* from its close relatives. Varying degrees of hybrid sterility, followed by predominately purifying selection within populations, together with possible group selection favoring populations with fewer introgressed genes, may finally have driven down non-*sapiens* genetic contributions. This certainly seems to be the case for Neanderthal introgression into living Europeans (Fu et al. 2016). Of course, since selective purification is hypothesized to act more quickly in populations with larger effective sizes, relatively smaller, often isolated, populations might tend to hold longer onto their genetic relics (perhaps the situation in Oceania), only to be swamped later by populations with fewer introgressed genes. Such a process might be accentuated, and causally confounded, if cultural/technological evolution was also more rapid in earlier occupied areas of denser population/larger population size.

Patterns of Introgression

It is worth emphasizing that while there is solid evidence of introgressed genes into *H. sapiens*, they are typically in small percentages, which validates the OA hypothesis of complete or nearly complete genetic replacement. There is now also evidence of numerous introgression events, including in Africa, but the fraction of retained non-*sapiens* genomes seems largely to have dropped to ~0–3%.

Reported introgressions between major lineages of *Homo* include multiple Neanderthal introgressions into non-Africans of 1.5–2.5% (average ~2%), multiple Denisovan into Melanesians, East Asians and possibly South Asians (0–5%), Neanderthal into the Denisovan (0–2%), an early diverging ghost lineage into the Denisovan (0–2%), *H. sapiens* or close pre-*sapiens* into the Altai Neanderthal (0–1.5%), and pre-*sapiens* African into African *H. sapiens* (~0–5%?). Interestingly, the limited nDNA data extracted from a ~430 kya pre-Neanderthal from Sima de los Huesos, Sierra de la Atapuerca (Meyer et al. 2016), has no signs of introgression as yet, barring unresolved discrepancies of the mtDNA tree.

Since multiple *Homo* lineages appear to have a low-surviving introgressed fraction, they should be considered separate species. This pattern contrasts with that within *H. sapiens*, with often large percentages of genomes mixing and genes surviving, despite documented cases of strong cultural pre-zygotic isolation, and even culturally influenced post-zygotic effects. This general theme does suggest that selection did have significant impact in evolution within the genus *Homo* in favoring a fairly coordinated gene complex that operated in spite of small-to-medium long-term effective population sizes (e.g., Neanderthals and

Denisovans <5,000 individuals, versus the ancestors of *H. sapiens*, ~15,000–30,000 individuals). The evidence for near-*sapiens* nuclear DNA introgression into the Altai Neanderthal suggests that the previously hypothesized Middle East hybrid zone had considerable long-range effects. However, a long-term hybrid zone need not equate with significant gene exchange if hybrids fail to successfully move into the ranges of parent species.

In summary, morphology and genetics taken together suggest first frequent introgression of *H. sapiens* genes into near relatives in contact zones, followed by strong selection against the majority of introgressed genes, with positive selection acting on only a few (Abi-Rached et al. 2011, Fu et al. 2016). Although this pattern has been referred to as "adaptive introgression" (e.g., Kuhlwilm et al. 2016), the term should be used with caution since the overall effect may be maladaptive in terms of the genetic load of all genes subject to purifying selection. Genomic data so far suggests that the pattern and outcome described above may have been a common characteristic of many lineages within *Homo*, and not just during the last OA event.

Limitations

The reviewers of an earlier version of this manuscript raised important points that need clarification. One is the validity of the method used here—OLS+with Procrustes distances—to infer a phylogeny based on skull shapes (Waddell 2014). This concern can be partly addressed by comparing this method to an emerging method of Maximum Likelihood (ML) estimation (Theobald and Wuttke 2008, Felsenstein 2015). An ML method being developed by Felsenstein and Bookstein (2016) aims to be a partial solution to the complexities of shape analysis of living forms that Bookstein (2009, 2015, 2016) discusses. Weber and Bookstein (2011) and Bookstein (2016) also address additional complications in current geometric morphometric analyses, pointing out the desirability of implicitly basing estimates of variance and covariance on mathematical equations that describe growth processes. This is a worthy goal in the long term, although phylogenetic systematics continues apace without this being achieved, for example, in virtually all DNA-based phylogenies. The danger, of course, is that many phylogenies will be misleading, at least in part, until the process is consistently well approximated (Swofford et al. 1996, Felsenstein 2004).

A Procrustes-based estimate of a phylogenetic pattern is effectively a sub-model of the ML method of Theobald and Wuttke (2008), or Felsenstein (2015), obtained by setting its variance-covariance matrix to the identity matrix. This simplification should run much faster, and may be a useful model for identifying "good" parts of tree-space. Other sub-models, such as a diagonal matrix only, may also offer advantages in practical analyses since they vastly reduce the need to estimate the values of a huge number of covariance parameters.

With regard to tree estimation, OLS+ and a Procrustes distance is expected to be less statistically efficient, and also perhaps less robust, than the equivalent ML method asymptotically, but it is computationally much faster (Swofford et al. 1996, Felsenstein 2004). As mentioned, a standard Procrustes distance assumes that the variance-covariance matrix between all landmarks is the identity matrix (zero variances and covariances). Felsenstein (2015) gives an elegant example in which some fish lineages have two different fins that grow either significantly more or less than other fins (larger variance) and are also closely correlated in size (non-zero covariances).

In Felsenstein's example, Procrustes distances may give a useful estimate of the relative distance between short-finned fish, but it may also slightly overestimate the distances between long-finned fish and more strongly overestimate the distance between short-finned and long-finned-fish. With four taxa, this problem translates to the inverse-Felsenstein effect, where taxa on long terminal edges may be repelled from each other (Waddell 1995). This problem is broadly analogous to using a Jukes Cantor distance to build at tree for a set of nucleotides that evolved by a fully general Markov (GM) model with divergent base compositions. Yet, a Jukes Cantor tree may be a useful estimate, and the GM tree may yield a more erratic estimate, if there is insufficient information to estimate the greater number of parameters and/or if the complications are not so pronounced. As noted earlier, since there is an interesting pattern of long edges in the tree of figure 8.1, it would be valuable, as ML software becomes available, to reexamine this. Another potentially robust approach is the use of median-based geometric morphometric distances (Rohlf and Slice 1990) with OLS+.

Clearly, phylogenetics and geometric morphometrics continue to wed in a variety of ways. However, as with marriage, some will be happier than others. There is a certainly a need to press on beyond Procrustes distances, but it is not yet clear what might work best in a given instance, or how the emphases of authors such as Gould (2002) and Hull (1988), who desire more process modeled in evolution, will be realized. Friston et al. (2012) provide one critique of least squares (e.g., Procrustes) when they are poorly equipped to handle unexpected changes, for example, in skull shape. Waddell et al. (2012) explore altering the fit function to reduce "surprise" or misfit human genome site patterns and, in a sense, this is what Theobald and Wuttke (2008) and Felsenstein and Bookstein (2016) are doing to move beyond Procrustes distances. A widespread transformation in this field will require development of different data/tools and a willingness to think of the problem in different ways (Fleck 1979). Bayesian platforms, such as Beast2, certainly offer still untapped tools for incorporating into a phylogenetic method not only expanded models of the evolution of shape, but also the ages, locations, associated technology, paleogeography, and so on of extinct lineages.

Concern has also been expressed regarding the conflicting conclusions of this and Templeton's (2017, this volume) chapter. In terms of the amount of non-*sapiens* introgression, Nested Clade Phylogeographic Analysis (NCPA, Templeton et al. 1995) estimated an ~10% contribution of non-*H. sapiens* to non-African forms (about four to five times greater than estimates herein based on the latest data). This method, however, has been shown to yield false positives in up to 75% of cases (e.g., Knowles and Maddison 2002, Knowles 2008, Petit 2008). A second issue is that, similar to the MR model, NCPA is biased towards inferring isolation by distance, even when the data lack structure (Beaumont and Panchal 2008) or there is a tree-like divergence pattern (Knowles and Maddison 2002, Knowles 2008). Amongst most modern populations, isolation-by-distance effects overlay some early *H. sapiens* migrations. In attempting to explain why some modern populations show more Neanderthal genes than others, these effects have been mixed with a structured ancestral African population (Eriksson and Manica 2012). However, these processes to explain the data are falsified by specific tests (Yang et al. 2012).

A second difference between these chapters comes from different definitions of the MR and OA models. Waddell and Penny (1996) defined the OA model and contrasted it with the MR model, as it was conceived prior to the work of Cann et al. (1987). Whole genome data has since shown that current parameter values are much closer to those of OA than to MR. Technically, if the MR model is treated as a general graph and OA a tree, the former is a superset and the latter a subset (Templeton 2017). However, this does not reflect the history of the debate, wherein non-*sapiens* components are rare and non-essential to the genetics of living *H. sapiens*. Further, considering the OA model as strictly a tree was never an essential component of the OA hypothesis. Not only have surviving non-*sapiens* elements been shown to be rare, they are now known to be biased towards less functional regions of the genome (e.g., Sankararaman et al. 2014). Further, some of the introgressed gene variants in *H. sapiens* that have medical implications show evidence of being selected against.

A final concern is going beyond trees to networks for understanding different kinds of information, morphology included. This is a difficult methodological problem that affects phylogenetics as a whole, and many current implementations towards networks have limitations; for example, Neighbor Net is limited to circular split systems. On the other hand, just as trees infer false edges when the data do not fit model expectations, it seems that networks are far more likely to infer false edges because they allow freer choice of splits/edges.

Even though no current method can extract all historical information from genomes, there seem to be real caps on the amount of introgression of all types between *H. sapiens* and non-African lineages of *Homo* that survived. If there are still-unidentified introgressions/migrants into *H. sapiens*, the best way to find them may be through careful examination of all regions of the thousands of available genomes that do not fit well (e.g., low posterior probability), in the highly parameterized models that are now being developed (e.g., Gronau et al. 2011, Schiffels and Durbin 2014).

A Struggling Genus?

It is now commonplace for articles to assert how tenuously lineages within *Homo* persisted, as though our lineage and others were at an unusually high risk of extinction. Given current trends, it may be necessary to sharply adjust "near-catastrophe" thinking (such as giant volcanic eruptions, ice ages, etc.) to explain the low-to-moderate effective population sizes of 5,000 to 30,000 individuals characteristic of *H. sapiens* and inferred for *H. neanderthalensis* and Denisovans. Rather than being disastrous, these population sizes apparently permitted the emergence of a diversity of novel adaptations and behaviors/cultures. During the last few hundred thousand years, there was a rich patchwork of distinct populations with many characteristics typical of species that overran each other at different times, apparently partly powered by selective advantage. Given the morphological (e.g., Trinkaus 2007, Waddell 2014) and DNA (Fu et al. 2016) evidence for increased numbers of introgressed hybrids where and when lineages meet, failure to swap large amounts of the genome was likely not due to pre-mating isolation, with hybrids likely being less fit than at least one source population.

A corollary of OA is that the greatest risk of extinction any ≤100 kya lineage within *Homo* faced was *H. sapiens*. Given the apparently low rates of genetic introgression across *Homo*, this risk of extinction by a co-generic was probably also a factor earlier in the history of the genus. This view of evolution in the genus *Homo*, from OA on, may be moving towards the antithesis of the MR model. This in turn may be evidence of co-adapted gene complexes related to development and behavior/culture that did not tolerate disruption.

The ongoing acquisition of genomic data of species of *Homo* may lead to testing one of the most interesting species concepts of all: The co-adapted genome. In the case of *Homo*, a tightly constrained co-adaptation might by crucial to the evolution of a socially networked thinking machine. Hints of this possibility come from the lack of introgressed genes on chromosome X between long-separated lineages: X is overrepresented in determinants of intelligence (Skuse 2005) and social networking (Startin et al. 2015). Additional support for this scenario comes from the observation that genes associated with specific brain regions, including the parietal lobes, that are greatly expanded in *H. sapiens*, are seriously lacking in the remnants of Neanderthal/Denisovan to *H. sapiens* introgression (Vernot and Akey 2014, 2015).

Phylogenetic Definitions of *Homo sapiens* and Species with Which They Interacted

Here I have argued, and only argued, for an extant-clade or crown-based definition of *H. sapiens*. A similar crown-based definition of *H. neanderthalensis* that includes specimens at or after an arbitrary but accessible date (e.g. ~100 kya, perhaps just prior to contact with

H. sapiens) may also be possible. As for other fossils, phylogenetic systematics cautions against basing species on grades (non-monophyletic groups; Hennig 1966). It is also unwise to assume that all members of a lineage that is closer to *H. sapiens* than to *H. neanderthalensis*, but which is not *H. sapiens*, is an ancestor (or sister taxon) of the extant species. There may well be considerable structure amongst these extinct hominids, as is suggested by the here-hypothesized Saldahna/LH18/Iwo Eleru clade. In order to pursue this, it will be essential to include more detail, including sample identity, and conceive a narrow definition of "species" or "morph" (Schwartz and Tattersal 2002), in order to have a better chance of understanding how these morphs or species interrelate and therefore also how their evolution may be traced. This also applies to estimating trees as well as more general networks.

Based on current genomic data for different specimens of *Homo*, it seems unwise, and obfuscating, to continue using the systematically meaningless, grab bag term *sensu lato* (see also discussion in Schwartz, this volume). Among other things, and in the context of this contribution, this modifier implies a kind of hybrid swarm. In contrast, the latest genomic results suggest that later lineages of *Homo* were pretty frugal about persistent gene flow between themselves. Further, *sensu lato* has outworn its utility (if it ever had one), inasmuch as it adds further to the taxonomic confusion that already exists: for example, currently in the literature one finds *Homo heidelbergensis*, *Homo heidelbergensis sensu stricto*, *Homo heidelbergensis sensu lato*, *Homo heidelbergensis sensu Stringer*, and so on, each with a different and often idiosyncratic definition. If one insists on being vague, it seems more useful to simply write "*Homo heidelbergensis*-like" as a way of letting the reader know that one is referring broadly to specimens that are grossly similar to one another, and not making a phylogenetic statement.

Future Directions

Africa is the key area for a better understanding of the origin, both in time and specific location, of *H. sapiens*. Robust reconstructions of pre-admixture genomes such as exist for the Hadza, Sandawe, and other exceptional African populations that are based on "pieces" of them that can be identified in living descendants represent an essential first step in furthering this field of inquiry (Pickrell et al. 2012). Genomes and morphology both suggest that the last ancestors of *H. sapiens* kept pretty much to themselves (e.g., witness the abrupt appearance of the inverted-T-shaped chin), and likely had a fairly long history in one part of Africa before expanding to overrun multiple neighboring species. aDNA from not very ancient specimens (e.g., Fish Hoek and Boskop) could provide currently missing information on major demographic shifts.

Other important contributions to the progress of studies of human origins include:

1. Examination of morphological characters useful to diagnosing *H. sapiens* in key populations including Khoisan, Pygmies, and Oceanians.

2. A well-characterized morphological matrix of discrete characters within *Homo*.

3. A freely accessible database of shape measures, such as landmarks, curves, and internal and external 3D scans of key fossils and recently living humans, as well as improved analyses that include networks of descent.

4. Critical reexamination of articles on non-*sapiens* genomes with regard to changing data (errors?), correspondence between established phylogenetic methods, and tracking changing parameter estimates via the communication of key data sets in compact forms.

5. Careful residual analysis of all genomes with a focus on regions that do not fit parameterized models; this may provide the best clues toward identifying currently undetected instances of introgressions and other avenues of gene flow.

In summary, the OA replacement model has continually been reinforced by new data and analyses. It also appears that near-complete (>90%) genetic replacement of independent lineages may have been a common and repeating feature of evolution within the genus *Homo*—which certainly contradicts some of the basic assumptions of the MR model and the chronospecies concept. Further, similar studies of other genera of mammals will help us to understand the fit of the general history of genus *Homo* with the rest of nature—a long-time but still unattained goal of paleoanthropology.

Acknowledgments

I am firstly very grateful to workshop organizer Jeff Schwartz for his invitation to this invigorating conference. For comment and critique of the manuscript I thank Fred Bookstein, Alan Templeton, and Joe Felsenstein. I also thank David Bryant, Mike Hendy, Hiro Kishino, David Penny, and David Swofford for years of intellectual stimulation. This manuscript was accepted November 8, 2016, and the research behind it partly supported by NIH grant 5R01LM008626.

References

Abi-Rached, L., M. J. Jobin, S. Kulkarni, A. McWhinnie, K. Dalva, L. Gragert, F. Babrzadeh, B. Gharizadeh, M. Luo, F. A. Plummer, J. Kimani, M. Carrington, D. Middleton, R. Rajalingam, M. Beksac, S. G. E. Marsh, M. Maiers, L. A. Guethlein, S. Tavoularis, A.-M. Little, R. E. Green, P. J. Norman, and P. Parham. 2011. "The Shaping of Modern Human Immune Systems by Multiregional Admixture with Archaic Humans." *Science* 334: 89–94.

Ackermanna, R. R., J. Rogers, and J. M. Cheverudc. 2006. "Identifying the Morphological Signatures of Hybridization in Primate and Human Evolution." *Journal of Human Evolution* 51:632–645.

Adams, D. C., F. J. Rohlf, and D. E. Slice. 2013. "A Field Comes of Age: Geometric Morphometrics in the 21st Century." *Hystrix* 24:7–14.

Bandelt, H.-J., and A. W. M. Dress. 1992. "Split Decomposition: A New and Useful Approach to Phylogenetic Analysis of Distance Data." *Molecular Phylogenetics and Evolution* 1:242–252.

Beaumont, M. A., and M. Panchal. 2008. "On the Validity of Nested Clade Phylogeographical Analysis." *Molecular Ecology* 17:2563–2565.

Blome, M.W., A.S. Cohen, C.A. Tryon, A.S. Brooks, and J. Russell. 2012. "The Environmental Context for the Origins of Modern Human Diversity: A Synthesis of Regional Variability in African Climate 150,000–30,000 Years Ago." *Journal of Human Evolution* 62:563–92.

Bookstein, F. L. 2009. "Measurement, Explanation, and Biology: Lessons from a Long Century." *Biological Theory* 4:6–20.

Bookstein, F. L. 2015. "Integration, Disintegration, and Self-Similarity: Characterizing the Scales of Shape Variation in Landmark Data." *Evolutionary Biology* 42:395–426.

Bookstein, F. L. 2016. "The Inappropriate Symmetries of Multi-variate Statistical Analysis in Geometric Morphometrics." *Evolutionary Biology* 43:277–313.

Bowler, J. M., H. Johnston, J. M. Olley, J. R. Prescott, R. G. Roberts, W. Shawcross, N. A. Spooner, O. Johnston, R. Prescott, and S. Shawcross. 2003. "New Ages for Human Occupation and Climatic Change at Lake Mungo, Australia." *Nature* 421:837–840.

Brothwell, D., and T. Shaw. 1971. "A late Upper Pleistocene Proto-West African Negro from Nigeria." *Man* 6:221–227

Buneman, P. 1971. "The Recovery of Trees from Measures of Dissimilarity." In *Mathematics in the Archaeological and Historical Sciences*, edited by F. R. Hodson, D. G. Kendall, and P. Tautu, 387–395. Edinburgh: Edinburgh University Press.

Cann, R. L., M. Stoneking, and A. C. Wilson. 1987. "Mitochondrial DNA and Human Evolution." *Nature* 325:31–36.

Caumul, R., and P. D. Polly. 2005. "Phylogenetic and Environmental Components of Morphological Variation: Skull, Mandible, and Molar Shape in Marmots Marmota, Rodentia." *Evolution* 59:2460–2472.

Cruciani, F., B. Trombetta, A. Massaia, G. Destro-Bisol, D. Sellitto, and R. Scozzari. 2011. "A Revised Root for the Human Y Chromosomal Phylogenetic Tree: The Origin of Patrilineal Diversity in Africa." *American Journal of Human Genetics* 88:814–818.

Curnoe, D., J. Xueping, A. I. R. Herries, B. Kanning, P. S. C. Taçon, B. Zhende, D. Fink, Z. Yunsheng, J. Hellstrom, L. Yun, G. Cassis, S. Bing, S. Wroe, H. Shi, W.C. Parr, H. Shengmin, and N. Rogers. 2012. "Human Remains from the Pleistocene-Holocene Transition of Southwest China Suggest a Complex Evolutionary History for East Asians." *PLoS ONE* 7(3):e31918.

Darwin, C. 1871. *The Descent of Man, and Selection in Relation to Sex*. London: John Murray.

Der Sarkissian, C., M. E. Allentoft, M. C. Avila-Arcos, R. Barnett, P. F. Campos, E. Cappellini, L. Ermini, R. Fernandez, R. da Fonseca, A. Ginolhac, A. J. Hansen, H. Jonsson, T. Korneliussen, A. Margaryan, M. D. Martin, J. V. Moreno-Mayar, M. Raghavan, M. Rasmussen, M. S. Velasco, H. Schroeder, M. Schubert, A. Seguin-Orlando, N. Wales, M. T. P. Gilbert, E. Willerslev, and L. Orlando. 2015. "Ancient Genomics." *Philosophical Transactions of the Royal Society B* 370: 20130387.

Eriksson, A., and A. Manica. 2012. "Effect of Ancient Population Structure on the Degree of Polymorphism Shared Between Modern Human Populations and Ancient Hominins." *Proceedings of the National Academy of Sciences* USA 109:13956–13960.

Felsenstein, J. 2002. "Quantitative Characters, Phylogenies, and Morphometrics." In *Morphology, Shape, and Phylogenetics*, edited by N. MacLeod, 27–44. Systematics Association Special Volume Series 64. London: Taylor and Francis.

Felsenstein, J. 2004. *Inferring Phylogenies*. Sunderland, MA: Sinauer Associates.

Felsenstein, J. 2015. "Morphometric Methods that Use Phylogenies, NIMBioS, 15 August 2015: Tutorial on Evolutionary Quantitative Genetics." Accessed May 28, 2016. http://citeseerx.ist.psu.edu/viewdoc/download?doi =10.1.1.717.7187&rep=rep1&type=pdf.

Felsenstein, J., and Bookstein, F. L. n.d. "Morphometrics on Phylogenies," unpublished manuscript.

Fleck, L. 1979. *The Genesis and Development of a Scientific Fact* (ed. T.J. Trenn and R.K. Merton, foreword T. Kuhn). Chicago: Chicago University Press.

Friston, K., C. Thornton, and A. Clark. 2012. "Free Energy Minimization and the Dark-Room Problem." *Frontiers in Psychology* 3(130):1–7.

Fu, Q., M. Hajdinjak, O. T. Moldovan, S. Constantin, S. Mallick, P. Skoglund, N. Patterson, N. Rohland, I. Lazaridis, B. Nickel, B. Viola, K. Prüfer, M. Meyer, J. Kelso, D. Reich, and S. Pääbo. 2015. "An Early Modern Human from Romania with a Recent Neanderthal Ancestor." *Nature* 524:216–219.

Fu, Q., C. Posth, M. Hajdinjak, M. Petr, S. Mallick, D. Fernandes, A. Furtwängler, W. Haak, M. Meyer, A. Mittnik, B. Nickel, A. Peltzer, N. Rohland, V. Slon, S. Talamo, I. Lazaridis, M. Lipson, I. Mathieson, S. Schiffels, P. Skoglund, A. P. Derevianko, N. Drozdov, V. Slavinsky, A. Tsybankov, R. G. Cremonesi, F. Mallegni, B. Gély, E. Vacca, M. R. G. Morales, L. G. Straus, C. Neugebauer-Maresch, M. Teschler-Nicola, S. Constantin, O. T. Moldovan, S. Benazzi, M. Peresani, D. Coppola, M. Lari, S. Ricci, A. Ronchitelli, F. Valentin, C. Thevenet, K. Wehrberger, D. Grigorescu, H. Rougier, I. Crevecoeur, D. Flas, P. Semal, M. A. Mannino, C. Cupillard, H. Bocherens, N. J. Conard, K. Harvati, V. Moiseyev, D. G. Drucker, J. Svoboda, M. P. Richards, D. Caramelli, R. Pinhasi, J. Kelso, N. Patterson, J. Krause, S. Pääbo, and D. Reich. 2016. "The Genetic History of Ice Age Europe." *Nature* 534:200–205.

Gascuel, O., and M. Steel. 2006. "Neighbor-Joining Revealed." *Molecular Biology and Evolution* 23:1997–2000.

Green, R. E., J. Krause, A. W. Briggs, T. Maricic, U. Stenzel, M. Kircher, N. Patterson, H. Li, W. Zhai, M. H. Y. Fritz, N. F. Hansen, E. Y. Durand, A. S. Malaspinas, J. D. Jensen, T. Marques-Bonet, C. Alkan, K. Prufer, M. Meyer, H. A. Burbano, J. M. Good, R. Schultz, A. Aximu-Petri, A. Butthof, B. Hober, B. Hoffner, M. Siege-mund, A. Weihmann, C. Nusbaum, E. S. Lander, C. Russ, N. Novod, J. Affourtit, M. Egholm, C. Verna, P. Rudan, D. Brajkovic, Z. Kucan, I. Gusic, V. B. Doronichev, L. V. Golovanova, C. Lalueza-Fox, M. de la Rasilla, J. Fortea, A. Rosas, R. W. Schmitz, P. L. F. Johnson, E. E. Eichler, D. Falush, E. Birney, J. C. Mullikin, M. Slatkin, R. Nielsen, J. Kelso, M. Lachmann, D. Reich, and S. Paabo. 2010. "A Draft Sequence of the Neanderthal Genome." *Science* 328:710–715.

Gonder, M. K., H. M. Mortensen, F. A. Reed, A. de Sousa, and S. A. Tishkoff. 2007. "Whole-mtDNA Genome Sequence Analysis of Ancient African Lineages." *Molecular Biology and Evolution* 24:757–768.

Gould, S. J. 2002. *The Structure of Evolutionary Theory*. Cambridge, MA: Harvard University Press.

Gronau, I., M. J. Hubisz, B. Gulko, C. G. Danko, and A. Siepel. 2011. "Bayesian Inference of Ancient Human Demography from Individual Genome Sequences." *Nature Genetics* 43:1031–1034.

Hammer, M. F., A. E. Woerner, F. L. Mendez, J. C. Watkins, and J. D. Wall. 2011. "Genetic Evidence for Archaic Admixture in Africa." *Proceedings of the National Academy of Sciences USA* 108:15123–15128

Harvati, K., C. Stringer, R. Grün, M. Aubert, P. Allsworth-Jones, and C. A. Folorunso. 2011. "The Later Stone Age Calvaria from Iwo Eleru, Nigeria: Morphology and Chronology." *PLoS ONE* 69:e24024.

Hendy, M. D., and D. Penny. 1993. "Spectral Analysis of Phylogenetic Data." *Journal of Classification* 10:5–24.

Henn, B., C. R. Gignoux, and M. Jobin. 2011. "Hunter-Gatherer Genomic Diversity Suggests a Southern African Origin for Modern Humans." *Proceedings of the National Academy of Sciences USA* 108:5154–5162.

Hennig, W. 1966. *Phylogenetic Systematics*, translated by D. Dwight Davis and Rainer Znagerl. Urbana: University of Illinois Press.

Hillis, D. M., and J. J. Bull. 1993. "An Empirical Test of Bootstrapping as a Method for Assessing Confidence in Phylogenetic Analysis." *Systematic Biology* 42:182–192.

Hillis, D. M., and J. P. Huelsenbeck. 1992. "Signal, Noise, and Reliability in Molecular Phylogenetic Analyses." *Journal of Heredity* 83:189–195.

Hull, D. L. 1988. *An Evolutionary Account of the Social and Conceptual Development of Science*. Chicago, IL: University of Chicago Press.

Huson, D. H., and D. Bryant. 2006. "Application of Phylogenetic Networks in Evolutionary Studies." *Molecular Biology and Evolution* 23:254–267.

Jefferies, R. P. S. 1979. "The Origin of ChordatesA Methodological Essay." In *The Origin of Major Invertebrate Groups*, edited by M. R. House, 443–447. London; New York: Academic Press for The Systematics Association.

Kitazoe, Y., H. Kishino, P. J. Waddell, N. Nakajima, T. Okabayashi, T. Watabe, and Y. Okuhara. 2007. "Robust Time Estimation Reconciles Views of the Antiquity of Placental Mammals." *PLoS ONE*. 2007 2;e384:1–11.

Kim, H. L., A. Ratan, G. H. Perry, A. Montenegro, W. Miller, and S. C. Schuster. 2014. "Khoisan Hunter-Gatherers Have Been the Largest Population Throughout Most of Modern-Human Demographic History." *Nature Communications* 5, Article number: 5692.

Klein, R. G. 2009. *The Human Career: Human Biological and Cultural Origins*. 3rd edition. Chicago IL: University of Chicago Press.

Knowles, L. L. 2008. "Why Does a Method that Fails Continue to Be Used?" *Evolution* 62:2713–2717.

Knowles, L. L., and W. P. Maddison. 2002. "Statistical Phylogeography." *Molecular Ecology* 11:2623–2635.

Kuhlwilm, M., I. Gronau, M. J. Hubisz, C. de Filippo, J. Prado-Martinez, M. Kircher, Q. Fu, H. A. Burbano, C. Lalueza-Fox, M. de la Rasilla, A. Rosas, P. Rudan, D. Brajkovic, Ž. Kucan, I. Gušic, T. Marques-Bonet, A. M. Andrés, B. Viola, S. Pääbo, M. Meyer, A. Siepel, and S. Castellano. 2016. "Ancient Gene Flow from Early Modern Humans into Eastern Neanderthals." *Nature* 530:429–435.

Li, J. Z., D. M. Absher, H. Tang, A. M. Southwick, A. M. Casto, S. Ramachandran, H. M. Cann, G. S. Barsh, M. Feldman, L. L. Cavalli-Sforza, and R. M. Myers. 2008. "Worldwide Human Relationships Inferred from Genome-wide Patterns of Variation." *Science* 319:1100e1104.

Lieberman, D. E. 2011. *The Evolution of the Human Head*. Cambridge, MA: Belknap Press/Harvard University Press.

Liu, W., C.-Z. Jin, Y.-Q. Zhang, Y.-J. Cai, S. Xing, X.-J. Wu, H. Cheng, R. L. Edwards, W.-S. Pan, D.-G. Qin, Z.-S. An, E. Trinkaus, and X.-Z. Wu. 2010. "Human Remains from Zhirendong, South China, and Modern Human Emergence in East Asia." *Proceedings of the National Academy of Sciences USA* 107:19201–19206.

Lordkipanidze, D., M. S. Ponce de León, A. Margvelashvili, Y. Rak, G. P. Rightmire, A. Vekua, and C. P. E. Zollikofer. 2013. "A Complete Skull from Dmanisi, Georgia, and the Evolutionary Biology of Early *Homo*." *Science* 342:326–331.

Mendez, F. L., G. D. Poznik, S. Castellano, and C. D. Bustamante. 2016. "The Divergence of Neandertal and Modern Human Y Chromosomes." *American Journal of Human Genetics* 98:728–734.

Meyer, M., M. Kircher, M.-T. Gansauge, H. Li, F. Racimo, S. Mallick, J. G. Schraiber, F. Jay, K. Prufer, C. de Filippo, P. H. Sudmant, C. Alkan, Q. Fu, R. Do, N. Rohland, A. Tandon, M. Siebauer, R. E. Green, K. Bryc, A. W. Briggs, U. Stenzel, J. Dabney, J. Shendure, J. Kitzman, M. F. Hammer, M. V. Shunkov, A. P. Derevianko, N. Patterson, A. M. Andres, E. E. Eichler, M. Slatkin, D. Reich, J. Kelso, and S. Paabo. 2012. "A High-Coverage Genome Sequence from an Archaic Denisovan Individual." *Science* 338:222–226.

Meyer, M., J.-L. Arsuaga, C. de Filippo, S. Nagel, A. Aximu-Petri, B. Nickel, I. Martínez, A. Gracia, J. M. B. de Castro, E. Carbonell, B. Viola, J. Kelso, K. Prüfer, and S. Pääbo. 2016. "Nuclear DNA Sequences from the Middle Pleistocene Sima de los Huesos Hominins." *Nature* 531:504–507.

Mounier, A., A. Balzeau, M. Caparros, and D. Grimaud-Herve. 2016. "Brain, Calvarium, Cladistics: A New Approach to an Old Question, Who Are Modern Humans and Neandertals?" *Journal of Human Evolution* 92:22–36.

Mounier, A., S. Condemi, and G. Manzi. 2011. "The Stem Species of Our Species: A Place for the Archaic Human Cranium from Ceprano, Italy." *PLoS ONE* 6:e18821.

O'Leary, M. A., J. I. Bloch, J. J. Flynn, T. J. Gaudin, A. Giallombardo, and N. P. Giannini. 2013. "The Placental Mammal Ancestor and the Post–K-Pg Radiation of Placentals." *Science* 339:662–667.

Paabo, S. 2014. *Neanderthal Man: In Search of Lost Genomes.* New York, NY: Basic Books.

Peterson, G.I., and J. Masel. 2009. "Quantitative Prediction of Molecular Clock and Ka/Ks at Short Timescales." *Molecular Biology and Evolution* 26:2595–2603.

Petit, R. J. 2008. "The Coup de Grace for Nested Clade Phylogeographic Analysis?" *Molecular Ecology* 17:516–518.

Pickrell, J. K., N. Patterson, C. Barbieri, F. Berthold, L. Gerlach, T. Güldemann, B. Kure, S. W. Mpoloka, H. Nakagawa, C. Naumann, M. Lipson, P.-R. Loh, J. Lachance, J. Mountain, C. Bustamante, B. Berger, S. Tishkoff, B. Henn, M. Stoneking, D. Reich, and B. Pakendorf. 2012. "The Genetic Prehistory of Southern Africa." *Nature Communications* 2012;3:1143.

Profico, A., F. Di Vincenzo, L. Gagliardi1, M. Piperno, and G. Manzi. 2016. "Filling the Gap. Human Cranial Remains from Gombore II (Melka Kunture, Ethiopia; ca. 850 ka) and the Origin of *Homo heidelbergensis.*" *Journal of Anthropological Sciences* 94:1–24.

Prüfer, K., F. Racimo, N. Patterson, F. Jay, S. Sankararaman, S. Sawyer, A. Heinze, G. Renaud, P. H. Sudmant, C. de Filippo, H. Li, S. Mallick, M. Dannemann, Q. Fu, M. Kircher, M. Kuhlwilm, M. Lachmann, M. Meyer, M. Ongyerth, M. Siebauer, C. Theunert, A. Tandon, P. Moorjani, J. Pickrell, J. C. Mullikin, S. H. Vohr, R. E. Green, I. Hellmann, P. L. F. Johnson, H. Blanche, H. Cann, J. O. Kitzman, J. Shendure, E. E. Eichler, E. S. Lein, T. E. Bakken, L. V. Golovanova, V. B. Doronichev, M. V. Shunkov, A. P. Derevianko, B. Viola, M. Slatkin, D. Reich, J. Kelso, and S. Pääbo. 2014. "The Complete Genome Sequence of a Neanderthal from the Altai Mountains." *Nature* 505:43–49.

Rasmussen, M., X. Guo, Y. Wang, K. E. Lohmueller, S. Rasmussen, A. Albrechtsen, L. Skotte, S. Lindgreen, M. Metspalu, T. Jombart, T. Kivisild, W. Zhai, A. Eriksson, A. Manica, L. Orlando, F. M. De La Vega, S. Tridico, E. Metspalu, K. Nielsen, M. C. Avila-Arcos, J. V. Moreno-Mayar, C. Muller, J. Dortch, M. T. P. Gilbert, O. Lund, A. Wesolowska, M. Karmin, L. A. Weinert, B. Wang, J. Li, S. Tai, F. Xiao, T. Hanihara, G. van Driem, A. R. Jha, F.-X. Ricaut, P. de Knijff, A. B. Migliano, I. Gallego Romero, K. Kristiansen, D. M. Lambert, S. Brunak, P. Forster, B. Brinkmann, O. Nehlich, M. Bunce, M. Richards, R. Gupta, C. D. Bustamante, A. Krogh, R. A. Foley, M. M. Lahr, F. Balloux, T. Sicheritz-Ponten, R. Villems, R. Nielsen, J. Wang, and E. Willerslev. 2011. "An Aboriginal Australian Genome Reveals Separate Human Dispersals into Asia." *Science* 334:94–98.

Reich, D., R. E. Green, M. Kircher, J. Krause, N. Patterson, E. Y. Durand, B. Viola, A. W. Briggs, U. Stenzel, P. L. F. Johnson, T. Maricic, J. M. Good, T. Marques-Bonet, C. Alkan, Q. Fu, S. Mallick, H. Li, M. Meyer, E. E. Eichler, M. Stoneking, M. Richards, S. Talamo, M. V. Shunkov, A. P. Derevianko, J.-J. Hublin, J. Kelso, M. Slatkin, and S. Pääbo. 2010. "Genetic History of an Archaic Hominin Group from Denisova Cave in Siberia." *Nature* 468:1053–1060.

Rohlf, F. J. and D. E. Slice. 1990. "Extensions of the Procrustes Method for the Optimal Superimposition of Landmarks." *Systematic Zoology* 39:40–59.

Rightmire, G. P. 1998. "Human Evolution in the Middle Pleistocene: The Role of *Homo heidelbergensis.*" *Evolutionary Anthropology* 6:218–227.

Roberts, R. G., R. Jones, and M. A. Smith. 1990. "Thermoluminescence Dating of a 50,000-Year-Old Human Occupation Site in Northern Australia." *Nature* 345:153–156.

Sankararaman, S., S. Mallick, M. Dannemann, K. Prüfer, J. Kelso, S. Pääbo, N. Patterson and D. Reich. 2014. "The Genomic Landscape of Neanderthal Ancestry in Present-Day Humans." *Nature* 507:354–357.

Scally, A. 2016. "Mutation Rates and the Evolution of Germline Structure." *Philosophical Transactions of the Royal Society* 371:20150137.

Schiffels, S., and R. Durbin. 2014. "Inferring Human Population Size and Separation History from Multiple Genome Sequences." *Nature Genetics* 46:919–925.

Schwartz, J. H. 2016. "What Constitutes *Homo sapiens*? Morphology Versus Received Wisdom." *Journal of Anthropological Sciences* 94:1–16.

Schwartz, J. H., and I. Tattersall. 2000. "The Human Chin Revisited: What Is It and Who Has It?" *Journal of Human Evolution* 38:367–409.

Schwartz, J. H., and I. Tattersall. 2002. *The Human Fossil Record: Terminology and Craniodental Morphology of Genus* Homo *Europe,* Volume 1. Hoboken, NJ: Wiley-Liss.

Schwartz, J. H., and I. Tattersall. 2003. *The Human Fossil Record: Craniodental Morphology of Genus* Homo *Africa and Asia,* Volume 2. Hoboken, NJ: Wiley-Liss.

Schwartz, J. H., and I. Tattersall. 2005. *The Human Fossil Record: Craniodental Morphology of Early Hominids Genera* Australopithecus, Paranthropus, Orrorin *and Overview, Volume 4.* Hoboken, NJ: Wiley-Liss.

Schwartz, J. H., and I. Tattersall. 2010. "Fossil Evidence for the Origin of *Homo Sapiens*." *American Journal of Physical Anthropology* 143 (Suppl. 51):94–121.

Skuse, D. H. 2005. "X-linked Genes and Mental Functioning." *Human Molecular Genetics* 149 (Spec No 1):R27–32.

Startin, C. M., C. Fiorentini, M. de Haan, and D. H. Skuse. 2015. "Variation in the X-linked *EFHC2* Gene Is Associated with Social Cognitive Abilities in Males." *PLoS ONE* 10(6):e0131604.

Stojanowski, C.M. 2014. "Iwo Eleru's Place Among Late Pleistocene and Early Holocene Populations of North and East Africa." *Journal of Human Evolution* 75:80–89.

Stringer, C. 2016. "The Origin and Evolution of *Homo sapiens*." *Philosophical Transactions of the Royal Society B* 371:20150237.

Swofford, D. L. 2000. *Phylogenetic Analysis Using Parsimony *and Other Methods*, Version 4.0b10. Sunderland, MA: Sinauer Associates.

Swofford, D. L., G. J. Olsen, P. J. Waddell, and D. M. Hillis. 1996. "Phylogenetic Inference." In *Molecular Systematics*, 2nd edition, edited by D. M. Hillis and C. Moritz, 450–572. Sunderland, MA: Sinauer Associates.

Tajima, F. 1983. "Evolutionary Relationship of DNA Sequences in Finite Populations." *Genetics* 105:437–460.

Templeton, A. R. 1998. "Nested Clade Analyses of Phylogeographic Data: Testing Hypotheses about Gene Flow and Population History." *Molecular Ecology* 7:381–397.

Templeton, A. R., E. Routman, and C. A. Phillips. 1995. "Separating Population Structure from Population History: A Cladistic Analysis of the Geographical Distribution of Mitochondrial DNA Haplotypes in the Tiger Salamander, *Ambystoma tigrinum*." *Genetics* 140:767–782.

Tishkoff, S. A., M. K. Gonder, B. M. Henn, H. Mortensen, A. Knight, C. Gignoux, N. Fernandopulle, G. Lema, T. B. Nyambo, U. Ramakrishnan, F. A. Reed, and J. L. Mountain. 2007. "History of Click-Speaking Populations of Africa Inferred from mtDNA and Y Chromosome Genetic Variation." *Molecular Biology and Evolution* 24:2180–2195.

Tishkoff, S. A., and M. K. Gonder. 2006. "Human Origins within and OA." In *Anthropological Genetics*, M. Crawford, ed. Cambridge, UK: Cambridge University Press.

Theobald, D. L., and D. S. Wuttke. 2008. "Accurate Structural Correlations from Maximum Likelihood Super-position." *PLoS Computational Biology* 42(2):e43.

Trinkaus, E. 2007. "European Early Modern Humans and the Fate of the Neandertals." *Proceedings of the National Academy of Sciences USA* 104:7367–7372.

Vernot, B., and J. M. Akey. 2014. "Resurrecting Surviving Neandertal Lineages from Modern Human Genomes." *Science* 343:1017–1021.

Vernot, B., and J. M. Akey. 2015. "Complex History of Admixture between Modern Humans and Neandertals." *American Journal of Human Genetics* 96:448–453.

Vigilant, L., M. Stoneking, H. Harpending, K. Hawkes, and A. C. Wilson. 1991. "African Populations and the Evolution of Human Mitochondrial DNA." *Science* 253:1503–1507.

Waddell, P. J. 1995. "Statistical Methods of Phylogenetic Analysis, Including Hadamard Conjugations, Logdet Transforms, and Maximum Likelihood." PhD thesis. Palmerston North, New Zealand: Massey University.

Waddell, P. J. 2013. "Happy New Year *Homo erectus*? More Evidence for Interbreeding with Archaics Predating the Modern Human/Neanderthal Split." *arXiv Quantitative Biology* 1312.7749:1–29.

Waddell, P. J. 2014. "Extended Distance-based Phylogenetic Analyses Applied to 3D *Homo* Fossil Skull Evolu-tion." *arXiv Quantitative Biology* 1501.0019:1–42.

Waddell, P. J. 2015. "Expanded Distance-based Phylogenetic Analyses of Fossil *Homo* Skull Shape Evolution." *arXiv Quantitative Biology* 1512.09115:1–20.

Waddell, P. J., A. Azad, and I. Khan. 2011a. "Resampling Residuals on Phylogenetic Trees: Extended Results." *arXiv Quantitative Biology* 1101.0020:1–9.

Waddell, P. J., H. Kishino, and R. Ota. 2001. "A Phylogenetic Foundation for Comparative Mammalian Genom-ics." *Genome Informatics Series* 12:141–154.

Waddell, P. J., H. Kishino, and R. Ota. 2005. "Statistical Tests for SINE Data and Resolution of the Species Tree." *Technical Report* 216:1–21. Columbia: Department of Statistics, University of South Carolina.

Waddell, P. J., N. Okada, and M. Hasegawa. 1999. "Towards Resolving the Interordinal Relationships of Placental Mammals." *Systematic Biology* 48:1–5.

Waddell, P. J., and D. Penny. 1996. "Evolutionary Trees of Apes and Humans from DNA Sequences." In *Hand-book of Symbolic Evolution*, edited by A.J. Lock and C.R. Peters, 53–73. Oxford: Clarendon Press.

Waddell, P. J., J. Ramos, and X. Tan. 2011b. "*Homo denisova*, Correspondence Spectral Analysis, and Finite Sites Reticulate Hierarchical Coalescent Models." *arXiv Quantitative Biology* 1112.6424:1–43.

Waddell, P. J., and X. Tan. 2012. "New g%AIC, g%AICc, g%BIC, and Power Divergence Fit Statistics Expose Mating between Modern Humans, Neanderthals and Other Archaics." *arXiv Quantitative Biology* 1212.6820:1–23.

Wagner, G. P. 2001. *The Character Concept in Evolutionary Biology.* Lieden: Elsevier.

Weber, G. W., and F. Bookstein. 2011. *Virtual Anthropology. A Guide to a New Interdisiplinary Field.* Vienna: Springer-Verlag.

Wilkins, J. S. 2011. *Species: A History of the Idea.* Berkeley: University of California Press.

Wolpoff, M. H. 1989. "Multiregional Evolution: the Fossil Alternative to Eden." In *The Human Revolution: Behav-ioral and Biological Perspectives on the Origins of Modern Humans,* edited by P. Mellars and C. B. Stringer, 62–108. Edinburgh: Edinburgh University Press.

Wolpoff, M. H., J. Hawks, D. W. Frayer, and K. Hunley. 2001. "Modern Human Ancestry at the Peripheries: a Test of the Replacement Theory." *Science* 291:293–297.

Wolpoff, M., A. G. Thorne, J. Jelinek, and Z. Yinyun. 1994. "The Case for Sinking *Homo erectus*: 100 Years of *Pithecanthropus* Is Enough!" *Courier Forschungs Senckenburg* 171:341–361.

Yang, J., S. Grünewald, and X.-F. Wan. 2013. "Quartet-Net: A Quartet-Based Method to Reconstruct Phylogenetic Networks." *Molecular Biology and Evolution* 30:1206–1217.

Yang, M. A., A. S. Malaspinas, E. Y. Durand, and M. Slatkin. 2012. "Ancient Structure in Africa Unlikely to Explain Neanderthal and Non-African Genetic Similarity." *Molecular Biology and Evolution* 29:2987–2995.

9 "Like Fixing an Airplane in Flight": On Paleoanthropology as an Evolutionary Discipline, or, Paleoanthropology for What?

Fred L. Bookstein

It seems appropriate to convene a workshop that focuses on the disconnect between human evolutionary studies and the theoretical and methodological standards and practice that inform the rest of evolutionary biology.
—Jeff Schwartz (this volume, preface)

A scientific fact is a socially imposed constraint on speculative thought.
—Ludwik Fleck (1935/1979), tr. FLB

The professional duty of a scientist confronted with a new and exciting theory is to try to prove it wrong.
—Freeman Dyson (1988)

Statistics: the measurement of uncertainty.
—Stephen Stigler (1986)

Introduction

I write as a biostatistician and as the founder of a morphometric method that is widely used throughout paleoanthropology. My main interest these days is the relationship between arithmetic and understanding across the natural sciences. In my view, paleoanthropology stretches that connection past the breaking point. It is, indeed, as Jeff Schwartz put it in the preface to this volume, a "disconnect" (see my 2014 *Measuring and Reasoning* for explication of my philosophy of that connection in a variety of contexts). In this chapter, I review what I perceive as the ironies driving the "disconnect" that Schwartz identified, critique some of the currently fashionable biostatistical approaches that deny or obscure its significance, and suggest some remedies.

Intellectual Origins of the Critique

My support for the claim that paleoanthropology necessarily fails to articulate with evolutionary biology is based on the observation that for about the last seventy years evolutionary biology has manifested many of the characteristics of a consensus-based quantitative natural science. Here I sketch some highlights of the argument in themes from Ludwik Fleck (via Thomas Kuhn), C. S. Peirce (via Alexander Rosenberg), and E. O. Wilson. I will claim three separate types of "disconnect" between paleoanthropology and evolutionary biology in its quantitative (natural-scientific) voice: (1) the unavailability of any community consensus regarding substantive agreements of fact (in the words of my epigraph from Fleck, the absence of "constraints on speculative thought"); (2) a failure to model the fundamental quantity of evolutionary biology—the concept of fitness—in appropriate quantitative garb; and (3) a severe paucity of valid examples of abduction, whereby a new hypothesis quantitatively resolves a previously acknowledged quantitative discrepancy between all reigning theories, whatever they are, and data that the community already agrees are valid.

Fleck's Crucial Insight

Ludwik Fleck (1935/1979) emphasized that *a scientific fact is a socially imposed constraint on speculative thought*; in the original German, "*Können wir [die wissenschaftliche Tatsache] als 'denkkollektives Widerstandsaviso' bezeichnen*" (p. 129). The sociologist in me accepts Thomas Kuhn's approach to the role of measurement in the natural sciences: not the better-known schema (paradigm?) of "paradigms," as in his widely read chapter for the *International Encyclopedia of Unified Science* (Kuhn 1962, 1970), but rather the earlier essay of his for a 1958 conference on quantification reprinted in Kuhn (1961). Kuhn noted that in the physical sciences (a rubric I extend to the other natural sciences in order to include evolutionary biologists), quantification plays two distinct roles, confirmation and disruption, and that disruption succeeds only after a replacement explanation has been tested that the data better confirm. Apparently, Kuhn had not been aware of Fleck's ideas when he wrote this, but later he acknowledged the priority of the Polish microbiologist, and even oversaw the translation of Fleck's treatise from the German. In either variant, the adoption of a quantitative explanation is best understood as a version of C. S. Peirce's *abduction* (today more often called "inference to the best explanation"). As Peirce put it in lecture notes from Harvard originally dated 1903 (Peirce, 1934, p. 117):

The surprising fact, C, is observed;

But if A were true, C would be a matter of course,

Hence, there is reason to suspect that A is true.

(See Bookstein 2014 for examples of this specific logical tactic.) Richard Feynman phrased it memorably in his commencement address at Caltech in 1974. Speaking of experimental psychology, he commented wryly, "I call these things Cargo Cult Science, because they follow all the apparent precepts and forms of scientific investigation, but they're missing something essential, because the planes don't land" (p. 11).

Following Feynman, my critique here can be captured in six words: In paleoanthropology, the planes don't land. In this context, quantities such as those scattered in Waddell's (2017, this volume) diagrams permit novel inferences only if they uniquely resolve discrepancies of data with previously posited theories. I take this prototype as the model for persuasion by numbers all across the natural sciences. But I am unable to find any examples of it in today's paleoanthropology, not if by "resolution" of a quantitative discrepancy one means the actual numerical agreement of the value of some covariance (not merely its plus or minus sign) with some theoretically specified path coefficient, or the agreement of some squared morphological or genetic "distance" with some diffusion constant multiplied by an elapsed time. It is not only that, as far as I can tell, the consensus processes posited by Fleck or Kuhn are missing, but also that, owing to the general flight from sophisticated mathematical-statistical forms of reasoning in this field, there are no norms about what patterns would count as "surprising" in Peirce's syllogism. Null-hypothesis p-values certainly cannot do the job. This particular disconnect is one general thrust of this essay.

Evolutionary Biology, Standing on One Foot

Whereas the preceding theme dealt with the role of arithmetic (manipulation of numerical data) in the "disconnect" that is our subject, my second concerns the role of evolutionary biology. Here I deal solely with the focus on *fitness*. I am no expert on this topic, but instead defer to Alexander Rosenberg (1984, p. 143), who comments that the principle of survival of the fittest (or, actually, "expansion of the fitter") is the culminating axiom of the entire theory. To paraphrase: if a subclan D' is sufficiently higher in fitness than the rest of clan D for sufficiently many generations, then the proportion of D' in D will increase.

As Rosenberg (1984, p. 218) notes in subchapter 7.8, "The Statistical Character of Evolutionary Theory," this axiom has real bite:

The axiom's appeals to "sufficiently many generations" and "sufficiently large differences" in fitness are not admissions of vagueness or uncertainty about how much is enough. They are reflections of the interrelationship between evolutionary variables and nonevolutionary ones on which they are supervenient. Failure to recognize the statistical character of this claim can lead to mistaken charges that the theory is false or unfalsifiable.

It is this second prong that applies to a critique of the paleoanthropological context: no quantitative theory here would seem to be based on enough data to support it against a principled skepticism on statistical grounds. (Rejecting null hypotheses of patternlessness is not sufficient [see Bookstein 2014, section 4.5].) Bluntly put, if we translate "sufficiently many generations" to mean "in the long run," then paleoanthropology does not have any runs that are long enough. Indeed, paleoanthropology's subject matter is nowhere near the domain of punctuated equilibrium within which we can expect the standard models of theoretical population genetics to operate (Felsenstein 2016).

This problem of fitness is difficult enough as applied to any ordinary taxon, let alone to *Homo sapiens*. As Millstein and Skipper (2007, p. 38) stated, "Biologists and philosophers have yet to provide an adequate interpretation of fitness." In the same volume, Francisco Ayala (2007, p. 241) highlights what he calls the "three frontiers of human biology": "the egg-to-adult transformation, the brain-to-mind transformation, and the ape-to-human transformation." It is fair to say that paleoanthropology is making very little progress on the first two frontiers. But, as E. O. Wilson (2012) elegantly explains, the issue of fitness in human evolution is considerably more difficult than in most taxa, because we are one of the only two truly eusocial taxa (the other being the social insects). Although the essential focus for studying a eusocial species must be to understand group selection across its kinship groups, this information, which is crucial to understanding human uniqueness, does not fossilize. As Wilson (2012, p. 47) points out, "The human condition is a singularity," and the scientific method has great difficulty dealing with singularities. Furthermore, "the fundamentals of the mammalian life cycle and population structure" prevent the "overpowering of individual selection by group selection" that characterized the evolution of social insects. Wilson (*ibid.*) concludes: "The human condition is an endemic turmoil rooted in the evolution processes that created us." While the study of turbulence in hydrodynamics can be described with Wilson's metaphors, this arena of inquiry is far from evolutionary biology in both methodology and rhetoric, especially with regard to causation. (Explanations of turbulence, for instance, typically do not model a particular system of eddies and whirlpools, which are not fixed, nor require these attributes to be at any particular position, such as their averages. Further, the correlations that characterize turbulence are interpreted not by their individual specific values, but by the trend according to which they fall off with distance. In other words, they are generic correlations that are rooted not in specific data sets but in physical models [see, e.g., Davidson (2015)]. The implications of this style of mathematical modeling for understanding evolutionary "turmoil" require more explication than can be provided here.)

Correlations, Covariances, and Matrices

A third and final theme of my epistemological argument likewise can be traced back to Peirce: the distinction between the roles of the Gaussian distribution (and thus the statistics of covariances summarized in this section) in physics and in biology. In a popular essay, "The Fixation of Belief," Peirce (1877) wrote:

Mr. Darwin proposed to apply the statistical method to biology. The same thing has been done in a widely different branch of science, the theory of gases. Though unable to say what the movements of any particular molecule of gas would be on a certain hypothesis regarding the constitution of this class of bodies, Clausius and Maxwell were yet able, eight years before the publication of Darwin's immortal work, by the application of the doctrine of probabilities, to predict that in the long run such and such a proportion of the molecules would, under given circumstances, acquire such and such velocities; that there would take place, every second, such and such a relative number of collisions, etc.; and from these propositions were able to deduce certain properties of gases, especially in regard to their heat-relations. In like manner, Darwin, while unable to say what the operation of variation and natural selection in any individual case will be, demonstrates that in the long run they will, or would, adapt animals to their circumstances. Whether or not existing animal forms are due to such action, or what position the theory ought to take, forms the subject of a discussion in which questions of fact and questions of logic are curiously interlaced. (Quoted in Rosenberg [1984, p. 218])

"Curiously interlaced," indeed, inasmuch as this prescient statement predated the discovery of the correlation coefficient and of the gene. It was the physicists Clausius and Maxwell, as Peirce noted, who brought the Gaussian distribution to the attention of the natural sciences. But evolutionary biology does not use the form of disorder represented there: the disorder of the gas laws, an entropy-maximizing chaos without pattern. Rather, biology uses the bell curve by attending to the *covariances* of any single measured variable with other measured variables. (Later, I will revisit the thermodynamic alternative, touching on its current resurgence in biology via its role in the explanation of consciousness, which is probably the most crucial component of the fitness of *H. sapiens*.)

Since 1928, when John Wishart wrote down the general formula for the probability distribution of a covariance matrix of arbitrarily many arbitrarily correlated Gaussian measurements, we have understood at least the rudiments of data analysis for multivariable data sets gathered under effective experimental control. But this is not remotely the situation in paleoanthropology. First of all, note that the Wishart distribution deals with matrices of *covariances*, not correlations. Algebraically speaking, covariance is the more accessible quantity, because it is an average, not a ratio. But the way in which covariance matrices are validly handled in statistics is oblique to the needs of evolutionary biology, where explanations are driven by path coefficients *sensu* Sewall Wright (1968), not covariances. (For a disquisition on the difference see Bookstein 2017, Chapter 2, and for Wishart's actual formula, which is much simpler than its derivation, see Chapter 4.)

Covariances Have a Picture

The covariance between two variables is the average of their product minus the product of their averages, or, what amounts to the same thing, the average product of their deviations from their own separate averages (figure 9.1, left panel). Why would a biologist want to compute a number answering to this description? After all, there is no reason to expect the product in question to be in scientific units having any sort of obvious biological meaning (e.g., for studies of ecophenotypy, the unit might be something like "degree-kilometers"),

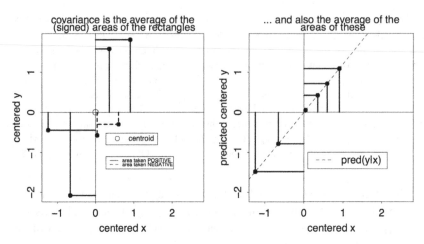

Figure 9.1
Two geometries of the covariance coefficient.

and nothing in physiology or biomechanics need authorize the multiplication. In fact, the usefulness of the covariance, beyond its appearance in Wishart's theory, arises mainly from the fact that it supplies the numerator of the coefficient of the predictor in the *linear regression,* which is the line of least-vertical-squares fit to the points in the scatterplot. This is the dashed line in the right-hand panel (which could have been drawn as well in the left panel) of figure 9.1. The denominator of this slope formula is the variance of that same predictor. The covariance is the same regardless of which variable is doing the predicting. That is, *covariances are acausal,* as is obvious from the formulation in Figure 9.1, the average of x times y is the same as the average of y times x.

The Covariance Matrix

To the modern statistician, the preceding text has missed the point. For a pair of variables x and y, it's not that we have "two variances and one covariance." Rather, we have a *list of variables,* here a set of pairs of measurements (x, y), and a *covariance matrix,* a 2×2 array of covariances of each variable of the list by every other. (The covariance of a variable with itself is just its variance.) In other words, the *mathematically* natural way of notating what we were just talking about is the table

$$\begin{pmatrix} \sigma_{xx} & \sigma_{xy} \\ \sigma_{xy} & \sigma_{yy} \end{pmatrix}.$$

Here I've taken advantage of the usual statistician's shorthand, "σ" for "covariance" and a subscript for the pair of variables in question. (Why the letter σ? Because it is the lower-case version of the Greek capital letter Σ, a symbol in every associated formula.) For the variances, it is common to write $\sigma_{xx} \equiv \sigma_x^2$ where σ_x, the *standard deviation* of x, is the square root of the variance.

Correlations modify the algebra of covariances. The easiest way to get to a correlation from a covariance is through the regression formula again, the slope of the line at the right in figure 9.1. The correlation is the square root of the product of two of those slopes, the one for x on y and the one for y on x, as in figure 9.2. This is one of at least *thirteen* different characterizations of this ostensibly useful number—see Rodgers and Nicewander, 1988—but it inherits the acausal character of the covariance on which it is based. And there emerges another problem: Why would we multiply two different prediction coefficients? Alternatively, why would we divide a covariance by the product of the standard deviations? Yes, we have cancelled out those pesky products of units, but does the arithmetic have any necessary evolutionary meaning?

Although the Victoriana polymath Francis Galton first intuited that quantities such as these had something to do with heredity (see Stigler 1986), correlations first entered into the dogmas of a quantitative evolutionary biology through the work of his disciple Karl Pearson. For instance, in a lay lecture of 1903 Pearson designed a method for demonstrating that the regression line sometimes actually fit the class means for scatters of physical measurements (such as comparisons of arm span between parents and children or between siblings), so that the relative constancy of the correlation coefficients around 0.5 for different pairs of such measurements hinted at a common hereditary process, just as his colleague W. F. R. Weldon was discovering for the crabs at the bottom of the Thames. When Pearson computed the same analyses for "psychical characters" such as "vivacity" or "intelligence," he achieved a similar concentration of values around 0.5. But then, in a leap of faith not too different from those of today's evolutionary psychologists, Pearson (1903, p. 204) took the argument too far:

We are forced, I think literally forced, to the general conclusion that the physical and psychical characters in man are inherited within broad lines in the same manner, and with the same intensity . . . . That sameness surely involves something additional. It involves a like heritage from parents . . . . We inherit our parents' tempers, our parents' conscientiousness, shyness and ability, even as we inherit their stature, forearm and span.

Note the triple repetition of the trope of physical force, to conceal the lack of all inferential force. The remainder of his lecture drew inferences about eugenics that today are offensive enough to demolish most of our scholarly respect for Pearson as statistician (Bookstein 2014, section 7.8.2). In fairness, analysis of variance had not been invented yet, nor the notation of $G \times E$ that drives modern nature-nurture analyses. Nevertheless one can trace this error forward through the twentieth century—the endless confusion between correlation and

squared correlation is the product of these slopes

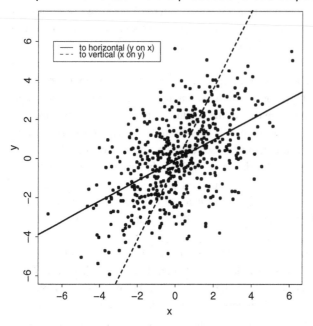

Figure 9.2
The correlation is the square root of the product of the slopes of the two regression lines, one (y on x) taken with respect to the horizontal direction, the other (x on y) with respect to the vertical. Does this product of slopes have any evolutionary meaning?

causation familiar to anybody who has ever taken a course in biometry. The only rhetoric that is persuasive is the rhetoric of path coefficients, regression slopes that are consilient across multiple studies and multiple mathematical models (Wright 1968).

Due Regard for Uncertainty

An epistemological defense of arithmetical inferences becomes more difficult when we generalize from a pair of measurements to a whole list $x_1, x_2, \ldots, x_p$. The idea of the covariance matrix itself is not affected, and so, from the Wishart probability distribution model for this matrix entity, mathematical statisticians have derived distributions of most of the quantities still relied upon in multivariate paleoanthropology, such as principal components loadings, factor analysis formulas, eigenvalues of "explained variance"—indeed, all the machinery of linear multivariate analysis. But these are just numbers; their application to evolutionary questions requires that they be used to differentiate between hypotheses. After all, statistics is a tool not for exploration, but for skepticism. As David

Cox (2006, p. 7) put it, statistics "takes the family of models as given and aims to give intervals or general sets of values within which ψ [the parameter] is likely to lie, [and] to assess the consistency of the data with a particular parameter value ψ_0." But nearly every application of statistics in paleoanthropology omits that assessment of "the consistency of the data with a particular parameter value" that is required for biostatistical interpretation of the value even after "tak[ing] the family of models as given."

When the "parameter" is a multivariate descriptor, like a principal component (PC), the "intervals" to which Cox refers are likewise multidimensional. Two of these intervals are particularly relevant in studies that use principal components (such as the computation of relative warps, or principal components of Procrustes shape). Although the derivation of these formulas is difficult, it is not difficult to set them down. The uncertainty of the *direction* of a principal component, for instance, is a fairly straightforward function of all the other terms in the eigendecomposition. Asymptotically, the variation in each direction PCk ($k > 1$) of the "end" of PC1—the tangent plane for representing variations in PC1—is Gaussian with variance $\lambda_1 \lambda_k / ((N-1)(\lambda_1 - \lambda_k)^2)$ independent of all the other axes. For instance, if the eigenvalue of PC2 is much larger than that of PC3, the uncertainty of PC1 is principally in the direction of PC2, with a spherical standard error roughly proportional to the reciprocal square root of $N\lambda_1 / \lambda_2$. (For analogous formulas see, e.g., Jolliffe [2002]; Bookstein [2014, p. 359] provides is a relatively elementary discussion.)

Continuing this line of reasoning, there is also a convenient formula for the *interpretability of components after the first*. Roughly, the ratio of consecutive eigenvalues necessary for any meaningful description of the direction of the one with the larger variance—a "$p \sim 0.50$" threshold, so to speak—is approximately $1 + \sqrt{8/N}$. (See Bookstein 2014, section 5.2.3.4, especially figure 5.12.) If this criterion is not satisfied, there is no basis for reifying either the formula or the scores of the corresponding component. Most of the claimed two- and three-dimensional ordinations in today's paleoanthropology would likely fail to satisfy this requirement. When relevant articles do not report the first ignored eigenvalue, which is usually the case in our field, neither the reader nor the reviewer has the information necessary to make an informed judgment about the reification.

Organized Skepticism

Needless to say, paleoanthropologists rarely invoke formulas such as these. If they did, many of their more positivistic assertions would probably evaporate. But our colleagues, and especially observers in the media, generally relish the affected certainty of paleoanthropological claims. The scientifically literate public enjoys the simulation of confidence, and academic tenuring committees prefer that candidates for promotion come with "discoveries," not "suggestions." Yet, as Feynman (1974) emphasized, it is mandatory to focus on the evidence that *disagrees* with one's claim. No one has put this better than Dyson (1988;

see Epigraphs above): A scientist's first responsibility is skepticism toward his or her own most beloved conjectures. If paleoanthropologists behaved in accordance with this rule, the field would be much less agonistic, but also much less disconnected from evolutionary biology than it is now. To state this more sociologically: It appears that paleoanthropology is particularly weak in "organized skepticism," the "OS" of the acronym CUDOS (Greek $\kappa v \delta o \varsigma$, "honor"), which, according to one of the founders of this branch of sociology, Robert Merton (1942), is one of the fundamental norms of normative scientific behavior.

What Is Wrong with Paleoanthropological Quantifications Today

In this section I touch lightly on six different themes that are common to many of the discussions in this volume.

There is no epistemology of covariance structures in evolutionary biology

I have not seen any demonstration that a covariance matrix *per se* is biologically real. By this I do not mean that such a matrix cannot be computed by using the embryo as its own measuring instrument, that is, by watching a number of specimens grow in order to record what covariances are actually manifest in idealized environments. Rather, I am not convinced that such a Platonic object is actually biochemically *encoded* anywhere in either the genome or the physiome in a way that can be measured without "running the organism" as a simulation of its own biochemical program. The pathways (and associated path coefficients) that control morphogenesis or physiology are surely not quantifiable via dissection, tomography, or genome-wide association studies. The problem begins with the roster of variables indexing the matrix, that is, the labels of the rows and columns. There *must* be a constitutive theory of this list somewhere, a justification of the entities included and, even more importantly, the entities that are *not* included in the analysis. (In morphometrics, this translates to the list of landmarks or semilandmarks driving the analysis. Different lists have entirely different covariance structures.) Other sciences provide constructive examples. For soils, it is discrete mineral components that add to 100 percent, while for functional data analysis, it is timelike or spacelike slices. If the field of morphology cannot claim something similar, our methods must accommodate this particularly inconvenient fact and acknowledge the nonexistence of any finite-dimensional standard for morphospace and the corresponding uncertainty that applies to *any* morphological inference owing to the ineluctably arbitrary selection of the features or landmarks to be measured or located.

"Distances" are not distances

The use of dissimilarities as measures on pairs of specimens has a long and honorable history in the psychological sciences (cf. Coombs 1956, Borg and Groenen 2005). However, the act of calling them "distances" does not render them acceptable to evolutionary biology.

Indeed, distances are functions of two forms, not one. Their reifications result in a riot of different algebraic formulas that lead to essentially different ordinations. If there is an epistemology of such abstract "distances," it would need to be based on a more fundamental quantity, such as the way in which expected squared distances in Brownian motion are the product of a diffusion coefficient by time (Bookstein 2013a). But apparently genes do not mutate, nor does morphology change, in that way. Consequently, following Coombs, it would be better to treat such dissimilarities as raw data, not as the reflection of any underlying physicalizable process. In short, "distances" are metaphors. They exist in the mind of the metaphor-maker or in the silicon of his computer, not in the bodies or physiologies of the organisms under study.

Procrustes shape coordinates

The preceding critique applies verbatim to "Procrustes distance," which is the fundamental quantity of geometric morphometrics. There are three essential problems here. First, the formula for Procrustes distance itself tacitly encodes geometric maneuvers that are surely unrelated to actual evolutionary processes. (For instance, landmarks at far ends of the longest diameter tend toward loadings on PC1 that are roughly equal and opposite, regardless of the actual morphogenetic processes that regulated their position.) Second, the Procrustes method systematically misrepresents synapomorphies, especially when they involve shape changes of relatively large magnitude (Bookstein 2016). The methods fail at a scale of variation less than that separating *Gorilla* from *Homo,* for instance, and so are unlikely to be reliable when applied to the emergence of hominids. And third, and perhaps most important, is the argument against studying shape at all. By an accident of intellectual history, the more familiar Procrustes formula deals purely with shape, which is the impact on one's retina of an equivalence class with respect to position, rotation, and scale. From an evolutionary perspective, however, (1) rotation could well be an important descriptive feature (in any world with gravity or with flowing fluids), (2) position (in any species that locomotes) is related to potential energy and thereby to biomechanics, and (3) scale is the fundamental quantity of allometry. Dryden and Mardia (1998, 2016) extended the Procrustes method to "size-shape space" (see also the Vienna construal of form space by Mitteroecker et al. 2004). But in an instance of the lock-in effect, it is only the first version, the size-free version, that is taught to our graduate students (see my more elaborate critique in Bookstein 2016).

Assuming that earlier criticisms (e.g., the relevance of landmarks to fitness, the availability of adequate fossil samples) have been dealt with, the two applications of Procrustes analysis I hold to be valid for evolutionary biology are (1) the identification of directions in these spaces with biomechanical indices deemed Darwinianly relevant on prior functional grounds (Bookstein 2015b), and (2) the explicit study of allometry (for which principal components in size-shape space are more appropriate than principal components of shape per se) and its methodological cousin, heterochrony (Mitteroecker et al. 2004). But these themes are absent from most of today's paleoanthropological literature.

Principal components

Above, I reviewed several problems with principal components using a rubric that acknowledges statistical uncertainty. Here I address a more serious concern: Regardless of sampling variation, the figure of merit that a principal component analysis is maximizing makes no biological sense. The first principal component has the maximum of variance for all linear combinations whose coefficients sum in square to 1. *But that formula—the sum of the squares of the coefficients—is devoid of biological content.* There is an alternative version of principal components that construes them as *relative eigenvectors bearing* extrema of the ratio of variance of linear combinations between two different covariance matrices (for instance, arising in two different groups). This is more reasonable, since now both of the matrices involved have been observed. This is not the case in ordinary PCA, for which the gauge metric of sphericity (completely uncorrelated sets of variables) never arises in nature, and for good reason. (Why, one may ask, would a reasonable person measure such a data set?) In particular, there is never any reason to believe that the uncorrelated PCs of any data set constitute a list of independently selectable or manipulable organismal variables of state. Linear combinations of quantities that are meaningful or selectable individually need not themselves be selectable or even meaningful; the inevitability of negative coefficients makes this difficulty even worse (see extended discussion in Bookstein and Mitteroecker 2014).

Likelihood ratios for point hypotheses

A common trope in paleoanthropology is the computation of the standardized difference between a particular measurement for one specimen and the distribution of the measurement in a reference group. Since half the square of such a ratio is the exponent of e in the Gaussian formula for log likelihood of that item under the hypothesis of membership in that group, the difference between two of these is the main component of the log-likelihood ratio for the hypothesis of membership in one group versus the other. Because this quantity is a function of specimen sampling in the reference groups, it has a standard error that can be approximated by a simple formula. To oversimplify slightly, the coefficient of variation of such a log-likelihood ratio will always be at least of order $\sqrt{2/(N-1)}$, where N is the sample size of the reference group at greater distance. If $N=9$, which is not an untypically low value in paleoanthropological studies, this puts the possibility of equipoise—identical likelihood of the data for both hypotheses—right at the $p \sim 0.05$ significance boundary. Whenever a likelihood ratio is wielded to summarize a particular comparison (e.g., "odds of XX to 1 that this form is not a Neanderthal") the reported odds ratio is entitled to its own estimate of uncertainty (see discussion in Bookstein 2018, section 3.6.3).

Uncertainties for trees and networks

However broad the uncertainties of likelihood-based classification, they are even broader for the composite hypotheses conveyed by trees and networks. For estimated trees, Felsenstein (2004) presents the machinery for the corresponding uncertainties, and his principles ought to be honored. For networks, we statisticians do not yet have any standard statistical methods at all—which raises the question of why paleoanthropologists nonetheless continue to extract these from fallible, spotty data. I suggest that trees and networks are metaphors at least to the same extent that Procrustes distance is, and should be regarded as expressions of poetic license to the same degree. There is, however, no place in evolutionary biology *sensu stricto* for metaphors. (Incidentally, neither "distances" nor networks appear to conform to the models of Gaussian variability that drive our customary language of correlation-based inferences.)

Some Suggested Remedies

The title of this essay can be taken as a guide to what I see as fundamental to a constructive transformation of paleoanthropology. The simile of the title, "fixing an airplane in flight," refers to repairing a complex system even as it is required to continue functioning competently in an open context. The analogy can be taken on several levels. A first interpretation of that "airplane in flight" is our species' conception of itself. Anthropology is part of our collective understanding of what it means to be human, and paleoanthropology attempts to reconstruct how we came to be that way. The evolutionary origins of a variety of contested behaviors, such as tribalism, warfare, rape, or burial of the dead, are notorious for their eruption beyond the polite bounds of scientific conferences into the mass media. A second aspect of the simile is the description of the medium through which we are flying, to wit, the boundary conditions that keep paleoanthropology aligned, however loosely, with the technical minutiae of evolutionary biology (e.g., the genomics of fossil DNA, the implications of which for the human tree are the topic of Pääbo's [2015] successful popularization of the probable origin of the Neanderthals).

But there is a third aspect of the simile that troubles me most. To repair something, one must look at the original design, by which I mean the design of the modern quantitative sciences. There, I think, the work to be done most closely approaches what the typical mathophobic paleoanthropology graduate student or postdoc can master. This work will entail abandoning the metaphors of correlation, covariance, and principal components for alternatives that are much better aligned with the realities of paleoanthropological data bases.

A paleoanthropology that is independent of correlations, regressions, "distances" and principal components might have not only better statistics, but also a more intimate relationship with contemporary evolutionary biology. It would be well to acknowledge that we are bound to the haphazard and fugitive nature of fossil human samples; to restrict the rhetoric of discovery to what is strongly supported by data and also admit where support is not particularly strong; to always provide errors of estimate for any numerical or geometrical quantity that purports to be relevant to the understanding of human evolution. Regardless of whether our conception of human evolution is privileged for species-specific reasons, there is no justification for privileging it on evolutionary grounds. Let us at least apply the same standards of skepticism to claims in this domain that we have come to apply across the rest of experimental evolutionary biology.

Among other possible advances are two that emerged during recent investigations of mine into approaches to quantitative morphology that are applicable to small samples of bony forms (i.e., to typical fossil data). One involves a new method for studies of "integration," a currently fashionable term, that is based on geometrical rather than correlational considerations. This would be more appropriate for the modestly sized samples that paleoanthroplogists are used to inasmuch as it matches better the language of systematic biology to which paleoanthropologists may be attempting to contribute. The other is a new, explicit morphometric method for relating morphological variability to one particular hypothetical functional effect, the strain consequent on a specific stress, that can be computed from same imaging data that drives the morphometrics. Here, the strain quantifications that relate cause to effect could, in principle at least, be path coefficients, which are true assessments of a selective effect on a specific indicator of fitness. A third approach offering potential insight into the coevolution of individuals and culture uses a novel and more compelling biomathematical methodology—the thermodynamics of information—to model the emergence of consciousness.

Integration is a matter of geometry, not correlation

As I have discussed elsewhere (Bookstein 2015a), the main thing we know about the growth and evolution of shape is that variations in position of the different key points of a form (the locations usually called "landmarks") are far from uncorrelated. This observation covers not only the obvious kinds of correlation, such as left-right symmetry, but also growth gradients, as, for example, the hypermorphosis of the hominid splanchnocranium relative to the neurocranium. We also know that there are subdivisions or *modules* of this structure that respond to specific genomic, epigenetic, or eco-phenotypic challenges. Consequently there is a pressing need for methods that assort the dominant pattern of overall correlation among aspects of shape in such a way as to separate the global from the local and all levels in-between (e.g., Windhager et al. 2017).

The approach to this topic that I currently favor for evolutionary applications is a manipulation of the mathematics of the thin-plate spline, which is the graphic most often used to display the findings of correlated dependencies among Procrustes shape coordinates (Bookstein 2015a). The key to this new methodology of integration is the replace-

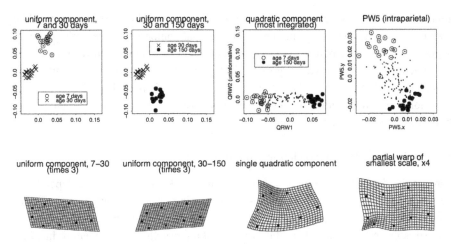

Figure 9.3
Example of an integration analysis. The growth of 18 experimental rodent neurocrania is decomposed as the superposition of four processes. Left to right, early-phase and late-phase uniform components; best-fitting quadratic term for the residual; and partial warp for the uniquely local phenomenon (departure from a smoothly integrated grid). Of these four components, only one (the quadratic term) derives from a covariance analysis. After Bookstein, 2017, figure 5.99.

ment of one null hypothesis by another. The hypothesis that is replaced is the null of total noncorrelation, which is the hypothesis tested by the usual likelihood ratios (see, for example, Mardia et al. 1979, section 5.3.2a). But this is a meaningless null for paleoanthropological work because no form this uncanalized could ever have existed. In contrast, the new null hypothesis appears to correspond better to one major goal of systematics: the detection and reporting of effects at all geometric scales in a manner that is formally independent of scale. This means that up to uniform transformations (shears) of the type shown in the left half of figure 9.3, changes of shape involving the same ratio of changes of distance will be reported equivalently regardless of the size, position, or orientation of the square sustaining those changes. The method is based on a specific transformation of morphospace that derives from a deep mathematical property of the thin-plate spline interpolation driving the usual diagrams.

Figure 9.3 presents an example of this analysis of integration (Bookstein 2015a). The data are the positions of eight midsagittal cranial landmarks collected on 18 laboratory rats at eight ages for each rat (the morphometrics textbooks call these the *Vilmann neurocranial octagons*). The four features in figure 9.3 emerge as potentially the most appropriate for specific explanation (developmental, in this case; selective, in applications to evolutionary processes per se). These features are clearly at three different scales. Two are uniform shears, first backward, then forward, combined with a monotonic decline in height-width ratio for this octagon as a whole. The third summarizes a clear quadratic growth-gradient (Bookstein 1991, section 7.6) over the form as a whole, while the fourth represents a singular effect local to one closely spaced pair of landmarks in the parietal region. While the quadratic growth gradient is computed as a principal component under stringent constraints, the three other components are independent of any correlations. Instead they summarize the geometry of the observed growth changes by its highlights over the full range of potential feature scales. I have been exploring the implications of this method for evolutionary biology through studies at higher taxonomic levels (Bookstein 2016); soon, I hope, somebody will apply it to hominid data.

Morphometrics from the biomechanical point of view

An approach that has rapidly gained attention recently seeks to visualize the dependence of strain upon starting form or upon species by computing principal components of skull shape for various anthropoid primates and then carrying out finite element analyses of the effects of stresses on forms at or near the extremes of the principal component ordination. (A *finite element analysis* divides a complicated form into a collection of simpler geometrical components, such as tetrahedra, and analyzes the effect of some mechanical load on the original object as the set of displacements it produces at the corners of these elements according to an algebraic re-expression of the laws of engineering.) Results are often displayed in colored diagrams of the magnitude of strain (percentage extension or compression of the material in directions of greatest or least such ratio) over the surface of the organismal form under study. Except in the simplest scenarios, computations of this type require elaborate computer software that is typically quite expensive to purchase and quite opaque in its operation.

I have argued (Bookstein 2013b) that this is not a good approach for visualizing the hypothetical selective effects under discussion, because while the morphometric step in the analysis cannot properly account for small-scale aspects of the stress-strain relationship, the biomechanical analysis demands that small-scale features be treated as no less important than features at a large scale. Although the extrema of strain are typically a function of small-scale features, such as cavity depth, the variation of features like these usually has little effect on the first few relative warps (see Weber et al. 2011, Weber and Bookstein 2011, section 2.6), and is thus typically overlooked by principal components analyses and allied methods.

Elsewhere (Bookstein 2013b) I advocated a systematic study of aspects of form that specifically predict strain under conditions of realistically modeled loads. (When applied to data from real living or fossil organisms, these would need to be numerical studies using finite-element software.) Responses to the imposed stress would be algebraic, not stochastic. This approach would rightly shift the focus from principal components of form to strain as a (numerically) predictable function of specific dimensions of form extracted for that particular explanatory purpose. In formulating the prototype of this approach, I took advantage of two helpful characteristics of the example. First, the morphospace was constrained to only

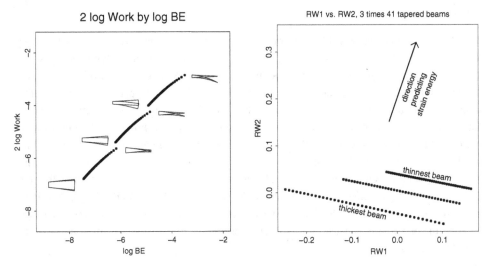

Figure 9.4
Structure of the relation between form and a particular aspect of function, namely, strain under fixed biomechanical load. Left: Design of the example: three families of 41 tapered beams each, clamped at the left end at one of three different thicknesses and subjected to a constant simulated load at the free (right) end. In each family the thinnest form tapers nearly to a point while the thickest form expands to 150% of the thickness at the clamped end. The little sketches show unstrained and strained forms for the thinnest and thickest members of each of the three families. Right: achieved strain in response to a constant end-load, according to the textbook formula, can be predicted very accurately by aspects of the shape coordinates, but the prediction is not aligned with the first Procrustes principal component. After Bookstein, 2014, figure 7.30.

two dimensions of variation, and, second, the strains induced by the particular load I chose were already known in closed algebraic form, thereby not requiring any high-dimensional data meshing or the black box of a commercial finite-element package.

The example I used (see figure 9.4) was the variation of a linearly tapered cantilevered beam of constant material properties under an end load. The computation showed that there was nearly exact proportionality between the bending energy of the ordinary thin-plate spline and the actual strain energy of the loaded beam. However, the geometric aspect of morphology that drove this relationship—the predictor of strain from form at constant load—was neither of the principal components of the distribution of the beams in morphospace. Rather, it was oblique to them, with the last relative warp (the one that explained the least Procrustes variance) accounting for the greater proportion of explained variation in the biomechanical response. Once again, implications for paleoanthropology are clear: Analyses that explain morphological changes by their apparent effects on fitness as assessed by integrated strain or maximum strain need to be aligned with the dimensions of morphology that are actually predicting those strain scores, not the principal components according to the Procrustes metric (or any other metric that is irrelevant to function and thereby to fitness).

Consciousness

Briefly, the biomathematical literature on consciousness is veering strongly away from any technology that, like psychometrics, is grounded in Gaussian correlation-type concerns, and toward an embrace of statistical thermodynamics, which, around the turn of the twentieth century, split from Maxwellian kinetic theory to encompass concepts of entropy and free energy as espoused by Willard Gibbs and others. To the extent that these new approaches have implications for conceptions of cultural evolution, by directing our attention to quite different kinds of measurement within and among multiple members of a human group, they are likely to open up the brain-to-mind aspect of human evolution through more promising forms of quantification than are current. The heavy lifting of this approach has thus far been in the hands of a mathematical neuroscientist, Karl Friston (2010), and his frequent collaborator, neuro-philosopher Andy Clark (2016). I have high hopes for this new methodology, in spite of (or perhaps because of) its circumvention of most of the assumptions of multivariate data analysis.

Relying on the Social Aspects of Our Science:
Consensus-Driven Paleoanthropology

In addition to these preliminary suggestions for a twenty-first-century paleomorphological methodology, I offer others that deal purely with matters of intellectual style. Do **not** name a new species (e.g., the *"Homo iwoeleruenesis"* referred to Prof. Waddell's chapter, this volume) prior to its confirmation by multiple sources of evidence. The rewards of such a claim are transient and the possibility of ensuing confusion extremely high. Within evolutionary biology, the announcement of a species would count as a particularly visible example of inference to a best explanation. As such, it must meet all of Peirce's general criteria of abduction. The morphological finding must be *surprising* on competing models (i.e., not ecotypes or diseased specimens); it must be "a matter of course" on the new hypothesis; and, most importantly, it must lead to confirmatory studies. Echoing Sherwood Washburn's effort 65 years ago to establish a "new" physical anthropology, a "new" taxonomy must be the best explanation, not only of the new data, but also of whatever old data are still regarded as valid. Hopefully, this new discipline will help bring paleoanthropology more in line with method, practice, and theory in the rest of evolutionary biology, and in particular will help winnow the current chaos of putative labels of paleospecies.

This concern for consensus is a special case of the more general methodological rule that any quantitative natural-scientific claim—a phylogeny, a network, a speciation event—requires an auxiliary rhetoric of uncertainty, just as any sample average needs to be accompanied by an estimated standard error. If the topic appears to be a historically conditioned one, such as a fossil-based description of a phylogenetic network, the requisite quantitative rhetoric does not yet exist; then the paleoanthropologist should simply stop talking about *p*-values and the like until the statisticians can cobble together a preliminary language for that purpose. Recall Wittgenstein's aphorism: "*Wovon man nicht sprechen kann, darüber muss man schweigen*" (whereof one cannot speak, thereof must one be silent). While such a rhetorical rule already exists for a range of more familiar styles of scientific reporting, it is largely ignored by paleoanthropologists. This is a matter of pedagogy, not philosophy. Claims about principal components, for instance, should *always* be accompanied by reports of their uncertainty. Paleoanthropology needs to become competent in reporting the uncertainty of principal components before it is entitled to commission a methodology for uncertainty of phylogenies.

Paleoanthropology for What?

In 1939 the American sociologist Robert M. Lynd published an underappreciated book-length essay, *Knowledge for What?*, that, according to its subtitle, attempted to rework "the place of social science in American culture." I would argue that an encomium for paleoanthropology should be similarly couched: "the place of bioanthropology in human culture."

After more than a decade of involvement in paleoanthropological disputes (e.g., Gunz et al. [2009] on the Out-of-Africa hypothesis, or Fornai et al. [2015] on the number of species of *Australopithecus*), and as a disciple of a different scientific *Denkkollektiv* (the statisticians rather than the anthropologists), I can say that my original point of view has changed little (if at all). Tools for resolving disputes over details of "the descent of man" are much weaker than the corresponding tools for resolving disagreements among evolutionary biologists. The primary problems in paleoanthropology remain essentially those identified by Ayala (egg-to-adult, brain-to-mind, ape-to-human), and the third of these appears not to interact much with either "egg-to-adult," which has turned into "Evo-Devo" (evolutionary developmental biology), or "brain-to-mind," which has taken off under the rubric of "consciousness studies" across a wide range of animals, not just mammals. There is no greater discrepancy of analytic notation than, for instance, between classic correlation-based biometry, the algebra of least-squares, and the mathematical models that drive E. O. Wilson's models of eusociality (Nowak et al. 2010) or Friston's (2010) thermodynamic approach to conscious intention. As I understand it, it was aspects of social behavior that made us human, and these just do not fossilize; nor is consciousness describable at all well by covariances among a small set of discrete measurements.

Compounding these methodological problems is the fact that *Homo sapiens* is a terrible model species. Generation times are too long. Experimentation is ethically prohibited. And, perhaps most bothersome to the natural scientist, virtually every study of humans is, to some degree, reflexive. People in groups study other people in groups for all sorts of reasons, including advertising, cosmetic plastic surgery, and the psychology of propaganda. That human evolution research fails to quantify fitness except by default makes it nearly impossible for paleoanthropological research to articulate with the core methodologies of evolutionary biology. In other words, the detailed study of our own species is unlikely to result in knowledge about how evolutionary processes might go "in general." We are too specialized a special case, especially as regards behavior. If paleoanthropology is to maintain a place in the academy, its practitioners will need to better defend why it should be classed with the sciences instead of the humanities. An interest in one's own species' evolutionary history does not translate into doing good science. The more sensible position would be to admit that certain questions may never be answered.

Schwartz's call for this workshop was shrewd: There is indeed a disconnect between paleoanthropology and evolutionary biology. This is mainly because paleoanthropology is not really a biological science at all. It is a testimonial to "man's conception of himself," to quote the title of my first course in anthropology in 1964, at the University of Michigan with Professor Robert MacLeod. After fifteen years as professor of anthropology, I conclude that we pursue the topic of human evolution not for the principal reason one engages in the evolutionary sciences, which is to gain control over our bodies and our environment, but for the same reason we pursue any other branch of the humanities: to understand better how we have come to be what we are both in body and in mind. Since it is primarily culture that gives meaning to our lives (*pace* Lieberman 2013), the database of paleoanthropology is mostly oblique to this fundamental question.

One crucial aspect of this restorative process would be straightforward intellectual honesty. It is time for paleoanthropologists to admit that most of the questions that evolutionary biologists ask of their laboratory species are unlikely ever to be answered for prehistoric *H. sapiens* (much less for other hominids). With this admission, the field would then be able to focus on those aspects of this evolutionary history about which we *do* have things to say. The raw material for this narrative must combine aspects of fossil form with information from the many channels that bear evidence of group structure. As Wilson (2012, p. 56) put it, "Most of culture, including especially the content of the creative arts, has arisen from the inevitable clash between individual selection and group selection." If there is to be a paleoanthropology of enduring validity, one whose facts do not change from year to year, it will have to incorporate evidence of group selection. For *H. sapiens*, such evidence is more often produced by the methods of the humanities than the methods of experimental evolutionary biology.

From this outsider's point of view, the other fields that risk abusing arithmetic in order to "answer" unanswerable questions lie within the humanistic side of the academy: for example, psychiatry, management science, evolutionary psychology (Bolhuis et al. 2011), psephology, and Biblical hermeneutics. As I perceive the state of affairs, paleoanthropology remains mired in specific methodological problems from which modern evolutionary biology is largely spared. If, as "gross national product" or "criminality" are metaphors, Procrustes distance is a metaphor, then the disciplines that rely on it need to rely on the rules for using numbers as metaphors. These are not the rules of correlation and regression.

Human evolution was necessarily chaotic, an "endemic turmoil," as Wilson (2012) said. But the quantitative methods of today's biometry and evolutionary biology hardly ever apply to chaotic systems. Yet this was our history, and the study of such processes is fascinating. If, as I believe, paleoanthropology is disconnected from the evolutionary sciences, it deserves to be attached to something else. That new academic home would be the *Geisteswissenschaften,* the historical sciences of human culture: "humanities" in the root etymological sense. Imagine how much fun it would be to teach those ancient departments how to use our expensive machines.

Acknowledgments

I am grateful to workshop organizer Jeff Schwartz for the invitation to contribute this chapter to a proceedings volume for a conference I had not actually attended. Many of the points here have been sharpened over the course of years of conversation with Philipp Mitteroecker, University of Vienna, and Joe Felsenstein, University of Washington. The method of strain analysis so briefly summarized in the discussion leading up to Figure 9.4 was originally conceived during discussions with Gerhard Weber and Paul O'Higgins. Figures 9.3 and 9.4 derive from research pursued and publications produced with the support of United States National Science Foundation grant DEB–1019583 to Joe Felsenstein and me.

This essay is dedicated to the memory of Werner Callebaut, 1952–2014, late managing director and resident intellectual of the KLI Institute, Klosterneuburg, Austria, whose breadth of insight into the role of philosophy of science in biological inquiry launched my own serious pursuit of the logical foundations of biological science in its statistical context.

References

Ayala, F. J. 2007. "Human Evolution: The Three Grand Challenges of Human Biology." In *The Cambridge Companion to the Philosophy of Biology*, edited by D. L. Hull and M. Ruse, 233–254. New York: Cambridge University Press.

Bolhuis, J. J., G. R. Brown, R. C. Richardson, and K. N. Laland. 2011. "Darwin in Mind: New Opportunities for Evolutionary Psychology." *PLoS Biology* 9:e1001109.

Bookstein, F. L. 1991. *Morphometric Tools for Landmark Data: Geometry and Biology.* New York: Cambridge University Press.

Bookstein, F. L. 2013a. "Random Walk as a Null Model for Geometric Morphometrics of Fossil Series." *Paleobiology* 39:52–74.

Bookstein, F. L. 2013b. "Allometry for the Twenty-First Century." *Biological Theory* 7:10–25.

Bookstein, F. L. 2014. *Measuring and Reasoning: Numerical Inference in the Sciences.* New York: Cambridge University Press.

Bookstein, F. L. 2015a. "The Relation between Geometric Morphometrics and Functional Morphology, as Explored by Procrustes Interpretation of Individual Shape Measures Pertinent to Function." *Anatomical Record* 298:314–327.

Bookstein, F. L. 2015b. "Integration, Disintegration, and Self-Similarity: Characterizing the Scales of Shape Variation in Landmark Data." *Evolutionary Biology* 42:395–426.

Bookstein, F. L. 2016. "The Inappropriate Symmetries of Multivariate Statistical Analysis in Geometric Morphometrics." *Evolutionary Biology* 43:277–313.

Bookstein, F. L. 2018 (to appear). *A Course in Morphometrics for Biologists.* New York: Cambridge University Press.

Bookstein, F. L., and P. M. Mitteroecker. 2014. "Comparing Covariance Matrices by Relative Eigenanalysis, with Applications to Organismal Biology." *Evolutionary Biology* 41:336–350.

Borg, I., and P. J. F. Groenen. 2005. *Modern Multidimensional Scaling*, 2nd ed. New York: Springer Science + Business Media.

Clark, A. 2016. *Surfing Uncertainty: Prediction, Action and the Embodied Mind.* New York: Oxford University Press.

Coombs, C. H. 1956. *A Theory of Data.* New York: John Wiley & Sons.

Cox, D. R. 2006. *Principles of Statistical Inference.* Cambridge, UK: Cambridge University Press.

Davidson, P. 2015. *Turbulence: An Introduction for Scientists and Engineers*, 2nd ed. Oxford, UK: Oxford University Press.

Dryden, I. L., and K. V. Mardia. 2016. *Statistical Shape Analysis*, 2nd ed., Chichester, UK: John Wiley & Sons.

Dyson, F. J. 1988. *Infinite in All Directions.* New York: Harper & Row.

Felsenstein, J. 2004. *Inferring Phylogenies.* Sunderland, MA: Sinauer Associates.

Felsenstein, J. 2016. *Theoretical Population Genetics.* Self-published textbook.

Feynman, R. P. 1974. "Cargo Cult Science." *Engineering and Science* June:10–13.

Fleck, L. 1935/1979. *Entstehung und Entwicklung einer wissenschaftlichen Tatsache.* Schwabe. Translated as *Genesis and Development of a Scientific Fact*, Chicago: University of Chicago Press.

Fornai, C., F. L. Bookstein, and G. W. Weber. 2015. "Variability of *Australopithecus* Second Maxillary Molars from Sterkfontein Member 4." *Journal of Human Evolution* 85:181–192.

Friston, K. 201. "The Free-Energy Principle: A Unified Brain Theory." *Nature Reviews: Neuroscience* 11:127–138.

Gunz, P., F. L. Bookstein, P. Mitteroecker, A. Stadlmayr, H. Seidler, and G. W. Weber. 2009. "Early Modern Human Diversity Suggests Subdivided Population Structure and a Complex Out-of-Africa Scenario." *PNAS* 106:6094–9098.

Hull, D. L., and M. Ruse, eds. 2007. *The Cambridge Companion to the Philosophy of Biology.* New York: Cambridge University Press.

Jolliffe, I. T. 2002. *Principal Component Analysis*, 2nd ed. New York: Springer Verlag.

Kuhn, T. S. 1961. The Function of Measurement in Modern Physical Science. *ISIS* 52:161–193, 1961. Also, pp. 31–63 in Woolf, H., ed., *Quantification: A History of the Meaning of Measurement in the Natural and Social Sciences,* 1961 Bobbs-Merrill.

Kuhn, T. S. 1962. *The Structure of Scientific Revolutions.* University of Chicago Press. Second edition, 1970.

Lieberman, D. E. 2013. *The Story of the Human Body: Evolution, Health, and Disease.* New York: Pantheon.

Lynd, R. M. 1939. *Knowledge for What? The Place of Social Science in American Culture.* Princeton, NJ: Princeton University Press.

Mardia, K. V., J. T. Kent, and J. Bibby. 1979. *Multivariate Analysis.* Chichester, UK: John Wiley & Sons.

Merton, R. K. 1942. "Science and Technology in a Democratic Order." *Journal of Legal and Political Sociology* 1:115–126. Reprinted in R. K. Merton, 1973. *The Sociology of Science: Theoretical and Empirical Investigations.* Chicago: University of Chicago Press.

Millstein, R. L., and R. A. Skipper, Jr. 2007. "Population Genetics." In *The Cambridge Companion to the Philosophy of Biology*, edited by D. L. Hull and M. Ruse, 22–43. New York: Cambridge University Press.

Mitteroecker, P., P. Gunz, M. Bernhard, K. Schaefer, and F. L. Bookstein. 2004. "Comparison of Cranial Ontogenetic Trajectories among Great Apes and Humans." *Journal of Human Evolution* 46:679–698.

Nowak, M. A., C. E. Tarnita, and E. O. Wilson. 2010. "The Evolution of Eusociality." *Nature* 466:1057–1062.

Pearson, K. 1903. "On the Inheritance of the Mental and Moral Characters in Man, and Its Comparison with the Inheritance of the Physical Characters." *Journal of the Anthropological Institute of Great Britain and Ireland* 33:179–237.

Peirce, C. S. 1934. "Pragmatism and Abduction: Lecture VII. (Lectures given at Harvard University, March 26 to May 17, 1903)." *Collected Papers of Charles Sanders Peirce,* Volume V: *Pragmatism and Pragmaticism.* Cambridge, MA: Belknap Press of Harvard University Press.

Peirce, C. S. 1877. "The Fixation of Belief." *Popular Science Monthly* 12(November):1–15. Reprinted in various places, including N. Hauser and C. Sloesel, eds. 1992. *The Essential Peirce, volume 1, 1867–1893.* Bloomington, IN: Indiana University Press.

Rodgers, J. L., and W. A. Nicewander. 1988. "Thirteen Ways to Look at the Correlation Coefficient." *The American Statistician* 42(1):59–66.

Rosenberg, A. 1985. *The Structure of Biological Science.* New York: Cambridge University Press.

Schwartz, J. H. 2015. [This volume, Preface.]

Stigler, S. M. 1986. *The History of Statistics: the Measurement of Uncertainty before 1900.* Cambridge, MA: Harvard University Press.

Torgerson, W. S. 1958. *Theory and Methods of Scaling.* New York: John Wiley & Sons.

Waddell, P. 2017 [This volume.]

Weber, G. W., and F. L. Bookstein. 2011. *Virtual Anthropology: A Guide to a New Interdisciplinary Field.* Berlin: Springer Verlag.

Weber, G. W., F. L. Bookstein, and D. S. Strait. 2011. "Virtual Anthropology Meets Biomechanics." *Journal of Biomechanics* 44:1429–1432.

Wilson, E. O. 2012. *The Social Conquest of Earth.* New York: Liveright.

Windhager, S., F. L. Bookstein, B. Wallner, E. Millesi, and K. Schaefer. 2017. "Patterns of Correlation of Facial Shape with Physiological Measurements Are More Integrated Than Patterns of Correlation with Ratings." *Nature: Scientific Reports* 7: 45340; doi: 10.1038/srep45340.

Wishart, J. 1928. "The Generalized Product Moment Distribution in Samples from a Normal Multivariate Population." *Biometrika* 290A: 32–52.

Wright, S. 1968. *Evolution and the Genetics of Populations.* Volume 1, *Genetic and Biometric Foundations.* Chicago: University of Chicago Press.

10 Back to Basics: Morphological Analysis in Paleoanthropology

Markus Bastir

Introduction

Paleoanthropology investigates human evolution through study of fossils (Macho 2007), which provide different kinds of information. Molecular information is increasingly gaining importance for investigating phylogenetic relatedness, but paleogenetic research is also restricted chronologically (Arsuaga et al. 2014, Green et al. 2010, Krause et al. 2007). Consequently, in general, it is the morphology of hominin fossils that remains the principal source of information about hominin evolution: its patterns, processes, and modes.

Fossil-based morphological analyses underlie evolutionary models about form, function, and, most frequently, hypotheses of phylogenetic relatedness. Because different fossils present different morphologies, paleoanthropologists are faced with a range of morphological variation that Riedl (1975) characterized as gradually changing fields of morphological overlapping similarity. Fields of morphological similarity offer difficulties in attempts to order them, to set boundaries, and to delimit or even define groups such as populations, demes, morphs, and species, or higher taxonomic units (Schwartz and Tattersal 2002, Tattersall and Mowbray 2005, Schwartz and Tattersall 2015). Hence, the justification for hypothesizing distinct groups on the basis of morphology requires knowledge about the biological causes of these features. Indeed, morphologies differ for different reasons, including the historical, functional, and/or developmental (Macho 2007). Thus, evolutionary paleoanthropologists should try to understand the factors of morphological variation in order to improve their biological interpretations and, in particular, to permit hypotheses of relatedness.

Prior to the use of molecular data, study of the comparative anatomy of recent and fossil vertebrates yielded a grossly correct tree of phylogenetic relationships (Riedl 1975, Müller 2005). At smaller scales, however, such as hominins, interpretation of morphological evidence has proven problematic. Problems in morphological research include the well-known discrepancies of morphological and genetic affinities, as is illustrated in the conflict involving African ape relationship: that is, interpretation of molecular data united humans and chimpanzees, while the study of morphology has long demonstrated a suite of synapomorphies seen only in chimpanzees and gorillas (Shea 1985, Ruvolo 1997).

The history of paleoanthropology also documents difficulties in standardization of taxonomic categories. Sometimes the taxonomic "pendulum" favors lumping fossils into few taxa, while, at others, it suggests allocating them to many taxa. Clearly, this "schizophrenia" should call attention to matter interpretative neutrality and objectivity in paleoanthropology. Further, these "swings," from one extreme to another, are at base due to different approaches to interpreting the underlying causes of morphological similarity or difference, which, in turn, redound on one's interpretation of ranges of morphological variation (Tattersall and Mowbray 2005).

The 1980s witnessed at times belligerent disputes between proponents of disparate scenarios regarding the emergence of "anatomically modern" *Homo sapiens* (AMS) as is demonstrated by the heated debates between advocates of a "remote multiregional" versus a "recent Out-of-Africa" model (Henke and Rothe 1994). Also central to this debate was the phylogenetic relatedness of Neanderthals and AMS (ancestor-descendent versus sister taxa). Prior to the 1980s, multiregional models espousing the gradual emergence models of anatomically modern humans were dominant. However, with the introduction of molecular approaches to systematics, the Out-of-Africa model replaced the linear, multiregional model (Stringer and Andrews 1988, Tattersall 1997, Wolpoff et al. 2001). During the last twenty years, *H. neanderthalsensis* has almost universally been accepted as a distinct species, the problematic and, if true, rare cases of hybridization with *H. sapiens* notwithstanding. Nevertheless, problematic or not, the increasing weight and rapidly instituted received wisdom of *H. sapiens*-Neanderthal hybridization has moved the scenario of later human evolution from "Out of Africa" to something yet to be properly articulated (Krause et al. 200, Green et al. 2010).

A central question in paleoanthropology thus remains: To what degrees can morphological similarity be understood in terms of phylogenetic relatedness rather than other biological factors that lend themselves to discussions of homology and analogy in morphological analysis (MA; Riedl 1975, Laland et al. 2013)? In this chapter, I will address these difficulties, problems, and questions by providing two examples that I hope will provide a framework for testing hypotheses and provoking positive steps forward in the "doing" of morphological analysis and systematics in paleoanthropology.

Craniofacial Variation in Early *Homo*

The hominin fossils from Dmanisi constitute an excellent example for illustrating these concerns. The stratigraphic conditions at Dmanisi suggest that several hominins were deposited in a relatively short period of time (Gabunia et al. 2000, Vekua 2002, Lordkipanidze et al. 2007, Lordkipanidze et al. 2013). Morphological differences between these fossils have been interpreted in the light of various taxonomic hypotheses (*H. erectus*, *H. ergaster*, *H. habilis*, *H. georgicus*; Rosas and Bermudez de Castro 1998, Skinner, Gordon, and Collard 2006, Lordkipanidze et al. 2007, Lordkipanidze et al. 2013 Bermudez de Castro et al. 2014).

The most radical interpretation of the Dmanisi hominins emerged with the latest discovery, Skull 5. In spite of the fact that this specimen is unlike any other from the site—an observation that had been suggested by the large and also morphologically unusual mandible D2600 with which Skull 5 appears to be associated—Lordkipanidze and collaborators (2013) maintained that all Dmanisi hominins were merely a variant of *H. erectus sensu lato*. Further, because all specimens came from the same geographically and stratigraphically constrained sediments, they must all belong to a paleodeme, analogous to those suggested for extant humans and chimpanzees. Lordkipanidze and colleagues (2013) concluded that early hominins from Africa and Dmanisi should be grouped (lumped) together and seen as representing a remarkably variable species. Not unexpectedly, some paleoanthropologists rejected such an unfettered inclusion of specimens into a species, *Homo erectus*, with whose type specimen— the Trinil 2 calvaria—they had nothing morphologically in common beyond comprising bones and teeth (Spoor 2013, Schwartz et al. 2014).

Schwartz et al. (2014) commented that, by accepting a priori that the Dmanisi sample constitutes a paleodeme, their analysis excludes from the start considering the possibility that differences might be taxically relevant. One may ask: "What is the relationship between intraspecific variation and interspecific difference?" Does this mean that investigating intraspecific factors of morphological variation is less important?

I argue that ranges of variation of intraspecific features (as can be demonstrated in recent species) are important because they constitute the natural axes of one expression of morphological difference, and because the patterns derived from them can be interpreted as representing the "lines" of least evolutionary resistance (Marroig and Cheverud 2005, Gunz 2012). When assessed quantitatively in recent species, intraspecific ranges of variation can permit theoretically useful null hypotheses against which potential taxonomically relevant ranges of variation, which underlie morphological dissimilarity, can be assessed (Skinner et al. 2006, Spoor et al. 2015). But this is only one kind of morphological enquiry.

Late Pleistocene *Homo*: Why Do Neanderthals and Modern Humans Differ in Skeletal Morphology?

Another problem is how one comes to interpret variation in skeletal morphology as reflecting adaptation, functional morphology, or random drift. Weaver (2009) provided a comprehensive review of such problems in comparing Neanderthals and AMS. For Weaver, the crucial question was: What are the evolutionary causes underlying skeletal anatomical difference?

After reviewing various studies on biomechanical as well as climatic adaptation (Rak 1986, Trinkaus 1986, 1987, Demes and Creel 1988, Spencer and Demes 1993, O'Connor, et al. 2005, Holton and Franciscus, 2008), Weaver (2009) concluded that the differences between Neanderthals and extant humans are not correlated with facial biomechanics or respiratory adaptation to climate. Rather, he asserted, Neanderthal cranial morphology, which is indisputably different from that of AMS (Schwartz and Tattersall 2010), is the result of "random drift" (but see Bastir in revision, Bastir and Rosas 2016). In contrast, Neanderthal postcranial morphology presents evidence of adaptation to increased levels of physical activity life in harsh climatic conditions (Churchill 1998, 2006, Weaver 2009). Consequently, since the cranial and postcranial morphological differences between Neanderthals and AMS are not correlated, they must result via different evolutionary processes.

While Weaver's (2009) paper was meant to be a critical review of studies that constituted "*forced attempts*" to fit any skeletal feature into the dominant Darwinian scenario, steeped in scenarios of functional adaptation and selection (Weaver 2009, Weaver pers. comm.), he did not consider other factors that could impact morphology. For example, Weaver (2009) all but does not consider internally induced variation and covariation, and their consequent organismal integration.

Consideration of early hominin craniofacial and cranial-postcranial-linked variation suggests different causal considerations for different expressions of morphology. This, in turn, echoes Mayr's proposed division of systematic labor: There are evolutionary morphologists who investigate proximate (functional) causation, and others who seek to identify ultimate (historical) causation (Laland et al. 2013, although the latter may not be the best explanatory framework for paleoanthropology).

The Conceptual Reference Framework

Recently Laland et al. (2015, 1813) argued that, in order "to make progress, scientists must specify phenomena that require explanation, identify causes and decide on what methods, data and analyses are explanatorily sufficient." In doing so, they also created a "conceptual framework" that reflects what Popper termed the hypothetico-deductive method for problem solving: (1) problem identification, (2) problem solutions recurring in hypotheses and theories, and (3) (probability-related) elimination of processes not belonging to the problem solution (by statistical hypothesis falsification; Popper 1968, 1994).

If we import these concepts and concerns into paleoanthropology, we may ask two questions, which relate to earlier discussion: "What is the meaning of craniofacial morphology in the Dmanisi sample?" and "Is Neanderthal skeletal morphology solely a matter of function?"

Both of these questions have taxonomic implications for hypothesizing species on the basis of morphology and/or the occupation of a new ecological niche (Muller and Wagner 1991, Macho 2007). While, in the first instance, comparative morphology is foregrounded, in the second, functional morphology and performance are relevant. Both questions, however, can be addressed via morphological data and hypothesis testing. Therefore, following Laland et al. (2015), testing morphological hypotheses requires deciding which methods,

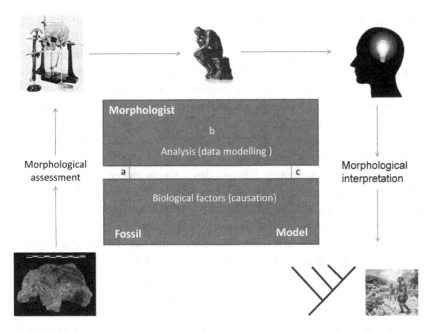

Figure 10.1
The knowledge-generating process in morphological analysis comparing/explaining two morphological patterns. The fossil belongs to nature, the hypothesis to the morphologist. While in nature the fossil relates to the evolutionary model by a natural process, in the hypothesis, this process is "modeled" by the morphologist handling the data within its conceptual framework. Vector "a" shows the morphological assessment, measuring the fossil translates its morphology into a numerical representation of the fossil. These numbers are then "modeled" to give rise to another representation, a result (vector "b"), this representation is then projected (interpreted) back to nature and the hypothesis is contrasted (vector "c"). This scheme shows that the better the representations of the fossils by the data the more verisimilar the interpretations.

data, and analyses are explanatorily sufficient (which means being nonreductionist) and, in turn, defining "sufficiency." I suggest that, by advancing in methodological and theoretical "sufficiency," paleoanthropology can prove itself an evolutionary science. The question is, however: How do paleoanthropologists use morphological methods, comparative data, and statistical analyses to extract and interpret morphological information?

Figure 10.1 illustrates what could be called the "knowledge-generating process" using morphological data in paleoanthropology. At the beginning, there is a fossil with certain morphologies; at the end, a specific evolutionary model is proposed to explain its specific morphologies. Between these two ends lies the paleoanthropologist, extracting, analyzing, and modeling data, and then interpreting the results in order to construct hypotheses. Ideally, this process includes quantitative, causal-analytical, and hypothetico-deductive methods. While trivial, figure 10.1 visualizes possible pitfalls in the knowledge-generating process: (1) when quantifying morphology (morphological and morphometric analyses) and (2) when inferring causation (concepts of morphological variation).

Morphological and Morphometric Analyses

Descriptive morphology comprises two elements of analysis: (1) the actual description of morphological detail (e.g., a hypocone that lies directly opposite the protocone versus a hypocone that extends farther lingually than the protocone; a bipartite versus a continuous supraorbital/glabellar region; a medially versus posteriorly directed lesser trochanter) that is based on standardized terminology (*nomina anatomica*; Feneis 1982, Berry et al. 1995), and (2) an assessment of the degree to which that feature is expressed (e.g., a large versus moderately large hypocone; a markedly versus slightly swollen glabellar region; a small versus moderately small lesser trochanter). Through descriptive anatomy, one is not limited to what measurements alone can convey, but is free to describe every morphological detail using as many words and terms as required. The use of descriptive morphology is particularly important because there are aspects of specimens that cannot be measured, especially when dealing with incomplete specimens where description is the only way to convey information. Further, because description is fossil/specimen specific, it is not necessary to force (reduce) the "morphology-extraction process" to numbers. Consequently, each description becomes a "unit of information" that can be compared to other, equally detailed, "units of information."

A description is a matter of fact. If a morphologist describes a feature incorrectly, that is an error. However, an assessment of size or overall shape (e.g., a very low versus moderately low frontal cranial profile, ovoid versus rounded posterior cranial profile, robust versus gracile zygomatic arch, blunt versus pointed lesser trochanter, bulbous versus conical hypocone) is more a function of one's accrued experience in comparative morphology. Indeed, since descriptions of relative terms such as "robust/gracile," "large/small," "sloping smoothly," or "projecting slightly" may be misinterpreted, particularly when used as "morphological characters" (Lieberman 1995, 1999, Wagner 1989), the goal must be to make clear the distinction between morphological detail itself, and relative terms of size and overall shape. If this distinction is not recognized—which, unfortunately, is often the case in studies that seek to code characters on the basis of size, shape, and relative position alone (e.g. Collard and Wood 2000, Gibbs, Collard, and Wood 2000, Gibbs, Collard, and Wood 2002, Shoshani et al. 1996, Groves 1986, Strait and Grine 2004)—vectors "a" and "c" (figure 10.1) are not readily accessible to, and their assessment and analysis not repeatable by, others.

In contrast to descriptive morphology, quantitative morphology, or morphometrics, is theoretically less likely to introduce confusion, especially when dealing with relative size and relative and overall shape. Because measurements are usually standardized (e.g., Martin and Saller 1959), they are typically reproducible. Nevertheless, one may ask: What is the quality of the information that measurements contain? Although measurements quantify ideas in order to generate and test hypotheses (Richtsmeier 2002), it is imperative that one be mindful of how closely measurements actually do translate a specimen's morphology into an appropriate ("sufficient" *sensu* Laland et al. [2015]) numerical representation. For, while the "representative relevance of a measurement" is a matter of methodology, the "biological relevance of a measurement" is a question for biological theory (Hacking 1983).

These relations are also visualized in figure 10.1. Vector "a" indicates the relevance of measuring and the similarity of "b" (analysis) with the actual (natural) process, that is, the biological relevance of the model. Vector "c" is a modeled version of vector "a," pointing in the opposite direction. Vector "c," which represents the *researcher's projection* of the model into a reality that is contrasted with it, is constrained by the nature/quality of vector "a" (the "extraction" of morphology [measurement]). Conclusions based on these measurements, as well as on statistical analyses, also lend themselves to reproducibility.

Traditional morphometrics is based on linear distances, or areas and angles or indices. Distances and areas usually express different notions of size, while angles and indices indicate some kind of relation and proportion, and thus shape (Bookstein 1991). However, traditional morphometrics suffers from a version of reductionism precisely because it reduces complex anatomical forms into abstract numbers. Thus, while seemingly resolving problems attendant to some aspects of descriptive morphology, traditional morphometrics suffers from difficulties in communication because numerical results must be reinterpreted as, or translated into morphological concepts. The process of data collection (observation, measurement) is operationally different from the analyses (statistics), but the process of interpretation is abstract and prone to bias.

Currently, the most advanced methods are three-dimensional (3D) geometric morphometrics (GM; Bookstein 1991, Richtsmeier et al. 1992, Rohlf and Marcus 1993, O'Higgins 2000, Richtsmeier 2002, Mitteroecker and Gunz 2009, Zelditch et al. 2012). GM does not measure morphology in terms of distances, but as triplets of distances (Cartesian coordinates in 3D space), wherein the goal is to describe the spatial (geometric) relation between points of measurements (landmarks) on anatomical structure (Zelditch et al. 2012).

Because landmarks used in GM are methodologically comparable, they are considered to be biologically "homologous" and meaningful (Oxnard and O'Higgins 2009) in the sense that "biological homology" (cf. Wagner [1989]) applies to structures that are comparable with respect to the same biological "reference process" and level of biological significance (Laubichler 2000, Laubichler and Wagner 2000). GM is superior to linear morphometrics because, through every step of the knowledge-generating process, the 3D properties of anatomical form are maintained (Rohlf and Marcus 1993, Zelditch et al. 2012). GM also independently calculates size and shape for a given data set, which can then be analyzed separately or simultaneously.

3D-GM converts complex anatomical form into numbers. In contrast to their use in traditional morphometrics, these numbers constitute a geometry that reflects a certain morphological *Gestalt*. Further, the statistical results are directly represented as forms (geometries). Nevertheless, although these results are subject to interpretation, other researchers can visualize them via 3D virtual and interactive manipulation using specialized software packages (O'Higgins and Jones 1998, EVAN-Society 2010, Halazonetis 2013). However,

GM is currently limited because these methods only capture general geometry. Analysis of morphological detail is complicated, especially in the absence of clear and/or comparable landmarks that would allow characterization of this information. Conceptual problems exist also regarding comparisons between evolutionarily novel morphological structures (when new *novel* landmarks appear in derived states). Further, because features relating to surface texture cannot be analyzed, which is complicated by the frequent incompleteness of specimens, the absence of relevant landmarks compromises sophisticated analyses (Gunz et al. 2009, Spoor et al. 2015). Nevertheless, with GM, the extraction and quantitative modeling of morphological data are generally reproducible, thereby diminishing problems of communication. This having been said, we must consider how information is interpreted.

Holistic Concepts of Morphology: Variation, Integration, Systems Models, and Organisms

Evolution is a historical process that is based on genetic continuity and the developmental translation of genotypes into performing phenotypes (Alberch 1990). Although the links between genes and morphology are far from clear, the evolutionary, developmental framework featuring "morphological integration" that has emerged over the past two decades is relevant to discussion here (Olson and Miller 1958, Riedl 1975, Wagner 1989, Raff 1996, Wagner 1996, Chernoff and Magwene 1999, Wagner and Laubichler 2000, Klingenberg et al. 2001, Klingenberg et al. 2003, Klingenberg 2009a, Martínez-Abadías et al. 2012, Esteve-Altava et al. 2013).

In paleoanthropology, the seminal review paper of Liebermann, Ross, and Ravosa (2000) stimulated research into morphological integration (Strait 2001, Hallgrimsson, et al. 2002, Rosas et al. 2002, Bookstein et al. 2003, Bastir et al. 2004, Bastir and Rosas 2005, Gunz and Harvati 2007, Mitteroecker and Bookstein 2008, Klingenberg 2009b, Mitteroecker et al. 2012, Bastir and Rosas 2013, 2016). There, Lieberman et al. took issue with phylogenetic systematics (cladistics), arguing that this approach to hypothesizing relatedness is based on the comparison of discrete morphological characters (morphological units) (Lieberman 1999). In this regard, assessment of morphological integration (or modularity in its absence [Bastir 2004]) as the study of covariation between morphological features helps clarify the relative independence in morphological structures (Chernoff and Magwene 1999).

In human craniofacial biology (Enlow 1990), analysis of integration and interdependence of morphological structures provides a reasonable platform from which to understand the process of biological reference and assess potential homology (Laubichler 2000). Further, a high degree of integration is apparently associated with the lack of "evolvability," which, in turn, may reflect biological constraints (Riedl 1977, Maynard Smith et al. 1985, Alberch 1989, Esteve-Altava et al. 2013). In contrast, lower levels of integration can lead to decoupled *co-variability* between organismal parts (modularity) and the possible introduction of more isolated ontogenetic pathways that would permit variation, and "evolvability" (Wagner and Altenberg 1996, Kirschner and Gerhart 1998). Most importantly, however, consideration of morphological integration can inform avenues of research that take into account variation of structure in relation to other structures and thus, lead to *an organismal perspective* that considers the relationships between "the whole" and its parts (Riedl 1975, Laubichler and Wagner 2000, Bastir 2004, Laubichler 2005, Rosas et al. 2006, Callebaut et al. 2007, Bastir 2008, Drack 2015).

An organismal perspective on morphology also redounds on the question of the biological *sufficiency* of morphological interpretation (Laland et al. 2015). This is illustrated in figure 10.2, which proposes a formal hierarchical organismal model for morphological research (Bastir 2004, 2008) and is based on previous work by Bertalanffy (1956, see review by Drack, 2015), Lorenz (1973), and Riedl (1975, 1977, 1981). From their epistemic perspective, these authors argued that all phenomena (i.e., the morphology of a structure and its "causes") are embedded in specific levels of complexity. Explanatorily, *sufficient* conceptual frameworks should include elements of neighboring levels of complexity that embody the actual level of complexity. In a vertical concept, for example, explaining the evolution of the hominin skull requires considering, on one level, the evolution of its different, functional components (Moss and Young 1960, Moss-Salentijn 1997) and, at another, the integration of the ("integrated units" of the) skull in the context of the entire

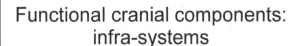

Functional cranial components: infra-systems

Skull: investigated system

Full organism: super-system

Figure 10.2
A systems model for morphological analysis schematizes hierarchical order and integration of the parts and the whole. A sufficient model of explanation of evolutionary morphology of the skull (investigated system) should consider and analyze factors related to interactions among craniofacial functional components (infra-systems) and those related to internal integration among functional elements of the entire organism (super-system).

organism, that is, investigate "the parts within the part," the parts and the wholes and their interrelationships.

Although this conception is simplistic, when it is applied exclusively to morphological analysis, the theory of integration becomes a mental construct that reflects a biologically real compartmentalization of structure, as would be the case, for example, with the morphologies of hominins, extant and extinct. In terms of evolutionary biology, integration and modularity could thus be seen as a "spatial" theoretical counterpart (heterotopy) to a temporal one ("heterochrony") (Zelditch and Fink 1996). With regard to the topic at hand, analyzing integration with an approach such as geometric morphometrics (Mitteroecker and Gunz 2009, Zelditch et al. 2012) allows one to quantify spatial relations and, therefore, provide a methodologically and theoretically sufficient conceptual framework.

Craniofacial Integration: Populations, Demes, Morphs, Species, and Genera

Enlow's (1990) school of human craniofacial biology recognized three different kinds of variation: growth (child-adulthood differences), sex (male-female differences), and head-form patterns (doli-brachycephaly or faciality; Enlow 1990). Growth and sex are related to allometry (size-shape correlations), and head-form patterns to a developmental integration between the brain, cranial base, and face (Enlow 1990, Lieberman et al. 2000a, 200b, Bastir et al. 2006, Richtsmeier et al. 2006, Bastir and Rosas 2016). With GM, it is possible to detect these factors, and principal axes of craniofacial variation, in nonhuman primates as well as hominins, both extinct and extant (Zollikofer and Ponce de León 2002, Bastir and Rosas 2004b, Bastir et al. 2005, Mitteroecker and Bookstein 2008).

An important question here is the interplay between "patterns of variability" and how these patterns interact with taxon-specific morphological features. For example, allometric variation in the mandibles of Middle and Late Pleistocene hominins differs from that in anatomically modern humans (Rosas 1997, Bastir et al. 2007). Allometric shape variation in Middle and Late Pleistocene hominins produces variations that are related to anteroposterior length. Because of allometry, larger mandibles from the Atapuerca Sima de los Huesos hominin (SH) paleodeme present Neanderthal-like features, e.g., a retromolar space (Rosas 1997, Rosas and Bastir 2004). In contrast, allometry in *H. sapiens* produces variation in superoinferior height (Rosas and Bastir 2004, Bastir et al. 2007). Consequently, while the biological reasons for allometric variation are likely similar (namely, being male or female and/or having grown more or less), shape variation differs. That is, the morphological context (i.e., the *Bauplan*) of a smaller brained and more prognathic, or a bigger brained and more orthognathic head apparently induces differences in allometric variation (Rosas and Bastir 2004, Bastir et al. 2007). This basic factor of intraspecific allometric variation can be confounded by taxon-specific features, for example, the Neanderthal retromolar space (Franciscus and Trinkaus 1995, Rosas 1997, Rosas and Bastir 2004, Bastir et al. 2007).

In Enlow's scheme, neurocranial globularity interacts (and coincides) with the expression of morphological characters such as the gonial angle, pre-angular notch (at the corpus-ramus juncture), and projection of menton, or with traits of robustness (Bastir et al. 2005, 2008, Alarcón et al. 2014). Skulls with narrow cranial bases and faces tend toward increased anterior facial lengths, including the anterior symphysis, rather than toward shorter and wider skulls (Enlow and McNamara 1973, Enlow 1990, Zollikofer and Ponce de León 2002, Bastir and Rosas 2004a, Bastir et al. 2005, Mitteroecker and Bookstein 2008). Thus, while "*height of the anterior mandible*" has recently been discussed in the context of defining the genus *Homo* (Schwartz and Tattersall 2015), study of head-form patterns suggests that *all aspects of anterior mandible height* may not be taxonomically relevant. Further, long faces also tend toward greater pre-angular notches, which may be intra- *and* interspecifically variable (Schwartz and Tattersall 2000, Rosas et al. 2002, Bastir et al. 2005). Another element that is relevant to considering Enlow's head-form patterns is robustness. For example, the purported robusticity of the occipital region and of the mandibles of the Neanderthal specimens from El Sidrón is related to their brachyfacial head-form pattern (Rosas et al. 2006, Bastir et al. 2010).

Clearly, the previous discussion serves to illustrate how difficult it is to find a consensus on the biological homology and systematic significance of some characters. Thus, it is important to take into account intraspecific factors of variation since they could serve as null models for hypothesizing of morphs, or in building testable hypotheses of relatedness. In this regard, Lordkipanidze et al.'s (2013) interpretation of craniofacial variation in the Dmanisi hominins—with regard to facial projection, facial length, and neurocranial shape—as being intraspecific could provide a valid null-hypothesis against which further research on the evolutionary significance of craniofacial traits can be carried out.

Because of the above-mentioned craniofacial interactions, it is not surprising that while some have proposed that morphological differences between the Dmanisi hominins represent demic variation (Lordkipanidze et al. 2013), others regard them as taxically relevant (Schwartz et al. 2014, Spoor et al. 2015). Since, evolution is the phylogenetic result of change in *development*, it should be expected that *developmentally caused* variation will be expressed to some degree intra- as well as interspecifically because of shared integrated trends and path of least evolutionary resistance (Marroig and Cheverud 2005, Gunz and Harvati 2007, Bastir and Rosas 2016).

This, in turn, suggests that, although "species are defined by specific morphological features, not by overall cranial shape" (Spoor et al. 2015, 453), both aspects of morphological analysis are important, because of the interrelatedness of overall shape and specific morphology. This kind of interaction (that is, character expression and standard features of intraspecific developmental variation) is certainly in need of further investigation—which highlights the importance of integrating 3D-GM with rigorous traditional phylogenetic analyses.

Neanderthal Skeletal Anatomy and Evolution: Craniofacial Biology
Needs Organismal Biology

Studies in craniofacial biology have demonstrated that major patterns of morphological variation can be explained by the covariation of functional cranial elements, that is, in their development, growth, and spatial relationships (Enlow 1990). In other words, craniofacial variation follows a developmental and constructional order. This view of *order* contrasts with Weaver's (2009) claims that craniofacial evolution in Neanderthals and modern humans is *random*, and that the randomness of craniofacial development and evolution is dissociated from nonrandom postcranial adaptations related to increased physical activity and climate.

Here, I propose an alternative interpretation. Namely, the different patterns of brain and body evolution in Neanderthals and modern humans (Bruner et al. 2003, Bruner 2004, Gunz et al. 2010, Bastir et al. 2011) can be considered an adaptive response to increasingly complex physical and behavioral interactions between these highly encephalized hominin species and their environments (Evans et al. 2005, 2006). Because of ontogenetic and morphological integration between the brain, base, and face (Bastir and Rosas 2016), different brain morphologies correspond to, and interact with, different basicranial and facial configurations. Specifically, for example, midfacial prognathism in Neanderthals may be related to their different encephalization patterns.

It has recently been demonstrated that an anteroposterior position of the wings of the greater sphenoid (alisphenoids; Bastir et al. 2008, 2011) is highly correlated with the anteroposterior position of the zygomatic (malar) bones (Bastir and Rosas 2016). Because of this, one would expect that the truncated protrusion of Neanderthal temporal lobes (due to a different mode of encephalization) is tied to a more posterior position of the alisphenoid relative to the midline of the cranial base (e.g., sphenoid body). Because of integration, the retracted alisphenoids create the "*zygomatic retreat*" that, in turn, leads to more sagittally oriented zygomas and structurally underlies Neanderthal midfacial prognathism (Trinkaus 1987). On the other hand, a retracted middle cranial fossa places the mandibular rami posteriorly relative to the midface and the dental arch (Enlow 1990, Bastir and Rosas, 2005). In turn, this contributes to the formation of a retromolar space, which is typical of large-faced Neanderthals (Franciscus and Trinkaus 1995, Rosas 1997, Rosas and Bastir 2004), and which reflects a separation of the anterior and posterior parts of the face. These craniofacial features are the consequence of local integration effects between the brain, cranial base, and face (infra-systems, figure 10.2).

But facial morphology also interacts with the postcranial skeleton and features of the entire body. Since the size of the airways within the nasal cavity, and overall size of the face, are very highly correlated (Bastir and Rosas 2013), the large Neanderthal face may reflect greater respiratory needs, and not adaptation to facial biomechanics (O'Connor et al. 2005). For, indeed, the energetic demands of increased estimated Neanderthal body mass (Froehle and Churchill 2009, Churchill 2014), and possibly also higher levels of physical activity (Weaver, 2009), are likely to place respiratory constraints on large cranial airways, which must be large because large bodies (organism, super-system, figure 10.2) have higher energetic and thus oxygen demands (Stahl 1967, Rosas et al. 2006, Bastir 2008, Froehle and Churchill 2009, Yokley et al. 2009, Churchill 2014). Body and facial sizes are highly correlated not only in hominins, but in mammals in general (Bastir 2004, Cardini and Polly 2013). Therefore, rather than being *random*, the above relationships reflect causally linked (*nonrandom*) interactions between different craniofacial components, the central nervous system, and the respiratory apparatus. This challenges Weaver's (2009) proposed dissociation of adaptive evolutionary forces involving the skull (*random factors*) and the postcranium (*adaptive factors*). Indeed, a clear causal, functional relationship between the cranium and postcranium is apparent when viewed from an organism-related perspective: The large cranial airways in the large Neanderthal face (Bastir and Rosas 2016) correspond to central and lower parts of the thorax that are also significantly larger in the heavier Neanderthals than in *H. sapiens* and thus able to cope with increased demands for oxygen (Franciscus and Churchill 2002, Gómez-Olivencia et al. 2009, García-Martínez et al. 2014, 2017, Bastir et al. 2017; figure 10.3).

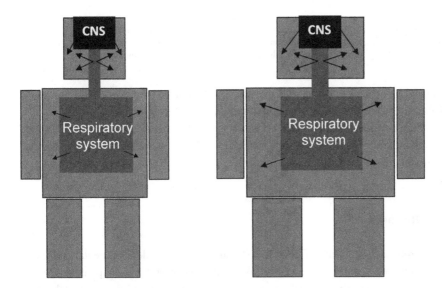

Figure 10.3
Integration at organismal levels. Central nervous system (CNS) and respiratory system. In anatomically modern humans (left), brain morphology (local factors) differs at a local level and energetic demands are reduced at an organismal level (general factor). This explains differences in the attachment of the face to the cranial base, reduced facial projection, and also smaller airways allowing for reduced facial size. Encephalization in Neanderthals leads to an endocranial pattern that differs from that of anatomically modern humans. Because of brain-base integration, the lateral parts of the cranial base and face are retracted relative to the midline. This produces midfacial prognathism and zygomatic retreat. Because of greater body mass and the resulting energetic demands, the airways and ribcages are larger in Neanderthals. These functional changes are coordinated because of organismal integration of the respiratory apparatus but can be seen (also) in the skull.

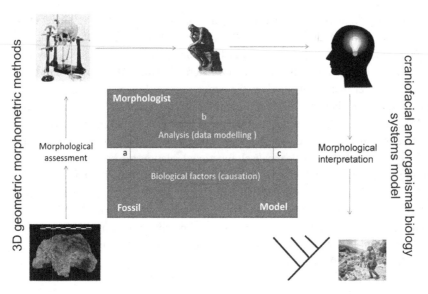

Figure 10.4
Morphological analysis in paleoanthropology. Improving conceptual frameworks at the level where morphological information is extracted (using 3D geometric morphometric methods) and at the level where morphometric data are interpreted (recurring to holistic systems models of craniofacial and organismal biology).

In summary, in order to understand craniofacial biology (infra-systems), one must consider organismal biology (i.e., the super-system) in order to provide a biologically sound and testable model of cranial evolution. Clearly, current evidence supports a model of organismal integration as an important factor in Neanderthal skeletal anatomy, and not Weaver's model of random drift affecting only the skull, that is dissociated from the postcranium.

Concluding Remarks

Here, I have proposed an outline of how 3D-GM and a theory of developmental and functional integration could contribute to a better understanding of problems in paleoanthropology with regard to human craniofacial biology and evolution. 3D-GM is currently the best available tool for investigating the spatial relationships of anatomical features *and should accompany descriptive morphological comparisons whenever possible*. An appreciation of the ontogenetic processes underlying craniofacial evolutionary biology can also lead to testable hypotheses that, in turn, and through study of morphological covariation between different functional units of the organism, can potentially provide insight into the factors that underlie variation. Although these factors are, at present, best understood in terms of intraspecific variation, they still provide useful (null) hypotheses to be tested via a causal analysis of evolutionary morphology. Indeed, models of organism integration based on developmental and functional principles are clearly needed in order to clarify how cranial and postcranial evolutionary trends may be linked. Thus, combining internal factors of *organismal integration* with existing models of externalist *organismal performance* may result in a "sufficient conceptual framework" (figure 10.4) (*sensu* Laland et al. 2015) that will lead to further refinement of morphological analyses in paleoanthropology.

Acknowledgments

I thank Jeff Schwartz for inviting me to this stimulating workshop, Eva Lackner, Isabella Sarto-Jackson, and Gerd Müller for their warm hospitality, and all workshop participants for the stimulating discussions. I thank Antonio Rosas and Daniel García Martínez for many discussions on skulls and ribs. Finally, I am grateful because this invitation made me remember the inspirations I had as an undergraduate student at the University of Vienna, where I shared many interesting discussions and seminars on morphology and evolution with the late Rupert Riedl, not only at the Biozentrum on the Wednesday evenings, but also at the KLI, then still in Altenberg. This research is funded by the Spanish Ministry of Economy and Competitiveness (CGL2012-37279, CGL2015-63648-P, MINECO).

References

Alarcón, J. A., M. Bastir, I. García-Espona, M. Menéndez-Núñez, and A. Rosas. 2014. "Morphological Integration of Mandible and Cranium: Orthodontic Implications." *Archives of Oral Biology* 59(1):22–29.

Alberch, P. 1989. "The Logic of Monsters: Evidence for Internal Constraint in Development and Evolution." *Geobios* 12:21–57.

Alberch, P. 1990. "Natural Selection and Developmental Constraints: External versus Internal Determinants of Order in Nature." In *Primate Life History and Evolution*, edited by C. J. DeRousseau, 15–35. New York: Wiley-Liss.

Arsuaga, J. L., I. Martinez, L. J. Arnold, A. Aranburu, A. Gracia-Tellez, W. D. Sharp, R. M. Quam, C. Falgueres, A. Pantoja-Perez, J. Bischoff, E. Poza-Rey, J. M. Pares, J. M. Carretero, M. Demuro, C. Lorenzo, N. Sala, M. Martinon-Torres, N. Garcia, A. Alcazar de Velasco, G. Cuenca-Bescos, A. Gomez-Olivencia, D. Moreno, A. Pablos, C.C. Shen, L. Rodriguez, A.I. Ortega, R. Garcia, A. Bonmati, J.M. Bermudez de Castro, E. Carbonell, 2014. Neandertal roots: Cranial and chronological evidence from Sima de los Huesos. Science 344, 1358–1363.2014. "Neandertal Roots: Cranial and Chronological Evidence from Sima de los Huesos." *Science* 344(6190):1358–1363.

Bastir, M. 2004. "A Geometric Morphometric Analysis of Integrative Morphology and Variation in Human Skulls with Implications for the Atapuerca-SH Hominids and the Evolution of Neandertals. Structural and Systemic Factors of Morphology in the Hominid Craniofacial System." [Doctoral Dissertation] Madrid: Autonoma University of Madrid.

Bastir, M. 2008. "A Systems-Model for the Morphological Analysis of Integration and Modularity in Human Craniofacial Evolution." *Journal of Anthropological Sciences* 86:37–58.

Bastir, M. In revision. "The Neanderthal Pharynx and Larynx; The Morphological, Functional and Structural Integration of the Upper Airways in Neanderthals." In *Neandertal Skeletal Anatomy: Form, Function and Paleobiology*, edited by S. Maddux and L. Cowgill. Cambridge: Cambridge University Press.

Bastir, M., D. García-Martínez, N. Torres-Tamayo, J. A. Sanchis-Gimeno, P. O'Higgins, C. Utrilla, I. Torres Sánchez, and F. García Río. 2017. "In Vivo 3D Analysis of Thoracic Kinematics: Changes in Size and Shape during Breathing and Their Implications for Respiratory Function in Recent Humans and Fossil Hominins." *The Anatomical Record* 300, 255–264: DOI: 10.1002/ar.23503.

Bastir, M., P. Godoy, and A. Rosas. 2011a. "Common Features of Sexual Dimorphism in the Cranial Airways of Different Human Populations." *American Journal of Physical Anthropology* 146(3):414–422.

Bastir, M., P. O'Higgins, and A. Rosas. 2007. "Facial Ontogeny in Neanderthals and Modern Humans." *Proceedings of the Royal Society of London B* 274:1125–1132.

Bastir, M., and A. Rosas. 2004a. "Geometric Morphometrics in Paleoanthropology: Mandibular Shape Variation, Allometry, and the Evolution of Modern Human Skull Morphology." In *Morphometrics: Applications in Biology and Paleontology*, edited by A. M. T. Elewa, 231–244. Springer Verlag: Berlin-Heidelberg.

Bastir, M., and A. Rosas. 2004b. "Facial Heights: Implications of Postnatal Ontogeny and Facial Orientation for Skull Morphology in Humans and Chimpanzees." *American Journal of Physical Anthropology* 123S:60–61.

Bastir, M., and A. Rosas. 2005. "Hierarchical Nature of Morphological Integration and Modularity in the Human Posterior Face." *American Journal of Physical Anthropology* 128(1):26–34.

Bastir, M., and A. Rosas. 2013. "Cranial Airways and the Integration between the Inner and Outer Facial Skeleton in Humans." *American Journal of Physical Anthropology* 152(2):287–293.

Bastir, M., and A. Rosas. 2016. "Cranial Base Topology and Basic Trends in the Facial Evolution of *Homo*." *Journal of Human Evolution* 91:26–35.

Bastir, M., A. Rosas, P. Gunz, A. Pena-Melian, G. Manzi, K. Harvati, R. Kruszynski, C. Stringer, and J-J. Hublin. 2011b. "Evolution of the Base of the Brain in Highly Encephalized Human Species." *Nature Communications* 2:588.

Bastir, M., A. Rosas, and K. Kuroe. 2004. "Petrosal Orientation and Mandibular Ramus Breadth: Evidence of a Developmental Integrated Petroso-mandibular Unit." *American Journal of Physical Anthropology* 123(4):340–350.

Bastir, M., A. Rosas, D. E. Lieberman, and P. O'Higgins. 2008a. "Middle Cranial Fossa Anatomy and the Origins of Modern Humans." *The Anatomical Record* 291(2):130–140.

Bastir, M., A. Rosas, and P. O'Higgins. 2006. "Craniofacial Levels and the Morphological Maturation of the Human Skull." *Journal of Anatomy* 209(5):637–654.

Bastir, M., A. Rosas, and D. H. Sheets. 2005. "The Morphological Integration of the Hominoid Skull: A Partial Least Squares and PC Analysis with Morphogenetic Implications for European Mid-Pleistocene Mandibles." In *Modern Morphometrics in Physical Anthropology*, edited by D. Slice, 265–284. New York: Kluwer Academic/ Plenum Publishers.

Bastir, M., A. Rosas, A. G. Tabernero, A. Peña-Melián, A. Estalrrich, M. de la Rasilla, and J. Fortea. 2010. "Comparative Morphology and Morphometric Assessment of the Neandertal Occipital Remains from the El Sidrón Site (Asturias, Spain: Years 2000–2008)." *Journal of Human Evolution* 58:68–78.

Bastir, M., P. G. Sobral, K. Kuroe, and A. Rosas. 2008b. "Human Craniofacial Sphericity: A Simultaneous Analysis of Frontal and Lateral Cephalograms Using Geometric Morphometrics and Partial Least Squares Analysis." *Archives of Oral Biology* 53(4):295–303.

Bermudez de Castro, J. M., M. Martinon-Torres, M. J. Sier, and L. Martin-Frances. 2014. "On the Variability of the Dmanisi Mandibles." *PLoS ONE* 9(2):e88212.

Berry, M., S. Standring, and L. Bannister. 1995. *Gray's Anatomy*. London: Harcourt Brace and Company Limited.

Bertalanffy, L. 1956. "A Biologist Looks at Human Nature." *The Scientific Monthly* 82(1):33–41.

Bookstein, F. L. 1991. *Morphometric Tools for Landmark Data*. Cambridge: Cambridge University Press.

Bookstein, F. L., P. Gunz, P. Mitteroecker, H. Prossinger, K. Schaefer, and H. Seidler. 2003. "Cranial Integration in *Homo*: Singular Warps Analysis of the Midsagittal Plane in Ontogeny and Evolution." *Journal of Human Evolution* 44(2):167–187.

Bruner, E. 2004. "Geometric Morphometrics and Paleoneurology: Brain Shape Evolution in the Genus *Homo*." *Journal of Human Evolution* 47(5):279–303.

Bruner, E., G. Manzi, and J-L. Arsuaga. 2003. "Encephalization and Allometric Trajectories in the Genus *Homo*. Evidence from the Neandertal and Modern Lineages." *Proceedings of the National Academy of Sciences* 100(26): 15335–15340.

Callebaut, W., G. B. Müller, and S. A. Newman. 2007. *The Organismic Systems Approach: Evo-Devo and the Streamlining of the Naturalistic Agenda. Integrating Development and Evolution from Theory to Praxis*. Cambridge, MA; London, UK: MIT Press.

Cardini, A., and D. Polly. 2013. "Larger Mammals Have Longer Faces Because of Size-Related Constraints on Skull Form." *Nature Communications* 4:2458 doi: 10.1038/ncomms3458 (2013)

Chernoff, B., and P. M. Magwene. 1999. "Afterword." In *Morphological Integration*, edited by E. C. Olson and P. L. Miller, 319–353. Chicago: University of Chicago.

Churchill, S. E. 1998. "Cold Adaptation, Heterochrony, and Neandertals." *Evolutionary Anthropology* 46:46–60.

Churchill, S. E. 2006. "Bioenergetic Perspectives on Neanderthal Thermoregulatory and Activity Budgets." In *Neanderthals Revisited*, edited by K. Harvati and T. Harrison, 113–156. New York City: Springer Verlag.

Churchill, S. E. 2014. *Thin on the Ground: Neandertal Biology, Archeology and Ecology*. Hoboken, NJ: Wiley Blackwell.

Collard, M., and B. Wood. 2000. "How Reliable Are Human Phylogenetic Hypotheses?" *Proceedings of the National Academy of Sciences of the USA* 97:5003–5006.

Demes, B., and N. Creel. 1988. "Bite Force, Diet, and Cranial Morphology of Fossil Hominids." *Journal of Human Evolution* 17:657–670.

Drack, M. 2015. "Ludwig von Bertalanffy's Organismic View on the Theory of Evolution." *Journal of Experimental Zoology. Part B, Molecular and Developmental Evolution* 324B:77–90.

Enlow, D. H. 1990. *Facial Growth*. Philadelphia: W. B. Saunders Company.

Enlow, D. H., and J. A. J. McNamara. 1973. "The Neurocranial Basis for Facial Form and Pattern." *American Journal of Orthodontics and Dentofacial Orthopedics* 43(3):256–270.

Esteve-Altava, B., J. Marugán-Lobón, H. Botella, M. Bastir, and D. Rasskin-Gutman. 2013. "Grist for Riedl's Mill: A Network Model Perspective on the Integration and Modularity of the Human Skull." *Journal of Experimental Zoology. Part B, Molecular and Developmental Evolution* 320(8):489–500.

EVAN-Society. 2010. *ET, Toolkit for Geometric Morphometric Analysis* (http://www.evan-society.org/).

Evans, P. D., S. L. Gilbert, N. Mekel-Bobrov, E. J. Vallender, J. R. Anderson, L. M. Vaez-Azizi, S. A. Tishkoff, R. R. Hudson, and B. T. Lahn. 2005. "Microcephalin, a Gene Regulating Brain Size, Continues to Evolve Adaptively in Humans." *Science* 309(5741):1717–1720.

Evans, P. D., N. Mekel-Bobrov, E. J. Vallender, R. R. Hudson, and B. T. Lahn. 2006. "Evidence that the Adaptive Allele of the Brain Size Gene Microcephalin Introgressed into *Homo Sapiens* from an Archaic *Homo* Lineage." *Proceedings of the National Academy of Sciences* 103(48):18178–18183.

Feneis, H. 1982. *Anatomisches Bildwörterbuch*. Stuttgart, New York: Georg Thieme Verlag.

Franciscus, R. G., and S. E. Churchill. 2002. "The Costal Skeleton of Shanidar 3 and a Reappraisal of Neandertal Thoracic Morphology." *Journal of Human Evolution* 42(3):303–356.

Franciscus, R. G., and E. Trinkaus. 1995. "Determinants of Retromolar Space Presence in Pleistocene *Homo* Mandibles." *Journal of Human Evolution* 28:577–595.

Froehle, A., and S. E. Churchill. 2009. "Energetic Competition between Neandertals and Anatomically Modern Humans." *PaleoAnthropology* 2009:96–116.

Gabunia, L., A. Vekua, D. Lordkipanidze, C. C. Swisher III, R. Ferring, A. Justus, M. Nioradze, M. Tvalchrelizde, S. C. Antón, G. Bosinsky O. Jöris, M.-A.d. Lumley, G. Majsuradze, A. Mouhkelishvili,. 2000. "Earliest Pleistocene Hominid Cranial Remains from Dmanisi, Republic of Georgia: Taxonomy, Geological Setting, and Age." *Science* 288:1019–1025.

García-Martínez, D., A. Barash, W. Recheis, C. Utrilla, I. Torres Sánchez, F. García Río, and M. Bastir. 2014. "On the Chest Size of Kebara 2." *Journal of Human Evolution* 70:69–72.

García-Martínez, D., M. Bastir, R. Huguet, A. Estalrrich, A. García-Tabernero, L. Ríos, E. Cunha, M. de la Rasilla, A. Rosas. 2017. "The costal remains of the El Sidrón Neanderthal site (Asturias, northern Spain) and its importance for understanding Neanderthal thorax morphology. *Journal of Human Evolution* 111:85–101.

Gibbs, S., M. Collard, and B. Wood. 2000. "Soft-Tissue Characters in Higher Primate Phylogenetics." *Proceedings of the National Academy of Science USA* 97:11130–11132.

Gibbs, S., M. Collard, and B. A. Wood. 2002. "Soft-Tissue Anatomy of the Extant Hominoids: A Review and Phylogenetic Analysis." *Journal of Anatomy* 200:3–49.

Gómez-Olivencia, A., K. L Eaves-Johnson, R. G. Franciscus, J. M. Carretero, and J. L. Arsuaga. 2009. "Kebara 2: New Insights Regarding the Most Complete Neandertal Thorax." *Journal of Human Evolution* 57(1):75–90.

Green, R. E., J. Krause, A. W. Briggs, T. Maricic, U. Stenzel, M. Kircher, N. Patterson, H. Li, W. Zhai, M. H-Y. Fritz, N.F. Hansen, E.Y. Durand, A.-S. Malaspinas, J.D Jensen, T. Marques-Bonet, C. Alkan, K. Prufer, M. Meyer, H.A Burbano, J.M. Good, R. Schultz, A. Aximu-Petri, A. Butthof, B. Hober, B Hoffner, M. Siegemund, A. Weihmann, C Nusbaum, E.S. Lander, C. Russ, N. Novod, J. Affourtit, M. Egholm, C. Verna, P. Rudan, D. Brajkovic, Z. Kucan, I. Gusic, V.B. Doronichev, L.V.Golovanova, C. Lalueza-Fox, M. de la Rasilla, J. Fortea, A. Rosas, R.W Schmitz, P.L.F. Johnson, E.E. Eichler, D. Falush, E. Birney, J.C. Mullikin, M. Slatkin, R. Nielsen, J. Kelso, M. Lachmann, D. Reich, S. Paabo, 2010. "A Draft Sequence of the Neandertal Genome." *Science* 328(5979):710–722.

Groves, C. 1986. "Systematics of the Great Apes." In *Comparative Primate Biology*, edited by D. R. Swindler and J. Erwin, 187–217. New York: Alan R. Liss.

Gunz, P. 2012. "Evolutionary Relationships Among Robust and Gracile Australopiths: An "Evo-Devo" Perspective." *BMC Evolutionary Biology* 39(4):472–487.

Gunz, P., and K. Harvati. 2007. "The Neanderthal 'Chignon': Variation, Integration, and Homology." *Journal of Human Evolution* 53(6):732–746.

Gunz, P., P. Mitteroecker, S. Neubauer, G. W. Weber, and F. L. Bookstein. 2009. "Principles for the Virtual Reconstruction of Hominin Crania." *Journal of Human Evolution* 57(1):48–62.

Gunz, P., S. Neubauer, B. Maureille, and J-J. Hublin. 2010. "Brain Development after Birth Differs between Neanderthals and Modern Humans." *Current Biology*: 20(21):R921–R922.

Hacking, I. 1983. *Representing and Intervening. Introductory Topics in the Philosophy of Science*. London, New York, New Rochelle, Melbourne, Sydney: Cambridge University Press.

Halazonetis, D. J. 2013. "ViewBox4."software, http://www.dhal.org.

Hallgrimsson, B., K. Willmore, and B. K. Hall. 2002. "Canalization, Developmental Stability, and Morphological Integration in Primate Limbs." *Yearbook of Physical Anthropology* 45:131–158.

Henke, W., and H. Rothe. 1994. *Paleoanthropologie*. Berlin, Heidelberg, New York: Springer Verlag.

Holton, N. E., and R. G. Franciscus. 2008. "The Paradox of a Wide Nasal Aperture in Cold-Adapted Neandertals: A Causal Assessment." *Journal of Human Evolution* 55(6):942–951.

Holton, N. E., T. R. Yokley, and R. G. Franciscus. 2011. "Climatic Adaptation and Neandertal Facial Evolution: A Comment on Rae et al." *Journal of Human Evolution* 61(5):624–627.

Kirschner, M., and J. Gerhart. 1998. "Evolvability." *Proceedings of the National Academy of Sciences* 95: 8420–8427.

Klingenberg, C. P. 2009a. "Morphological Integration and Developmental Modularity." *Annual Review of Ecology, Evolution, and Systematics* 39:115–132.

Klingenberg, C. P. 2009b. "Morphometric Integration and Modularity in Configurations of Landmarks: Tools for Evaluating a priori Hypotheses." *Evolution and Development* 11(4):405–421.

Klingenberg, C. P., A. V. Badyaev, S. M. Sowry, and N. J. Beckwith. 2001. "Inferring Developmental Modularity from Morphological Integration: Analysis of Individual Variation and Asymmetry in Bumblebee Wings." *The American Naturalist* 157(1):11–23.

Klingenberg, C. P., K. Mebus, and J-C. Auffray. 2003. "Developmental Integration in a Complex Morphological Structure: How Distinct Are the Modules in the Mouse Mandible?" *Evolution and Development* 5(5):522–531.

Krause, J., L. Orlando, D. Serre, B. Viola, K. Prufer, M. P. Richards, J-J. Hublin, C. Hanni, A. P. Derevianko, and S. Paabo. 2007. "Neanderthals in Central Asia and Siberia." *Nature* 449(7164):902–904.

Laland, K. N., J. Odling-Smee, W. Hoppitt, and T. Uller. 2013. "More on How and Why: Cause and Effect in Biology Revisited." *Biology and Philosophy* 28(5):719–745.

Laland, K. N., T. Uller, M. W. Feldman, K. Sterelny, G. B. Müller, A. Moczek, E. Jablonka, and J. Odling-Smee. 2015. "The Extended Evolutionary Synthesis: Its Structure, Assumptions and Predictions." *Proceedings of the Royal Society of London B*: 282(1813).

Laubichler, M. 2005. "Systemtheoretische Organismuskonzeptionen." In *Philosophie der Biologie*, edited by G. T. Ulrich Krohs, 109–124. Frankfurt am Main: Suhrkamp.

Laubichler, M., and G. P. Wagner. 2000. "Organisms and Character Decomposition: Steps Towards an Integrative Theory of Biology." *Philosophy of Science* 67:S289–S300.

Laubichler, M. D. 2000. "Homology in Development and the Development of the Homology Concept." *American Zoologist* 40:777–788.

Lieberman, D. E. 1995. "Testing Hypotheses about Recent Human Evolution from Skulls." *Current Anthropology* 36(2):159–197.

Lieberman, D. E. 1999. "Homology and Hominid Phylogeny: Problems and Potential Solutions." *Evolutionary Anthropology* 7(4):142–151.

Lieberman D. E., C. Ross, and M. J. Ravosa. 2000a. "The Primate Cranial Base: Ontogeny, Function, and Integration." *Yearbook of Physical Anthropology* 43:117–169.

Lieberman, D. E., O. M. Pearson, and K. M. Mowbray. 2000b. "Basicranial Influence on Overall Cranial Shape." *Journal of Human Evolution* 38:291–315.

Lordkipanidze, D., T. Jashashvili, A. Vekua, M. S. P. de Leon, C. P. E. Zollikofer, G. P. Rightmire, H. Pontzer, R. Ferring, O. Oms, M. Tappen M. Bukhsianidze, J. Agusti, R. Kahlke, G. Kiladze, B. Martinez-Navarro, A. Mouskhelishvili, M. Nioradze, Rook, L.,2007. "Postcranial Evidence from Early *Homo* from Dmanisi, Georgia." *Nature* 449(7160):305–310.

Lordkipanidze, D., M. S. Ponce de León, A. Margvelashvili, Y. Rak, G. P. Rightmire, A. Vekua, and C. P. E. Zollikofer. 2013. "A Complete Skull from Dmanisi, Georgia, and the Evol Biol of Early *Homo.*" *Science* 342(6156): 326–331.

Lorenz, K. 1973. *Die Rückseite des Spiegels. Versuch einer Naturgeschichte menschlichen Erkennens.* München: Deutscher Taschenbuch Verlag.

Macho, G. 2007. *General Principles of Evolutionary Morphology. Handbook of Paleoanthropology.* Berlin, Heidelberg: Springer-Verlag.

Marroig, G., and J. M. Cheverud. 2005. "Size as a Line of Least Evolutionary Resistance: Diet and Adaptive Morphological Radiation in New World Monkeys." *Evolution*: 59(5):1128–1142.

Martin, R., and K. Saller. 1959. *Lehrbuch der Anthropologie.* Stuttgart: Gustav Fischer.

Martínez-Abadías, N., M. Esparza, T. Sjøvold, R. González-José, M. Santos, M. Hernández, and C. P. Klingenberg. 2012. "Pervasive Genetic Integration Directs the Evolution of Human Skull Shape." *Evolution* 66(4): 1010–1023.

Maynard Smith, J., R. Burian, S. Kauffman, P. Alberch, J. Campbell, B. Goodwin, R. Lande, D. Raup, and L. Wolpert. 1985. "Developmental Constraints and Evolution." *The Quarterly Review of Biology* 60(3):265–287.

Mitteroecker, P., and F. Bookstein. 2008. "The Evolutionary Role of Modularity and Integration in the Hominoid Cranium." *Evolution* 62(4):943–958.

Mitteroecker, P., and P. Gunz. 2009. "Advances in Geometric Morphometrics." *BMC Evolutionary Biology* 36(2): 235–247.

Mitteroecker, P., P. Gunz, S. Neubauer, and G. Müller. 2012. "How to Explore Morphological Integration in Human Evolution and Development?" *BMC Evolutionary Biology* 39(4):536–553.

Moss-Salentijn, L. 1997. "Melvin L. Moss and the Functional Matrix." *Journal of Dental Research* 76(12): 1814–1817.

Moss, M., and R. W. Young. 1960. "A Functional Approach to Craniology." *American Journal of Physical Anthropology* 45:281–292.

Müller, G. 2005. "Evolutionary Developmental Biology." In *Handbook of Evolution*, edited by F. Wuketits and F. J. Ayala, 87–115. Weinheim: Wiley-VCH Verlag.

Muller, G. B., and G. P. Wagner. 1991. "Novelty in Evolution: Restructuring the Concept." *Annual Review of Ecology and Systematics* 22:229–256.

O'Connor, C. F., R. G. Franciscus, and N. E. Holton. 2005. "Bite Force Production Capability and Efficiency in Neandertals and Modern Humans." *American Journal of Physical Anthropology* 127(2):129–151.

O'Higgins, P. 2000. "The Study of Morphological Variation in the Hominid Fossil Record: Biology, Landmarks and Geometry." *Journal of Anatomy* 197:103–120.

O'Higgins, P., and N. Jones. 1998. "Facial Growth in Cercocebus Torquatus: An Application of Three-Dimensional Geometric Morphometric Techniques to the Study of Morphological Variation." *Journal of Anatomy* 193: 251–272.

Olson, E. C., and R. L. Miller. 1958. *Morphological Integration.* Chicago: University of Chicago Press.

Oxnard, C., and P. O'Higgins. 2009. "Biology Clearly Needs Morphometrics. Does Morphometrics Need Biology?" *Biological Theory* 4(1):84–97.

Popper, K. 1968. *The Logic of Scientific Discovery.* London: Hutchinson.

Popper, K. R. 1994. *Alles Leben ist Problemlösen. Über Erkentnis, Geschichte und Politik.* München: Piper.

Raff, R. A. 1996. *The Shape of Life. Genes, Development, and the Evolution of Animal Form*. Chicago and London: University of Chicago Press.

Rak, Y. 1986. "The Neandertal: A New Look at an Old Face." *Journal of Human Evolution* 15:151–164.

Richtsmeier, J. T. 2002. "The Promise of Geometric Morphometrics." *Yearbook of Physical Anthropology* 45:63–91.

Richtsmeier, J. T., K. Aldridge, V. B. DeLeon, J. Panchal, A. A. Kane, J. L. Marsh, P. Yan, and T.M. Cole III. 2006. "Phenotypic Integration of Neurocranium and Brain." *Journal of Experimental Zoology. Part B, Molecular and Developmental Evolution* 306(4):360–378.

Richtsmeier, J. T., J. M. Cheverud, and S. Lele. 1992. "Advances in Anthropological Morphometrics." *Annual Review of Anthropology* 21:283–305.

Riedl, R. 1975. *Die Ordnung des Lebendigen. Systembedingungen der Evolution*. Hamburg: Paul Parey Verlag.

Riedl, R. 1977. "A Systems-Analytical Approach to Macro-Evolutionary Phenomena." *The Quarterly Review of Biology* 52(4):351–370.

Riedl, R. 1981. *Biologie der Erkenntnis. Die stammesgeschichtlichen Grundlagen der Vernunft*. Berlin und Hamburg: Paul Parey.

Rohlf, F. J., and L. F. Marcus. 1993. "A Revolution in Morphometrics." *Tree* 8(4):129–132.

Rosas, A. 1997. "A Gradient of Size and Shape for the Atapuerca Sample and Middle Pleistocene Hominid Variability." *Journal of Human Evolution* 33:319–331.

Rosas, A., and M. Bastir. 2004. "Geometric Morphometric Analysis of Allometric Variation in the Mandibular Morphology from the Hominids of Atapuerca, Sima de los Huesos Site." *The Anatomical Record* 278A:551–560.

Rosas, A., M. Bastir, and C. Martinez-Maza. 2002. "Morphological Integration and Predictive Value of the Mandible in the Craniofacial System of Hominids: A Test with the Atapuerca SH Mandibular Sample." *Collegium Antropologicum* 26:171–172.

Rosas, A., and J. M. Bermudez de Castro. 1998. "On the Taxonomic Affinities of the Dmanisi Mandible (Georgia)." *American Journal of Physical Anthropology* 107:145–162.

Rosas, A, M. Bastir, C. Martínez-Maza, A. García-Tabernero, and C. Lalueza-Fox. 2006a. "Inquiries into Neanderthal Cranio-facial Development and Evolution: 'Accretion' vs. 'Organismic' Models." In *Neanderthals Revisited*, edited by T. Harrison and K. Harvati, 38–69. New York Springer Verlag.

Rosas, A., C. Martinez-Maza, M. Bastir, A. Garcia-Tabernero, C. Lalueza-Fox, R. Huguet, J. E. Ortiz, R. Julia, V. Soler, T. D. Torres, E. Martinez, J.C. Cañaveras, S. Sanchez-Moral, S. Cuezva, J. Lariol, D. Santamaria, M. de la Rasilla, J. Fortea. 2006b. "Paleobiology and Comparative Morphology of a Late Neandertal Sample from El Sidrón, Asturias, Spain." *Proceedings of the National Academy of Sciences* 103:19266–19271.

Ruvolo, M. 1997. "Molecular Phylogeny of the Hominoids: Inferences from Mulitple Independent CDNA Sequence Data Sets." *Molecular Biology and Evolution* 14:248–265.

Schwartz, J., and I. Tattersall. 2000. "The Human Chin Revisited: What Is It and Who Has It?" *Journal of Human Evolution* 38:367–409.

Schwartz, J., and I. Tattersal. 2002. *The Human Fossil Record. Terminology and Craniodental Morphology of Genus* Homo *(Europe)*. New York: Wiley-Liss.

Schwartz, J. H., and I. Tattersall. 2010. "Fossil Evidence for the Origin of *Homo Sapiens*." *American Journal of Physical Anthropology* 143(S51):94–121.

Schwartz, J. H., and I. Tattersall. 2015. "Defining the Genus *Homo*." *Science* 349(6251):931–932.

Schwartz, J. H., I. Tattersall, and Z. Chi. 2014. "Comment on 'A Complete Skull from Dmanisi, Georgia, and the Evol Biol of Early *Homo*.'" *Science* 344(6182):360.

Shea, B. T. 1985. "Ontogenetic Allometry and Scaling, a Discussion Based on the Growth and Form of the Skull in African Apes." In *Size and Scaling in Primate Biology*, edited by W. L. Jungers, 175–205. New York: Plenum Press.

Shoshani, J., C. Groves, E. L. Simons, and G. R. Gunnell. 1996. "Primate Phylogeny: Morphological vs. Molecular Results." *Molecular Phylogenetics and Evolution* 5:102–154.

Skinner, M. M., A. D. Gordon, and N. J. Collard. 2006. "Mandibular Size and Shape Variation in the Hominins at Dmanisi, Republic of Georgia." *Journal of Human Evolution* 51(1):36–49.

Spencer, M. A., and B. Demes. 1993. "Biomechanical Analysis of Masticatory System Configuration in Neandertals and Inuits." *American Journal of Physical Anthropology* 91:1–20.

Spoor, F. 2013. "Palaeoanthropology: Small-Brained and Big-Mouthed." *Nature* 502(7472):452–453.

Spoor, F., P. Gunz, S. Neubauer, S. Stelzer, N. Scott, A. Kwekason, and M. C. Dean. 2015. "Reconstructed *Homo Habilis* Type OH 7 Suggests Deep-Rooted Species Diversity in Early *Homo*." *Nature* 519(7541):83–86.

Stahl, W. R. 1967. "Scaling of Respiratory Variables in Mammals." *Journal of Applied Biology* 22:453–460.

Strait, D. S. 2001. "Integration, Phylogeny, and the Hominid Cranial Base." *American Journal of Physical Anthropology* 114:273–297.

Strait, D.S., and F.E. Grine. 2004. "Inferring Hominoid and Early Hominid Phylogeny Using Craniodental Characters: The Role of Fossil Taxa." *Journal of Human Evolution* 47:399–452.

Stringer, C., and P. Andrews. 1988. "Genetic and Fossil Evidence for the Origins of Modern Humans." *Science* 239:1263–1268.

Tattersall, I. 1997. "Out of Africa Again . . . and Again?" *Scientific American* (April):60–67.

Tattersall, I., and K. Mowbray. 2005. "Species and Paleoanthropology." *Theory in Biosciences* 123(4):371–379.

Trinkaus, E. 1986. "The Neandertals and Modern Human Origins." *Annual Review of Anthropology* 15(1):193–218.

Trinkaus, E. 1987. "The Neandertal Face: Evolutionary and Functional Perspectives on a Recent Hominid Face." *Journal of Human Evolution* 16:429–443.

Vekua, A. K. 2002. "A New Skull of Early *Homo* from Dmanisi, Georgia." *Science* 297:85–89.

Wagner, G. 1996. "Homologues, Natural Kinds and the Evolution of Modularity." *American Zoologist* 36:36–43.

Wagner, G. P. 1989. "The Origin of Morphological Characters and the Biological Basis of Homology." *Evolution* 43(6):1157–1171.

Wagner, G. P., and L. Altenberg. 1996. "Complex Adaptation and the Evolution of Evolvability." *Evolution* 50(3):967–976.

Wagner, G. P., and M. Laubichler. 2000. "Character Identification: The Role of the Organism." *Theory in Biosciences* 119:20–40.

Weaver, T. D. 2009. "The Meaning of Neandertal Skeletal Morphology." *Proceedings of the National Academy of Sciences* 106(38):16028–16033.

Wolpoff, M. H., J. Hawks, D. W. Frayer, and K. Hunley. 2001. "Modern Human Ancestry at the Peripheries: A Test of the Replacement Theory." *Science* 291:293–297.

Yokley, T., N. E. Holton, R. G. Franciscus, and S. E. Churchill. 2009. "The Role of Body Mass in the Evolution of the Modern Human Nasofacial Skeleton." *Paleoanthropology*.A39–40 (abs).

Zelditch, M. L., and W. L. Fink. 1996. "Heterochrony and Heterotopy: Stability and Innovation in the Evolution of Form." *Paleobiology* 22(2):241–254.

Zelditch, M. L., D. L. Swiderski, H. D. Sheets, and W. L. Fink. 2012. *Geometric Morphometrics for Biologists: A Primer*. San Diego: Elsevier Academic Press.

Zollikofer, C. P. E., and M. S. Ponce de León. 2002. "Visualizing Patterns of Craniofacial Shape Variation in *Homo Sapiens*." *Proceedings of the Royal Society of London B* 269(1493):801–807.

11 Where Evolutionary Biology Meets History: Ethno-nationalism and Modern Human Origins in East Asia

Robin Dennell

Introduction

In keeping with the theme of this volume, the focus of this paper is on whether paleoanthropology is an evolutionary science, with a focus on natural selection, biogeography, genetic change, and drift, or a historical discipline which seeks to establish and explain a sequence of events or developments. "History" here is used in two ways: first, as part of "deep history," in which the focus is on events and processes during the timespan of human evolution; and second, as part of modern history, during the lifetimes of those who study human origins. As I (Dennell 1990, 2001) and others (e.g., Hammond 1980, 1982) have previously argued, the two are interwoven and often inseparable: the study of our remote past is inextricably linked with our perceptions and concerns of the present. As examples, in British paleoanthropology before World War II, authors such as Sir Arthur Keith and Sir Grafton Eliot Smith focused on what they perceived as the deep antiquity of racial differences between whites and blacks in an era when Britain ruled most of sub-Saharan Africa; after World War II and the legacy of the Holocaust, American-led researchers emphasized instead the unity of humankind, the trivial nature of racial differences, and humankind's adaptability and diversity (Dennell 2001).

   Here, I explore the social and political context within which paleoanthropology has developed in China, and show how this has been used to promote a narrative about the antiquity and unity of the Chinese people. I begin with a quotation on the opening of the Olympic Games in Beijing in 2008 from the People's Daily (*Renmin Ribao*), the official newspaper of the Chinese government. It describes how the Olympic Torch had been taken through five continents by 20,000 torch bearers. The final leg of the Olympic Torch to the Bird's Nest Stadium was from Zhoukoudian, "*once inhabited by the ancestors of the Chinese* (my italics) . . . and marks a long awaited moment. This glorious historical moment congeals with the unswerving pursuit of a people; it records *the steadfastly progressive steps of a nation*; and is filled with true desires of the Chinese sons and daughters for friendship and peace with peoples of the world." The choice of Zhoukoudian was significant and deliberate in two ways: first, because Peking Man (a.k.a. *Sinanthropus pekinensis* and *Homo erectus*) was one of the first to use fire (according to Chinese archaeologists, and contra Weiner et al.

1998)—thus establishing a link with the Olympic Torch—and second, it is presented in China by the official media as the direct ancestor of the modern Chinese. It would be unimaginable for a Tanzanian today to regard Olduvai as the birthplace of modern Tanzanians, or an Indonesian to see Sangiran or Trinil in the same light, or for a German to feel the same way about Mauer. (It is however worth noting that Woodward [1948] entitled his book on Piltdown *The Earliest Englishman*). The challenge thus is to explain how and why in 2008, *Homo erectus* at Zhoukoudian is seen in China as providing an unbroken link not only to *Homo sapiens* but also to the modern Chinese nation. As a first step, it is necessary to survey developments in China over the previous century.

The Context of Paleoanthropology in Twentieth-Century China

Paleoanthropological research in China extends back almost 100 years to the early part of the twentieth century, and it has experienced three main phases (Shen et al. 2015). Each has been heavily dependent upon China's relations with the international community, particularly the USA and Western Europe. The first period was one in which Western influence was very strong in ideas, techniques, and researchers who often invested considerable parts of their careers in China. This period began in the last years of the Qing dynasty, and ended in 1941 when Japan attacked the United States and Britain.[1] For much of this period, China was weak in the face of foreign aggression, and often in turmoil. As landmarks, the Qing dynasty fell in 1912 and was followed by a short-lived first republic. The years between World War I and World War II were marked by the era of the warlords, constant local wars, attempts to unify China under the Nationalists, the outbreak of civil war in 1927, the Japanese takeover of Manchuria in 1931, and their invasion of northern China and the eastern seaboard in 1937, and their attack on Pearl Harbor in 1941. Despite this background of instability and turmoil, this period saw some remarkable collaboration in paleoanthropology between Chinese and non-Chinese researchers, most conspicuously at Chou-kou-tien (now known as Zhoukoudian). The second period began in 1949 with the founding of the Peoples' Republic of China (PRC), and is most associated with Mao Zhedong, who died in 1976. During this period, China was largely isolated from the wider international community (particularly after 1960 following the Sino-Soviet split), and also anti-Western. There was thus no meaningful contact with Chinese and Western researchers between 1949 and the 1970s. It was during this period that the idea was developed that Peking Man was the ancestor of the modern Chinese. The third period has no clear beginning but largely took shape after 1989 under Deng Xiao Ping, and takes us to the present. This period is marked above all by China's phenomenal economic development and re-engagement with the outside world, symbolized by the Olympic Games of 2008. In paleoanthropology, this has created new opportunities (particularly of collaboration with Western scientists) but also challenges to their own assumptions and beliefs. Conversely, for Western researchers, re-engagement with Chinese colleagues requires recognizing the social and political context in which their theories about human evolution have developed, as well as their own Western ones. To a far greater extent than in Europe and North America, ideas about human evolution that were prevalent in the 1930s are still evident in China 80 years later, and are also deeply politicized. These points are not appreciated as much as they should be by Western researchers.

To begin with, we can briefly review the wider background to the early study of paleo-anthropology in China.

Background: Late Nineteenth- and Early Twentieth-Century China

The nineteenth century had been disastrous for China: it had lost territory and sovereignty to Britain in two Opium Wars (1839–42 and 1856–1860), ceded Outer Manchuria to Russia in 1860, and Korea and Formosa (Taiwan) after its first war with Japan (1894–1895). Numerous indemnities almost bankrupted the Empire, which was further weakened by rebellions such as the Boxer Rebellion (1899–1901) and the Taiping Rebellion (1850–1864), which was the most lethal conflict of the nineteenth century. Most of its misfortune stemmed from the predatory behavior of the major European powers (Britain, France, Germany, and Russia) and in the late nineteenth century, Japan. On a more positive side, Western military and commercial intrusion was also accompanied by the influx of Western ideas. In the final years of the Qing dynasty before its end in 1912, China was like a sponge that soaked up the full spectrum of Western thinking, from liberalism, constitutional monarchy, and parliamentary democracy to republicanism, Marxism, and class warfare. In most cases, these ideas were developed to suit Chinese concerns and perceptions. A key institution in which many of these ideas were developed by Peking University ("Beida"), which was established in 1898 and largely replaced the traditional Confucian examination system that was by then glaringly inappropriate for China and finally abolished in 1905. Students from Beida were usually at the forefront of any mass movement, from the May 4th movement of 1919 (following the decision at the Versailles peace conference to transfer the German colony of Shandong to Japan instead of China) up to the Cultural Revolution and the events of 1989.

Darwinism and human evolution in China

In the case of paleoanthropology, there was an indigenous community of Chinese researchers at the turn of the twentieth century who were familiar with Western ideas of natural and social evolution. Darwin's theories on evolution, perhaps surprisingly to Western scholars, had an early and enthusiastic start in China (see Pusey 1983, Schmalzer 2008). However, the main interest was in social Darwinism, as developed by Huxley and Spencer. Huxley's (1893) *Evolution and Ethics* was translated into Chinese by a Westernizer, Yan Fu, to promote the necessity of social and political change in China (Schmalzer 2008, p. 20–21). The reformers and modernizers appreciated Darwin for three reasons. First, the implicit atheism of Darwinism—the absence of a Grand Design or a benevolent creator—fitted well with a materialist outlook that was hostile to the idea of gods and spirits controlling human destiny. Secondly, Darwin offered a way of countering superstition with science: Instead of dragons, there were instead dinosaurs and a demonstrable fossil record of extinct animals. Thirdly, when linked to Huxley's and Spencer's ideas of social evolution and the survival of the fittest, Darwin's ideas helped emerging Marxists legitimate a class struggle against a reactionary, elite (and foreign) dynasty. For non-Marxist modernizers, ideas about social evolution could be used to champion Chinese reformers against a conservative, non-Chinese rule under the Manchus. Although the main ideas that developed in paleoanthropological research in China before Japan invaded China in 1937 were Western-led, there was already a small but influential Chinese community that was comfortable with the idea of human evolution, and its usefulness in documenting the origins of races in general and the Chinese (i.e., the Han and the minorities within the former Empire) in particular. As example, when Black secured funding for the excavations of Chou-kou-tien, the excavations were under the direction of the Chinese. Three were particularly important: Zhongjian Yang (a.k.a. C. C. Young, 1897–1979), who graduated from Peking University in 1923 and took a Ph.D. in Munich in 1927; Pei Wenzhong (1904–1982), who also graduated from Peking University, in 1928; and Jia Lanpo (1908–2001), who graduated in geology from Huiwen Academy in 1929. As seen below, all three went on to be senior figures in human origins research in the PRC after 1949.

Nationalism and the Search for Chinese Origins

As with Europe, nationalism was a powerful force in China in the late nineteenth century for the modernizers who wished to see their country unified, strong enough to withstand foreign (Western and Japanese) interference, and freed from rule by a foreign, Manchu, dynasty. As with Europe, the past could be harnessed to reinforce a sense of national identity, and thus there was great interest among Chinese thinkers in the origins of the Chinese people. These ideas were initially developed by Europeans.

Chinese origins and the ancient Near East

One early theory was that the Chinese were derived from Egypt or Mesopotamia, on the basis of some spurious similarities between hieroglyphics or cuneiform and Chinese writing, as well as some similarities in ceramics (see Yen 2014). For a Chinese audience, this could be seen in a positive light, since both Chinese and European civilization could be seen as derived from Southwest Asia, and thus if they both shared the same roots, they should be of equal standing: China might therefore one day, it was thought, be treated as an equal. The negative of course was that Chinese culture was seen as derivative and not original. Debate over this was sharpened by the discovery of the Neolithic Yangshao culture in Northwest

China by the Swede John Gunar Andersson (1874–1960), whom we'll meet later at Zhouk-oudian. Initially, Andersson argued that the pottery style showed similarities with Southwest Asia. Later, Davidson Black (1884–1934)—one of the key figures at Zhoukoudian—argued that the skulls from Yangshao were identical to those of modern Chinese, and the Yangshao culture was thus indigenous. Andersson in time accepted this conclusion, and Yangshao thus gave the Chinese an ancestry that extended to the Neolithic.

Human Evolution and the Discovery of a Remote Past: Centers and Dispersal

Those who thought about human evolution in the late nineteenth and early twentieth centuries tended to think about (as now) centers in which humanity first developed, and from which they subsequently dispersed to other regions. As is well known, Darwin (1871, p. 161) cautiously hypothesized that as our nearest relatives, the chimpanzee and gorilla, lived in Africa, so it was "somewhat more probable that our early progenitors lived on the African continent than elsewhere." Less often remembered is that he immediately qualified this suggestion in the sentence that followed: "But it is useless to speculate on this subject: for . . . since so remote a period the earth has certainly undergone many great revolutions, and there has been ample time for migration on the largest scale."

Darwin, Haeckel, and Lemuria

Initially, the main rival to Darwin's cautious suggestion about an African center of human evolution was Ernst Haeckel's (1834–1919) fantastical theory that humanity originated on the lost continent of Lemuria, which sank somewhere in the Indian Ocean. This idea was expounded in his *Natürliche Schöpfungsgeschichte* in 1868, and translated into English in 1876 as *The History of Creation*. On Lemuria, the ancestors of chimpanzees and gorillas supposedly managed to reach Africa, while the ancestors of humans and orang-utans arrived in Southeast Asia. He even named this hypothetical human ancestor *Pithecanthropus allalus*, or speechless ape-man, on the grounds that it was language that most distinguished us from the apes. Perversely, it was a rare example of a theory about human evolution seemingly being proven correct (a false positive) because it inspired Eugene Dubois (1858–1940) to go to the Dutch East Indies (now Indonesia), where he found a hominin skull-cap and associated femur at Trinil, Java, in 1891 (see Theunissen 1989). In homage to Haeckel, he named it *Pithecanthropus*, but selected *erectus* as the specific name because of his conviction that it was associated with the skull, and indicated an upright posture. Lemuria had a surprisingly long inning as one of the most bizarre notions of human origins.[2] For example, A. C. Haddon (1912, p. 54) considered, "There is reason to believe that mankind did not originate in Africa, but that all the main races in that continent reached it from southern Asia." One of the last academic sightings of Lemuria in Western writing was in a ghastly racist book called *Savage Survivals* by J. H. Moore (1933, p. 60) in a series inappropriately named

The Thinkers Library: "It is believed that man evolved somewhere in southern Asia . . . in land now drowned by the Indian Ocean. This supposed land is called Lemuria." Strangely, however, and thanks to an impeccable Marxist pedigree, Lemuria enjoyed a long life in Chinese writing.

Lemuria and Engels

Less well-known by Western researchers is that Lemuria played an important, and long-lived role in Chinese paleoanthropology, thanks to Engel's (1876) essay, *"The Part Played by Labour in the Transition from Ape to Man."* This was translated into Chinese in 1928 (Schmalzer 2008, p. 60–61), and was widely quoted in communist writing for its message that "labour created humanity" through the liberation of the hands by bipedalism. (This theme later played a large part in the iconic significance of *Sinanthropus* in Marxist and Maoist writing). Engels also postulated that "a particularly highly-developed species of anthropoid apes lived somewhere in the tropical zone—probably on a great continent that has now sunk to the bottom of the Indian Ocean." The longevity of Engel's essay of 1876 is evident in that the quotation is taken from Jia Lanpo's *"Early Man in China,"* published in English in 1980.

Central Asia

A much more significant theory involving a center was that proposed by the American researcher William Dillinger Matthew (1871–1930), whose paper *"Climate and Evolution"* was published in 1915. Here, he firmly placed human origins on Central Asia.[3] In his model, the tectonic uplift that resulted in the Tibetan Plateau was the prime mover in mammalian (and not just human) evolution: tectonic instability and uplift prompted the appearance of new species that then dispersed outwards, thus displacing earlier and more primitive types. In his view, "The most advanced stages should be nearest the centre of dispersal, the most conservative stages furthest from it" (cited in Black 1925, p. 141). He effectively inverted Darwin's suggestion that humans evolved in Africa because that was where our closest cousins are found; for Matthew, the existence of the chimpanzee and gorilla in Africa indicated that they had been dispersed there by more advanced types (such as humans) that originated in Central Asia.

Matthew's Central Asian hypothesis had serious support in North America, particularly from fellow American Henry Osborn (1857–1935), William Gregory (1876–1970), and most importantly, the Canadian Davidson Black (1884–1934), who later discovered *Sinanthropus* at Chou-kou-tien (now Zhoukoudian) near Beijing. For Gregory (1927, p. 463) human origins lay "somewhere between western Europe and eastern Asia," with "the Tarim desert as the most likely place." For Osborn (1918, p. 19), "Western Europe is to be viewed as a peninsula, surrounded on all sides by the sea and stretching westward from the great land mass of eastern Europe and of Asia, which was the chief theatre of evolution both of animal and human life." In greater detail, "The most likely part of the world in which to discover these 'Dawn Men' . . . is the high plateau region of Asia embraced within the great prominences of Chinese Turkestan, of Tibet and of Mongolia" (Osborn 1927, p. 377). Osborn was a key figure at this time. He visited China, and lectured to a Chinese audience about the Central Asia hypothesis, and this was published in Chinese. Another influential figure at the time was the German geologist Walter Grabau, who taught at Beida, and argued to a Chinese audience, including his students, on the probability that humanity originated in Central Asia, in territory that included parts of China (Tibet, Inner Mongolia, and Chinese Turkestan). He also published these ideas in both English and Chinese (see Yen 2014).

As events showed, the most significant support came from Davidson Black, who put theory into action by seeking employment at the newly created Peking Medical College in order to be nearer this potential center of human evolution—just as Dubois had accepted a medical post in the Dutch East Indies in order to test Haeckel's theory about Lemuria. Black's (1925) paper was a thorough development of Matthew's (1915) paper, in which he highlighted the fossiliferous Siwaliks of British India, and the Yung-Ling and Tarim Basins of North China as areas worth prospecting. His 1934 paper to the Royal Society probably marks the apogee of this line of thinking.

Two other developments helped the Central Asian theory. The first was the remarkable reconnaissance in 1922 by the French missionary Émile Licent (1876–1952) and the Jesuit Teilhard de Chardin (1881–1955) through the Ordos Desert of North China (where they found the first Pleistocene remains of humans), to the Nihewan Basin (where there was a small Christian community) and on to Shuidonggou, where they excavated SDG1, which is still the classic Early Upper Paleolithic site in China (Boule et al., 1928, Licent and Chardin 1925). Second, the showman and explorer Ray Chapman Andrews (1884–1960), a later director of the American Museum of Natural History, raised lavish funding for expeditions from 1922–1928 into the Gobi Desert and Inner Mongolia to search for man's ancestors; he was unsuccessful, but did find important dinosaur remains. The preface to his book *On the Trail of the Ancient Man* (Andrews 1926) was written by none less than Osborne, who stated (Osborne 1926, p. ix) that he had predicted as early as 1900 (Osborne 1900) "that the home of the most remote ancestors of man, Primates, was placed in northern Asia."

Chou-kou-tien was thus on the edge of what was hypothesized as the likely cradle of humankind. Nevertheless, it was by far the nearest to it compared to any other site in Asia. Thus when Andersson found Locality 1 ("Dragon Bone Hill") at Chou-kou-tien in 1921 and Otto Zdansky (1894–1988) found two human teeth in 1924 in samples that had been sent back to Sweden, it seemed to confirm the Central Asian hypothesis. These were published in 1927 by Zdansky as *Homo* sp., but it was Black (1927) who announced the discovery of *Sinanthropus pekinensis* on the basis of a single left lower molar. Where Black was shrewd was in using this discovery to ask the Rockefeller Foundation (which was already responsible for running Peking Medical College[4]) for additional funding to continue excavations under the direction of Li Jie. The discovery in 1928 of more teeth, part of a juvenile mandible, and some skull fragments enabled Black to secure another $80,000 from the Rockefeller Foundation to set up the Cenozoic Research Laboratory of the Geological Survey of China in 1928. Black thus had at hand the most important paleoanthropological site in the world, and the laboratory in which to study the hominin remains that were found there. (The terms of the grant also insisted that Chinese researchers could study the animal remains, but that Black was solely responsible for any hominin material. As Jia (1980, p. 22) comments with understandable bitterness, "In recognition of a $80,000 donation to the project by the Rockefeller Foundation, the reactionary Chinese government at that time had gone to the length of relinquishing the right to study these human remains found on its own territory.")

Development, Not Origins: Weidenreich, 1937–1948 and Multiregionalism

After the sudden death of Davidson Black in 1934, his replacement by Franz Weidenreich (1873–1948) in 1935 as the senior anatomist studying the *Sinanthropus* specimens marked a significant shift of emphasis in the Chou-kou-tien project. In China, Weidenreich was more concerned with development than with origins[5] and came from a completely different intellectual framework from that of the North Americans such as Matthew, Osborne, Gregory, and Black. While not hostile to the idea that Central Asia was the cradle of mankind, he was more interested in how hominins such as *Sinanthropus* and *Pithecanthropus* fitted into the greater scheme of human evolution in Asia, Europe, and Africa. His background as an anatomist lay in Central Europe; he studied under Gustav Schwalbe (1844–1916), who in turn had studied the Neanderthal remains from Krapina, Croatia, and concluded (contrary to Boule) that Neanderthals had evolved into *H. sapiens*. As is well-known, it was during and after his time at Chou-kou-tien (1937–1941) that he developed his theory of multiregional evolution, best summarized by his famous (or infamous) trellis figure, published in 1946 (see figure 11.1). For Weidenreich, hominins such as *Sinanthropus* and *Pithecanthropus* were linked by gene exchange to each other and their counterparts in Europe and Africa, and in each continent, *H. sapiens* emerged as a result of gene exchange. *Pithecanthropus* seemed more primitive, as there was no evidence that it used stone tools or fire, unlike *Sinanthropus,* which could thus be regarded as one of the ancestors of *H. sapiens.*

In addition to Locality 1, the other important locality at Zhoukoudian was Upper Cave (now Locality 26), which Pei excavated in 1933–1934 and which dates to the late Upper Pleistocene. According to Weidenreich, the human remains showed affinities to *Sinanthropus* and also to modern Mongolians (with traits such as shovel-shaped incisors, and an Inca bone) as well as to Eskimos. This implied a deep ancestry of both peoples, and could be interpreted (or misinterpreted) as extending Chinese identity deep into the Pleistocene.

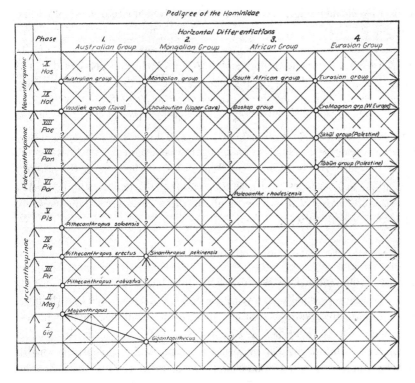

Figure 11.1
Weidenreich's model of multiregional evolution. Note first that he placed the Chinese *Gigantopithecus* at the base of the hominin lineage rather than the South African *Meganthropus*. Second, in his figure, the modern Mongolian group is shown as directly descended from Choukoutien Upper Cave and *Sinanthropus pekinensis*.
Source: Weidenreich, F. 1948. *Apes, Giants, and Man*, Fig. 30. Chicago: University of Chicago Press.

Human Origins Research in Asia before World War II

The field research in China in the years before 1937 can be assessed for both its intellectual and social context.

The intellectual context

Several points stand out when considering the ideas driving paleoanthropology in China in the years before World War II. The first is that it developed within an intellectual framework that could be considered as linked to evolutionary biology, particularly biogeography. Second, it was perhaps the only time when a theory about human origins preceded the discovery of data that was then used to confirm it. In contrast, the reverse has normally happened: unexpected discoveries in an area are usually then followed by theories that attempt to explain why the area might have been a center of evolution, or have played an important role within it. A clear example is East Africa, where the discovery of *Zinjanthropus, Homo habilis,* and *Australopithecus afarensis* was later followed by theories as to why East Africa might have been a cradle of humankind. A third feature of pre-World War II paleoanthropology is that most authority figures supported theories that gave Asia primacy over Africa as the cradle of humankind; no major writer of any significance at this time appears to have advocated Africa over Asia. This was due partly to the belief that a hallmark of early hominins was a large brain (as shown notoriously by Piltdown), and partly due to an implicit and sometimes explicit racial prejudice against envisaging black Africa as the place where white Europeans had originated (see Dennell 2001). The belief in the primacy of the brain and a bias against black Africa as a potential cradle doubly stymied Dart's small-brained *Australopithecus africanus* from southern Africa. However, the main driving forces behind the Asian paradigm were biogeographical.

The social context

In the years preceding the outbreak of World War II, most research into human origins took place in Asia, or more specifically, in three areas of Asia. These were in British India, in the Siwaliks and major river valleys such as the Narmada; in the Dutch East Indies, particularly Java; and in China. In the case of British India and the Dutch East Indies, this research was undertaken almost entirely by Western scientists. Examples from British India are the fieldwork in Kashmir, in the Soan and Narmada valleys by the Swiss-German Helmut de Terra, the British archaeologist Thomas Paterson, the French Jesuit Teilhard de Chardin, the Americans G. Edward Lewis at Haritayangar, and Barnum Brown, and members of the India Geological Survey such as Guy Pilgrim, who was one of the giants of Siwalik research. In Java, the main figures were staff of the Dutch Geological Survey, such as Duyfijes, Oppenorth, Koenigswald, and of course Dubois, who pioneered field research in the 1890s with his discovery of *Pithecanthropus*. In both British India and the Dutch East Indies, indigenous "natives" played only very minor roles as field assistants; they did not study or publish the artifacts and fossils that were collected, even though they were often invaluable in the field as collectors. When British and Dutch rule ended after World War II, there were no indigenous, qualified Indian or Indonesian scientists who could step into the shoes of their former rulers and continue research into paleoanthropology. There is no indication that research in the Siwaliks of British India or the Sangiran Basin played any role in developing national consciousness by the indigenous Indians and Javanese.

China stands out as significantly different from British India and the Dutch East Indies in that it was not part of a Western empire, and there was an emerging educated group of Chinese researchers who could and did work in collaboration with Western scientists. To a much greater extent than in India and the Dutch East Indies, Western ideas were assimilated, debated, and developed by local groups. Although the story of Chou-kou-tien has been told many times by Western writers, the emphasis is nearly always upon the key Western figures such as Andersson, Chardin, Black, and Weidenreich, and the significance of this research for the Chinese is usually overlooked, as is the importance of the Chinese scientists such as Yang, Pei, and Jia who were an integral part of this work. As seen below, Chinese involvement in the research at Chou-kou-tien played a major role in the development of ideas about how the Chinese define themselves.

Chinese Isolation (1937–c.1980) and the Development of Ethno-nationalism

Two developments killed off the international collaboration at Chou-kou-tien. The first was the Japanese invasion of 1937, when fieldwork became too dangerous for the Chinese workforce.[6] The second was the continuation of the civil war in China between the nationalists and communists after 1945, resulting in the proclamation of the PRC in 1949 and the defeat of the United States' main ally in the region, the nationalists. The ending of American influence in mainland China, and Mao's anti-imperialist foreign policy, resulted in the "bamboo curtain," and China's isolation from the West, with the USSR and later, North Korea and Albania as its only allies. Chinese isolation increased further following the Sino-Soviet split in 1960, and the turmoil created by the Great Leap Forward (and ensuing Great Famine) and the Cultural Revolution of the 1960s. The severance of contact between China and the outside world thus marooned a generation of Chinese researchers who could only warily tread their own path (Shen et al. 2015), and hope to avoid censure by whichever political storm blew their way.[7] The thawing of China's isolation from the wider international community, and the United States in particular began to end only after Nixon's visit to Beijing in 1972, but did not really begin to take effect until the Deng Xiao Ping era of the 1990s.

Nevertheless, although few in number, Chinese paleoanthropologists achieved a great deal after 1949. Excavations resumed at Locality 1, Zhoukoudian, from 1949–1951, and several times thereafter (1958–1960; 1966; and 1978–1982), resulting in the discovery of a further 100 hominin fossils from 40 individuals. Hominin fossils were found at several other locations, notably Dali, Lantian, and Yuanmou (Shen et al. 2015), and many important paleolithic sites were excavated, such as Dingcun, Xujiayao, and several sites in the Nihewan Basin. Institutionally, the Institute of Vertebrate Paleontology and Paleoanthropology (IVPP) in Beijing emerged as the leading research center, where the main researchers were Pei Wen-Zhong and his student Jia Lanpo, and C. C. Young, all of whom had learned their trade at Chou-kou-tien before World War II, and the physical anthropologist Wu Rukang (1916–2006).

For the generation of Chinese researchers who had begun their careers before World War II and who remained in mainland China, their main intellectual capital were the ideas and techniques that they had assimilated before 1937. The two main ideas that were carried forward in paleoanthropology were the primacy of Asia over Africa, and the idea of mul-tiregional evolution. Both ideas proved remarkably resilient. For adherents of the Central Asian model, the acceptance of the South African australopithecines (in 1948) was inter-esting, but did not demonstrate a direct ancestral link to the genus *Homo*; the unmasking of Piltdown as a fraud demonstrated the gullibility of Western scientists; and even the discovery of *Homo habilis* in 1960 could be seen as comparable to *Pithecanthropus*. It was only when the trickle of new discoveries in East Africa in the 1950s and 1960s became a flood that an African origin of hominins, including our own genus, became convincing to the likes of Pei Weizhong. However, even in the 1980s (as seen previously), Jia Lanpo adhered to the theory of an Asian origin of humanity.[8] As recently as 1998, the Chinese government granted 5 million yuan (ca. £500,000) to find australopithecines in China (Schmalzer 2008, p. 257), so far without success.

The shift from Asia-centricism to Chinese ethno-nationalism

The founding of the PRC in 1949 had direct consequences on paleoanthropology in China. First, the resumption of excavations at Zhoukoudian was made an immediate priority (unlike in India or Indonesia, where fieldwork in the Siwaliks or Sangiran was not resumed for several decades), and were restarted at only a few weeks' notice in 1949 (Jia and Huang 1990, p. 191). Second, Weidenreich's multiregionalism was reformulated in terms of ethno-nationalism, whereby *Sinanthropus* was officially portrayed as the direct ancestor of not only *H. sapiens* but also the Chinese people. This was not an original idea: accord-ing to Dikötter (2015, p. 85), Chinese archaeologist Lin Yan in 1940 cited *Sinanthropus* as proof that the Chinese race had inhabited the Middle Kingdom from the earliest stage of human history. This claim needs to be understood in the context of its time. In 1940, the existence of China was under severe threat from the Japanese, and in 1949, the newly founded PRC was faced with the problems of ruling a huge country that had been trauma-tized by eight years of Japanese occupation (14 years in the case of Manchuria), over two decades of civil war, and was also confronted by hostile Western powers. *Sinanthropus* thus became a very useful device for stressing the unity and deep historic roots of the Chinese people[9] (see Yen 2014, Sautman 2001, and Schmalzer 2008). Following Engel's essay of 1876, *Sinanthropus* also served socialism in showing the primacy of human labor, particularly (in the Mao era) of male strength, as *Sinanthropus* (unlike *Pithecanthropus*) had mastered the ability to make tools (Schmalzer 2008). These "messages" of the deep antiquity and unity of the Chinese people, and the primacy of human labor were disseminated to as many as possible through visits to the Zhoukoudian museum by mass groups from schools, army, and work units (Schmalzer 2008, p. 153). As seen from the quotation about the 2008 Olympic Games, this official view remains unchanged.

Hallam Movius (1907–1987) and the marginalization of China

An additional but external development that contributed greatly to severance of Old World paleolithic archaeology from China was Hallam Movius's publication in 1948 of his synthesis of the early paleolithic of East Asia. As is well known, Movius depicted the early paleolithic world as comprising two blocks: one in Africa, western Europe, and Southwest and South Asia defined by the presence of hand-axes, and the other in East and Southeast Asia (with an outlier in Northwest India, the Soanian) that was defined by a simple chopper-chopping tool technology. This dichotomy was seen in cognitive, or even implicitly, racial terms: those with hand-axes were seen as "progressive" and "dynamic"; those without as "conservative" and an example of "cultural retardation" (see Dennell 2014a, 2014b for a critique of these views). In the words of Movius, (1948, p. 411), in East and Southeast Asia the tools are "relatively monotonous and unimaginative assemblages of choppers, chopping tools, and hand-adzes . . . as early as Lower Palaeolithic times, southern and eastern Asia was a region of cultural retardation . . . it seems very unlikely that this vast area could ever have played a vital and dynamic role in early human evolution. . . . Very primitive forms of Early Man apparently persisted there long after types at a comparable stage of physical evolution had become extinct elsewhere." Movius, who had no prior experience of early palaeolithic archaeology, was undoubtedly heavily influenced by Teilhard de Chardin, who was the senior figure in the field expedition to Burma in 1938 in which he collaborated with Helmut de Terra (as he had already in Kashmir in 1935). Chardin was unambiguous about China's context in the wider paleolithic world: "Early Palaeolithic China was a quiet and conservative corner on account of its marginal geographical position . . . in contrast with the already 'steaming' West, Early Pleistocene Eastern Asia seems to have represented . . . a quiet and conservative corner amidst the fast advancing human world" (Chardin 1941, p. 60). He went on to state that "East Asia gives the impression of having acted (*just as historical China and in sharp contrast with the Mediterranean world*) *as an isolated and self-sufficient area*, closed to any major human migratory wave" (1941, p.86, 88; italics mine). Movius and Chardin effectively stripped East Asia of any paleolithic history, since it was seen as static and thus unchanging.

In effect, Movius extended the bamboo curtain from post-war China to the early Pleistocene: in his view, China is, and always had been isolated from the West. It provided an ideal excuse for Western scholars to ignore East Asia: they were unable to work there as they had done before World War II, but even if they had been able to, there was no point in wasting effort on a culturally "retarded" region that had played no significant role in human evolution. As Athreya (2010) has pointed out, between 1950 and 2003, there was not a single session on Asian paleoanthropology by the annual meetings of the American Association of Physical Anthropologists.

Post-1989: The Ending of China's Isolation

There is no clear turning point at which China re-joined the wider international community. Landmarks in this process are: Nixon's visit in 1972 and the PR display of table-tennis diplomacy; Howells et al.'s visit in the 1970s; visits—not always a meeting of minds—by Binford, Weiner, Boaz, Desmond Clark, Schick, and Toth working at Nihewan in the 1980s, and others such as Ciochon, Diane Gifford-Gonzalez, Pope, and Olsen. All these helped in that it was the first time for many Chinese to have met a Westerner.[10] Equally significant was that young Chinese researchers such as Gao Xing, Shen Chen, and Lipeng Zhou could now leave China to train for a Ph.D. or undertake post-doctoral research in the West. Likewise, Chinese and Western scholars met more frequently at international conferences (e.g., myself in Beijing in 1999, Huang Weiwen in Honolulu in 2001). Other changes were more gradual, notably younger Chinese researchers learning English and being bold enough to practize it by talking to foreigners, and now, frequently publishing their research in peer-reviewed Western journals.

For the study of human evolution, these developments were marked by the encounter of ideas that had been preserved since the 1930s on multiregional evolution with Western-led ideas. For the Chinese, this has been a very exciting time in which they have been exposed to Western ideas about the central importance of Africa, and the nature of lithic variability. They have also encountered techniques such as taphonomic analysis, 3D recording of archaeological excavations, GIS data manipulation, soil micromorphology, use wear analysis, and advanced dating techniques. On the whole, they have been very adept at learning these, and integrating them with their own field programs, as at Shuidonggou, for example. The post-1989 generation of Chinese researchers has been highly successful at assimilating and adapting new techniques of dating, analysis, recording, and so on. There has also been a subtle but important shift in Chinese perceptions of human evolution.

Continuity with Hybridization

The term "continuity with hybridization" is often used by Chinese researchers to summarize current thinking about the evolution of *Homo sapiens*. This model deserves to be examined for its academic content along with its rival, the Out-of-Africa model. The term "continuity with hybridization" also provides an accurate description of the political context in which modern human origins are studied in China, and this also needs examination.

"Continuity with hybridisation": the academic context

According to this view, the basic tenets of multiregional evolution still hold true: that *Homo erectus* at Zhoukoudian evolved into *H. sapiens*, and this process is marked by transitional specimens of "archaic *H. sapiens*." In their eyes, Chinese specimens such as Zhirendong that they attribute to "archaic *H. sapiens*" (Liu et al. 2010) are comparable to African specimens such as Herto, Ethiopia, that are also classified as "archaic *H. sapiens*" by some Western researchers. For most non-Chinese researchers, this practice is very confusing because the Out-of-Africa model requires only one source of "archaic *H. sapiens*" from which all other modern humans are derived.

For many years, debate over the origin of "modern" humans was highly polarized between those advocating that *H. sapiens* originated in Africa and then replaced all indigenous populations in Europe and Asia, such as Neanderthals, East Asian *H. erectus*, and *H. floresiensis* on Flores, and thus argued for multiregional evolution. For many years, there was little common ground between the two camps, as evident by the papers and not infrequent clashes by the likes of Stringer and Wolpoff and their supporters.

In recent years, the debate has become more nuanced and is no longer so bipolar. One important discovery has been from studies of ancient DNA (aDNA) of Neanderthals, showing that there is some (ca. 2–4%) of Neanderthal DNA in all modern non-Africans, that is, those in Europe and Asia and their descendent populations in the Americas and Australia (Prüfer et al. 2013). Some degree of hybridization has to therefore be accepted by those arguing for a model based on dispersal and replacement. Another development is the discovery—again from studies of aDNA—of Denisovans, a population first identified from the analysis of specimens from Denisova Cave, Siberia, of a previously unsuspected population that was neither Neanderthal nor *H. sapiens,* and likely a sister clade of Neanderthals (Krause et al. 2010, Reich et al. 2010, Meyer et al. 2012). As Denisovan DNA appears in modern populations in Melanesia but not on mainland or island Southeast Asia, there may therefore have been a dispersal event involving those of Denisovan ancestry, followed by their replacement in Southeast Asia by groups lacking Denisovan DNA (Reich et al. 2011). In other words, the emergence of *H. sapiens* in Southeast Asia is likely to have been a complex process rather than a single replacement event of *H. erectus* by *H. sapiens*, or unilinear evolution from one to the other. Some indication of the likely degree of complexity is indicated by the presence of Denisovan aDNA in a tibia from the Middle Pleistocene site of Sima de los Huesos, Atapuerca, Spain (Meyer et al. 2014). As this specimen (and the associated ones in the same cave) are usually referred to as early Neanderthal or as *H. heidelbergensis*, the presence of Denisovan aDNA was unexpected and provides an additional indicator of the emerging complexity of human evolution in Eurasia.

A similar pattern is emerging from recent detailed studies of Middle and Upper Pleistocene specimens in China, such as those at Dali, Panxian Dadong, and Xujiayao. On environmental grounds, it is likely that China was not always isolated during the Middle Pleistocene but open to immigration. One such interval may have been between MIS 15 and MIS 13 because MIS 14 (563−533 ka) is very weakly expressed in Siberia and North China, and this may have "provided favorable conditions for the second major dispersal episode of African hominins into Eurasia" (Hao et al. 2015, p.1).

The likelihood that such dispersals occurred is reflected in the Chinese fossil hominin record. Rightmire (2001), for example, classified the specimens from Dali and Jinnuishan as *H. heidelbergensis*, which has to imply that "[The] spread of some populations of *Homo heidelbergensis* into the Far East cannot be ruled out" (Rightmire 1998, p. 225). For others, this possibility is a near-certainty: "*Homo erectus* at first existed by itself in China; *Homo heidelbergensis* then entered and the two coexisted for a time; finally *H. erectus* became extinct there, and *H. heidelbergensis* persisted alone: an early 'replacement' event" (Groves and Lahr 1994, p. 3). Others have commented on the overlap between *H. erectus* and non-*erectus* Middle Pleistocene *Homo*. As example, Wu and Athreya (2013, p. 154) comment, "This overlap reflects the fact that the evolutionary trajectory from archaic to more modern forms was the result of highly variable patterns of population dynamics across the Old World between different regional groups." As Howell (1999, p. 223) remarked, "Such shifts are still ill-appreciated." Under the "continuity with hybridization model" of hominin evolution in East Asia (see e.g., Liu et al., 2010), "the suite of traits exhibited by Dali could be indicative of a local transition between *H. erectus* and *H. sapiens* that included some influence from western Eurasian populations during the Middle Pleistocene" (Wu and Athreya 2013, p. 154). As Bae (2010, p. 89) comments, "The question of whether eastern Asian archaic *H. sapiens* should be classified as *H. heidelbergensis* can also be viewed in the light of dispersing hominin populations. In particular, if *H. heidelbergensis* dispersed from the western Old World and into eastern Asia some time during the Middle Pleistocene, then it would support the hypothesis that a third major dispersal event out of Africa occurred." Bae (2010, p. 90) further comments "Even small amounts of gene flow from dispersing *H. heidelbergensis* groups into eastern Asia during the Middle Pleistocene is probably the most parsimonious explanation as to why similar morphological features occasionally appear among penecontemporaneous western and eastern Old World hominins."

The teeth from Panxian Dadong (Liu et al. 2013) that are ca. 300–130 ka show how far we are from a clear picture of later human evolution in China. They are described as non-Neanderthal but not obviously linked to *H. sapiens*. The same is true of the later site of Xujiayao, North China (Song et al. 2015, p. 224): ". . . Evinces the existence in China of a population of unclear taxonomic status with regard to other contemporary populations such as *H. sapiens* and *H. neanderthalensis*. The morphological and metric studies of the Xujiayao teeth expand the variability known for early Late Pleistocene hominin fossils and suggest the possibility that a primitive hominin lineage may have survived late into the Late Pleistocene in China."

"Continuity with hybridization": The social and political context

As the quotation at the start of this paper about the historic link between Zhoukoudian and the Olympic Games in Beijing in 2008 shows very clearly, the official line in China is that modern Chinese are the product of multiregional evolution from an ancestral population of *H. erectus* at Zhoukoudian. According to this view, the Chinese originated in China, and have always lived within China as a coherent grouping, and there is no reason to suppose that this view will change in the foreseeable future.

What we do have now is some degree of hybridization with foreign scientists, and an acceptance that some hybridization occurred in the deep past—though not enough to override the driving force of evolution directly from *Sinanthropus (H. erectus s.s)* to *H. sapiens*. So long as the debate is left open-ended, Chinese scientists are prepared to recognize that the Chinese fossil record is complex, and more evidence is required. Non-Chinese researchers for their part should recognize that human origins research in China is highly politicized, and there are well-defined limits to any debate. Those Western researchers who engage with Chinese researchers need to appreciate the deep background to Chinese perceptions of human evolution, and the role of Zhoukoudian within it. If there is to be a productive dialogue, there has to be respect and understanding of the way that Chinese paleoanthropology has developed.

Another way of looking at the "continuity with hybridization" model is to consider it from the viewpoint of Chinese researchers. At an individual level, two aspects stand out. The first is a desire that Western scientists should regard the Chinese fossil hominin and paleolithic record as important, and not marginal to a global paleoanthropology. "Hybridization" thus implies that the Chinese record is integrated into its wider Eurasian and African context, and not seen as separate and marginal. The second is the entirely understandable desire to be treated with respect as equal collaborators with non-Chinese researchers. Here, the memory of how Black and Weidenreich monopolized the study of the Zhoukoudian *Sinanthropus* fossils runs deep, and they (entirely reasonably) expect their views and analyses to be treated with as much respect and rigor as that given to Western researchers. "Hybridization" thus implies interacting with non-Chinese researchers on an equal footing: prestige and respect are integral aspects of discussions about the role of hybridization.

The Movius Line—a zombie that refuses to die?

The Movius Line (Movius 1948) was based on fieldwork in Burma in 1937–1938, and is probably the longest-accepted example of pre-World War II fieldwork that has still found adherents. In my own writing (Dennell 2014a, 2014b, 2015), I have argued that it was a house built on sand, has no validity, and needs to be forgotten. Bar-Yosef (2015) has expressed a similar opinion. This is one concept that could be dropped with equal benefits to Chinese and non-Chinese researchers.

Conclusion

From the above, it is obvious that the Chinese fossil record is viewed very differently by Chinese and (most) Western researchers: "In the West, scientists treat the Chinese fossil evidence as part of the broad picture of human evolution worldwide; in China, it is part of national history—an ancient and fragmentary part, it is true, but none the less one that is called upon to promote a unifying concept of unique origin and continuity within the Chinese nation" (Reader 1990, p. 111). Given the profound differences in the intellectual traditions and histories of China and the West, these differences of interpretation should not be surprising. "Human evolution is our origin narrative, and such narratives have cultural salience" (Marks 2015, p. 15) within specific cultural milieux. In China, both the Central Asian hypothesis and multiregionalism—the fundamental theories that initially drove the study of human evolution in China—were developed within a framework of evolutionary biology. The ethno-nationalism that was imposed upon the fossil hominin record after 1949 needs to be placed in its historical and political context, and is thus part of the study of modern history. As these ideas present a sequence of events that directly links Locality 1 at Zhoukoudian with the Birds' Nest Stadium and the 2008 Olympic Games, they are also historical in that they narrate a sequence of events—even if imperfectly understood—between the deep past and the present. Non-Chinese researchers in paleoanthropology have to understand that their Chinese counterparts operate within a very different social and political context from their own. At the same time, non-Chinese researchers need also to examine their own belief frameworks and academic traditions as critically as they examine those of their Chinese colleagues.

Acknowledgments

I wish to thank the Wenner-Gren Foundation and Jeff Schwartz for inviting me to the workshop underlying this volume, and to the KLI for their exemplary generous hospitality, which did so much to make this meeting a truly memorable occasion.

Notes

1. As a symbolic ending of this period in Chinese paleoanthropology, the *Sinanthropus* remains from Zhoukoudian were lost at this time when in transit from Beijing to the USA.

2. Lemuria still lurks on the Internet on what might be termed the lunatic fringes of mysticism and spiritualism, and is located variously in the Indian and Pacific Oceans.

3. Central Asia here was defined as Tibet, Mongolia, and Inner Mongolia, and not, as now, the "Five Stans" of Kazakstan, Kyrgyzstan, Tajikistan, Turkmenistan, and Uzbekistan.

4. Despite the phenomenal growth and modernization of Beijing in recent years, the Peking Medical College is still there in its original buildings. Somewhat incongruously for a Communist state, it contains a bust of J. D. Rockefeller, the ruthless capitalist who provided the key funding.

5. However, in his 1946 book *Apes, Giants, and Man*, Weidenreich did propose an evolutionary lineage from Asian *Gigantopithecus* to *Homo*. Although this now seems strange, Krogman (1947, p. 118) thought "The biological continuity is an uncontestable one."

6. At Chou-kou-tien, three workers were used for bayonet practice (Jia and Huang 1990, p. 153).

7. These political storms are incomprehensible for any post-war west European or American researcher. As example, in the Cultural Revolution, C. C. Young was forced to bend at right angles for long periods with a heavy placard around his neck, and was not allowed to eat until everyone else had finished (Schmalzer 2008, p. 124); Jia Lanpo was placed under house arrest (Schmalzer 2008, p. 124) and burnt the manuscript of his book on Chou-kou-tien because of its too-frequent references to Western researchers, and the fear that he might be accused of being an American spy (Jia and Huang, 1990, p. 2). At a time when I was enjoying undergraduate life at Cambridge, a Chinese friend and colleague spent three years planting trees and digging ditches in a labor camp as part of an Education through Labor program.

8. The rationale now had shifted to the argument that the Miocene *Ramapithecus*, known mainly from modern Pakistan, was a hominin. This notion gained traction in the 1960s through the work of David Pilbeam and Elwyn Simons, but it was invalidated by the discovery of GSP15000 in Pakistan, which showed the main affinities were with orang-utans. We should also note that Weidenreich (1946) postulated an evolutionary line from *Giganto-pithecus* to *Homo*, and Koenigswald was also sympathetic to East Asia as the cradle of humanity.

9. It is not unusual for the past to be used for propaganda purposes at times of national crisis. As example, see Boule's contrast of the creative (French) Cro-Magnons with the brutal (German) Neanderthals in WW1, or Jaquetta and Christopher Hawke's (1940) depiction in their popular bestseller *Prehistoric Britain*, of Britain as a beleaguered isle that developed its own specific identity.

10. I had the same experience working in Bulgaria in the early 1970s in the early phase of détente.

References

Andrews, R. C. 1926. *On the Trail of Ancient Man*. New York: Garden City Publishing Company, Inc.

Athreya, S. 2010. "Book review: the Palaeolithic Settlement of Asia. By Robin Dennell." *American Journal of Physical Anthropology* 142:501–502.

Bae, C. J. 2010. "The Late Middle Pleistocene Hominin Fossil Record of Eastern Asia: Synthesis and Review." *Yearbook of Physical Anthropology* 53:75–93.

Bar-Yosef. O. 2015. "Chinese Palaeolithic Challenges for Interpretations of Palaeolithic Archaeology." *L'Anthropologie* 53(1–2):77–92.

Black, D. 1925. "Asia and the Dispersal of Primates." *Bulletin of the Geological Society of China* 4:133–183.

Black, D. 1927. "Further Hominid Remains of Lower Quaternary Age from the Chou-kou-tien Deposit." *Nature* 120:954.

Black, D. 1934. "The Croonian Lecture—on the Discovery, Morphology, and Environment of *Sinanthropus Pekinensis*." *Philosophical Transactions of the Royal Society of London* 123:57–120.

Boule, M., H. Breuil, E. Licent, and P. Teilhard de Chardin. 1928. "Le Paléolithique de la Chine." *Archives de l'Institut de Paléontologie Humaine (Paris)* 4: 1–138.

Chardin, T. de 1941. "Early Man in China." *Institut de Géo-Biologie, Pékin* 7:1–100.

Darwin, C. 1871. *The Descent of Man and Selection in Relation to Sex*. London: John Murray.

Dennell, R. W. 1990. "Progressive Gradualism, Imperialism, and Academic Fashion: Lower Palaeolithic Archaeology in the Twentieth Century." *Antiquity* 64:549–558.

Dennell, R. W. 2001. "From Sangiran to Olduvai, 1937–1960: the Quest for 'centres' of Hominid Origins in Asia and Africa." In *Studying Human Origins: Disciplinary History and Epistemology*, edited by R. Corbey and W. Roebroeks, 45–66. Amsterdam: Amsterdam University Press.

Dennell, R. W. 2014a. "East Asia and Human Evolution: From Cradle of Mankind to Cul-de-sac." In *Southern Asia, Australia and the Search for Human Origins*, edited by R. W. Dennell and M. Porr, 8–20. Cambridge: Cambridge University Press.

Dennell, R. W. 2014b. "Hallam Movius, Helmut de Terra, and the Line that Never Was: Burma, 1938." In *Living in the Landscape: Essays in Honour of Graeme Barker*, edited by K. Boyle, R. J. Rabett, and C. Hunt, 11–34. Cambridge: McDonald Institute for Archaeological Research.

Dennell, R. W. 2016. "Life Without the Movius Line." *Quaternary International* 400: 14–22.

Dikötter, F. 2015. *The Discourse of Race in Modern China*. London: Hurst and Company.

Engels, F. 1950 (original edition 1876). *The Part Played by Labor in the Transition from Ape to Man*. New York: International Publishers.

Gregory, W. K. 1927. "The Origin of Man from the Anthropoid Stem—When and Where?" *Proceedings of the American Philosophical Society* 66:439–463.

Groves C. P., and M. M. Lahr. 1994. "A Bush Not a Ladder: Speciation and Replacement in Human Evolution." *Perspectives in Human Biology* 4:1–11.

Haddon, A. C. 1912. *The Wandering of Peoples*. Cambridge: Cambridge University Press.

Hammond, M. 1980. "Anthropology as a Weapon of Social Combat in Late Nineteenth Century France." *Journal of the History of the Behavioural Sciences* 16:118–132.

Hammond, M. 1982. "The Expulsion of the Neanderthals from Human Ancestry: Marcellin Boule and the Social Context of Scientific Research." *Social Studies of Science* 12:1–36.

Hao, Q., L. Wang, F. Oldfield, and Z. Guo. 2015. "Extra-Long Interglacial in Northern Hemisphere during MISs 15-13 Arising from Limited Extent of Arctic Ice Sheets in Glacial MIS 14." *Nature Scientific Reports* 5:12103. doi: 10.1038/srep12103.

Howell, F. C., 1999. "Paleo-demes, Species Clades, and Extinctions in the Pleistocene Hominin Record." *Journal of Anthropological Research* 55:191–243.

Howells, W. W., and P. J. Tsuchitani, P. J. (editors). 1977. *Palaeoanthropology in the People's Republic of China*. Washington, DC: National Academy of Sciences.

Jia, Lanpo. 1980. *Early Man in China*. Beijing: Foreign Languages Press.

Jia, Lanpo, and Huang Weiwen. 1990. *The Story of Peking Man: From Archaeology to Mystery*. Beijing: Foreign Languages Press/Hong Kong: Oxford University Press.

Krause, J., Q. Fu, J. M. Good, B. Viola, M. V. Shunkov, A. P. Derevianko, and S. Pääbo. 2010. "The Complete Mitochondrial DNA Genome of an Unknown Hominin from Southern Siberia." *Nature* 464:894–897. doi: 10.1038/nature08976.

Krogman, W. W. 1947. "Review of *Apes, Giants and Man* by F. Weidenreich." *American Anthropologist* 49: 115–118.

Licent, E., and P. Teilhard de Chardin. 1925. "Le paléolithique de la Chine." *L'Anthropologie* 25:201–234.

Liu, W., C. Jin, Y. Zhang, Y. Cai, S. Xing, J. Wu, H. Cheng, R. L. Edwards, W. Pan, D. Qin, Z. An, E. Trinkhaus, and X. Wu. 2010. "Human Remains from Zhirendong, South China, and Modern Human Emergence in East Asia" *Proceedings of the National Academy of Sciences USA* 107:19201–19206.

Liu, W., L. A. Schepartz, X. Song, S. Miller-Antonio, X. Wu, E. Trinkhaus, and M. Martinón-Torres. 2013. "Late Middle Pleistocene Hominin Teeth from Panxian Dadong, South China." *Journal of Human Evolution* 64: 337–355.

Marks, J. 2015. *Tales of the Ex-Apes: How We Think about Human Evolution*. Berkeley: University of California Press.

Matthew, W. D. 1915. "Climate and Evolution." *Annals of the New York Academy of Science* 24:171–318.

Meyer, M., Q. Fu, A. Aximu-Petri, I. Glocke, B. Nickel, J. L. Arsuaga, I. Martínez, A. Gracia, J. M. de Castro, E. Carbonell, and S. Pääbo. 2014. "A Mitochondrial Genome Sequence of a Hominin from Sima de los Huesos." *Nature* 505:403–406. doi: 10.1038/nature12788.

Meyer, M., M. Kircher, M. T. Gansauge, H. Li, F. Racimo, S. Mallick, J. G. Schraiber, F. Jay, K. Prüfer, C. de Filippo, P. H. Sudmant, C. Alkan, Q. Fu, R. Do, N. Rohland, A. Tandon, M. Siebauer, R. E. Green, K. Bryc, A. W. Briggs, U. Stenzel, J. Dabney, J. Shendure, J. Kitzman, M. F. Hammer, M. V. Shunkov, A. P. Derevianko, N. Patterson, A. M. Andrés, E. E. Eichler, M. Slatkin, D. Reich, J. Kelso, and S. Pääbo. 2012. "A High-Coverage Genome Sequence from an Archaic Denisovan Individual." *Science* 338:222–226. doi:10.1126/science.1224344.

Moore, J. H. 1933. *Savage Survivals: The Story of Race Told in Simple Language*. London: Watts and Co.

Movius, H. L. 1948. "The Lower Palaeolithic Cultures of Southern and Eastern Asia." *Transactions of the American Philosophical Society* 38(4):329–420.

Osborne, H. E. 1900. "The Geological and Faunal Relations of Europe and America during the Tertiary Period and the Theory of the Successive Invasions of an African Fauna." *Science* N.S. 11:561–574.

Osborne, H. E. 1918. *Men of the Old Stone Age*. New York: Charles Scribner's Sons.

Osborne, H. E. 1926. "Foreword." In *On the Trail of Ancient Man*, by R. C. Andrews, i–xi. New York: Garden City Publishing Company, Inc.

Osborne, H. E. 1927. "Recent Discoveries Relating to the Origin of Man." *Proceedings of the American Philosophical Society* 66:373–389.

Prüfer, K., F. Racimo, N. Patterson, F. Jay, S. Sankararaman, S. Sawyer, A. Heinze,G. Renaud, P. Sudmant, C. de Filippo, Li, H., S. Mallick, M. Dannemann, Q. Fu, M. Kircher, M. Kuhlwilm, M. Lachmann, M. Meyer, M. Ongyerth, M. Siebauer, C. Theunert, A. Tandon, P. Moorjani, J. Pickrell, J. Mullikin, S. Vohr, R. Green, I. Hellmann, P. Johnson, H. Blanche, H. Cann, J. Kitzman, J. Shendure, E. Eichler, E. Lein, T. Bakken, L. Golovanova, V. Doronichev, M. Shunkov, A. Derevianko, B. Viola, M. Slatkin, D. Reich, J. Kelso and S. Pääbo. 2013. "The Complete Genome Sequence of a Neanderthal from the Altai Mountains." *Nature* 505:-43–49. doi: 10.1038/nature12886.

Pusey, J. R. 1983. *China and Charles Darwin*. Cambridge, MA: Harvard University Press.

Reader, J. 1990. *Missing Links: The Hunt for Ancient Man*. London: Penguin.

Reich, D., R. E. Green, M. Kircher, J. Krause, N. Patterson, E. Y. Durand, B. Viola, A. W. Briggs, U. Stenzel, P. L. Johnson et al. 2010. "Genetic History of an Archaic Hominin Group from Denisova Cave in Siberia." *Nature* 468:1053–1060. doi: 10.1038/nature09710.

Rightmire G. P., 1998. "Human Evolution in the Middle Pleistocene: The Role of *Homo Heidelbergensis*." *Evolutionary Anthropology* 6:218–227.

Rightmire, G. P., 2001. "Comparison of Middle Pleistocene Hominids from Africa and Asia." In *Human Roots: Africa and Asia in the Middle Pleistocene*, edited by L. Barham and K. Robson-Brown, 123–133. Bristol: Western Academic and Specialist Press Ltd.

Sautman, B. 2001. "Peking Man and the Politics of Palaeoanthropological Nationalism in China." *Journal of Asian Studies* 60(1):95–124.

Schmalzer, S. 2008. *The People's Peking Man: Popular Science And Human Identity in Twentieth-Century China*. Chicago: University of Chicago Press.

Shen, C., X. Zhang, and X. Gao.2016. "Zhoukoudian in Transition: Research History, Lithic Technologies, and Transformation of Chinese Palaeolithic Archaeology." *Quaternary International* 400: 4–13.

Song, X., M. Martinón-Torres, J. M. Bermúdez de Castro, X. Wu, and L. Wu. 2015. "Hominin Teeth from the Early Late Pleistocene Site of Xujiayao, Northern China." *American Journal of Physical Anthropology* 156:224–240.

Theunissen, B. 1989. *Eugene Dubois and the Ape-Man from Java*. Dordrecht: Kluwer Academic Publications.

Weidenreich, F. 1946. *Apes, Giants, and Man*. Chicago: University of Chicago Press.

Weiner, S., W. Qinqi, P. Goldberg, J. Liu, and O. Bar-Yosef. 1998. "Evidence for the Use of Fire at Zhoukoudian, China" *Science* 281:251–253.

Woodward, A. S. 1948. *The Earliest Englishman*. London, Watts and Co.

Wu, X., and S. Athreya. 2013. "A Description of the Geological Context, Discrete Traits, and Linear Morphometrics of the Middle Pleistocene Hominin from Dali, Shaanxi Province, China." *American Journal of Physical Anthropology* 150:141–157.

Yen, H.-P. 2014. "Evolutionary Asiacentrism, Peking Man, and the Origins of Sinocentric Ethno-centrism." *Journal of the History of Biology* 47:585–625.

12 Referential Models for the Study of Hominin Evolution: How Many Do We Need?

Gabriele A. Macho

Introduction

Ever since Jolly's influential—albeit controversial—paper on baboon feeding ecology and its potential for assessing hominin niche separation (1970), baboons are regarded as good ecological models for early hominin evolution (Elton 2006). This has been fostered by long-term field studies on all aspects of baboon behavior, life history, and sociality (e.g., Alberts and Altmann 2012), which have highlighted many parallels between baboons and hominins. Notably, both groups are eclectic omnivores (Alberts et al. 2005, Macho 2015a) and live in large groups with complex modular, multilevel societies (Grueter et al. 2012, Swedell and Plummer 2012). With the exception of Japanese macaques, baboons, like hominins, are the only large-bodied primates that live(d) outside the tropics at higher latitudes where environments are harsher and more seasonal. In fact, both lineages have been synchron and sympatric throughout most of their early evolutionary histories, that is, until the emergence of *Homo* and its dispersal out of Africa to even more temperate zones during the Pleistocene. Their co-occurrence implies that competition between both clades was inevitable and almost certainly contributed to each group's evolutionary pathway (Hunt 2016). It is against this background that researchers draw on baboons when reconstructing habitat use by early hominins. Where questions other than ecology are concerned however, for example, those relating to morphological and cognitive evolution, hominins are usually compared with chimpanzees. This is considered justified on grounds that *Pan* is modern human's closest relative (e.g., Scally et al. 2012) and that the former's generalised morphology is likely to resemble that of the Last Common Ancestor (LCA). Recent theoretical and empirical studies, including fossil finds, have called into questions the validity of such propositions.

The Genetic and Fossil Evidence

The last common ancestor of hominins and chimpanzees is estimated to date back to about 7.42 (5.3–10) Ma, while the split of gorillas from the chimpanzee-human clade occurred some 9.18 (6.8–12.2) Ma ago, that is, only marginally earlier. This quick succession of splits, together with different rates of evolution within lineages, gene flow, and recombination, results in incomplete lineage sorting between gorillas, humans, and chimpanzees in about 30 percent of the genome (Scally et al. 2012). This is substantial and indicates that the distinction between taxa is not as clear-cut as commonly assumed. These uncertainties are unlikely to be resolved any time soon, even though the divergence dates continue to be refined through better incorporation of species' generation times (Langergraber et al. 2012) and more accurate sequencing and assembly of genomes (Gordon et al. 2016). The recently discovered *Chororapithecus abyssinicus* (Suwa et al. 2007), thought to be a basal member of the gorilla clade, constrains the divergence of the gorilla-hominin lineage further and places it to between 8.5 Ma and 7 Ma (Katoh et al. 2016). Shortly thereafter, the first putative hominins appear in the fossil record at the end of the Miocene: they are *Sahelanthropus tchadensis* at 6–7 Ma (Brunet et al. 2002) and *Orrorin tugenensis* at 6 Ma (Senut et al. 2001). Due to the fragmentary nature of these early hominin remains, as well as the possible confounding effects of homoplasy, these fossils and their attribution have been found difficult to assess (Wood and Harrison 2011). Interpretation is less problematic for the later *Ardipithecus ramidus* at about 4.4 Ma (White et al. 2009). Most researchers accept the hominin status of this taxon, although some dissenting opinions remain (Wood and Harrison 2011). Either way, the completeness of *A. ramidus* and the preservation of almost all parts of the body unequivocally proves that this Pliocene primate, whether a hominin or a hominid, combined a unique suite of primitive and derived features. Most importantly, *A. ramidus* does not resemble extant chimpanzees in its overall morphology (White et al. 2015). In some aspects, such as canine reduction, cranial base, and some hand and pelvis morphologies, the species is undoubtedly derived towards the modern human condition (or both may be plesiomorphic in some traits?), whereas in other features, like limb proportions and hand and foot morphologies, *A. ramidus* is more primitive than *Pan* (White et al. 2015). This mosaic of traits in an early hominin/hominid raises doubts as to whether chimpanzees are indeed good models for the LCA, and it strengthens arguments for chimpanzees being derived, at least with regard to positional behavior and the postcranial morphologies associated with such behaviors (Crompton 2016; Hunt 2016). This is not necessarily in conflict with genetic evidence. On the contrary, mutation rates in the chimpanzee lineage were found to be high, whereas the human lineage revealed surprisingly low mutation rates in comparison (Backwell and Zhang 2007; Scally and Durbin 2012). Hence, both genetic and morphological evidence refute propositions that chimpanzees have retained the primitive conditions and that this extant primate therefore resembles the ancestral condition of the hominin-panin LCA. This conclusion may spark a search for an alternative model among some paleoanthropologists, but the real significance of this finding lies elsewhere. Specifically, the fact that early hominins, like *A. ramidus*, were primitive relative to *Pan* and were generalized in key aspects of their morphology has implications for an understanding of evolutionary processes and pathways.

Undoubtedly, hominins underwent rapid anatomical modifications during the late Miocene to middle Pliocene. These changes affected key parts of their body, notably those that are associated with substrate use, such as the leg (Senut et al. 2001), the hand (Almécija and Alba 2014), the pelvis (Lovejoy et al. 2009a) and the shoulder girdle (Haile-Selassie et al. 2010). Moreover, hominins also changed their dietary ecology and broadened their dietary niche, which resulted in their orofacial skeletons becoming more robust and their teeth increasing in overall size and in enamel thickness (Teaford and Ungar 2000, Ward et al. 2001). The morphological changes in the cranium and the postcranium occurred more or less in tandem, rather than sequentially, and seem to have been the result of a common underlying cause: climate and habitat changes during the late Miocene to Pliocene (Levin 2015). It is *because* of their unspecialized (primitive) nature that hominins were able to adjust rapidly to the novel, more open environments. Morphological transitions to terrestrial bipedality and dietary adaptations were particularly fast. Although this is also the case for the hominin hand, these changes are seldom viewed as a direct response to more open habitats. Rather, the modern human (hominin) hand is regarded as specialized and capable of a number of unique grips that enabled our lineage to become skilled tool users/manufacturers (Napier 1993). However, it may have initially evolved as a feeding adaptation.

The early hominin hand was generalized with regard to its finger proportions (Almécija and Alba, 2014, Almécija et al. 2015) and primitive in carpo-metacarpal morphologies, that is, less elongated than those of chimpanzees, and capable of a greater range of movements (White et al. 2015). With the discovery of more hominin fossils and the deployment of sophisticated analytical tools, the evolution of the hominin hand is now becoming clearer. Whilst the carpus of *A. ramidus* and *A. anamensis* is primitive and suggests that these species engaged in arboreal behaviors, the carpal joint configurations of only slightly younger australopiths at 3.5 Ma had already changed significantly towards the modern human condition (Leakey et al. 1998, Ward et al. 1999); as a consequence, middle Pliocene hominin hands were probably more versatile and capable of transverse transmission of forces across the carpus (figure 12.1; Macho et al. 2010). Transverse forces are commonly associated with squeeze power grips and are thought to be a consequence—or be indicative—of tool use (Marzke and Marzke 2000, Tocheri et al. 2005). The link between tool use and transverse forces is undeniable, but it is noteworthy that squeeze power grips are also employed when extracting Underground Storage Organs (USOs). For example, baboons in the Drakensberg (South Africa) were observed to grab and pull out bundles of grasses in order to access the below-ground corms (Whiten et al. 1987); these foods are ubiquitous in the seemingly impoverished environments in which hominins lived, and they are high in nutrients and calories (Macho 2014a, 2015a). Such USOs would therefore have been valuable food sources for hominins too. Before consumption USOs need to be cleaned and peeled (Altmann 1988), and this requires (or benefits from) fine motor control of the fingers and the ability of precision grips (of some sort). Anatomically, squeeze power and precision grips are aided by a continuous, more distally orientated MCII articulation at the capitate (figure 12.1A), thus allowing direct load transfer from

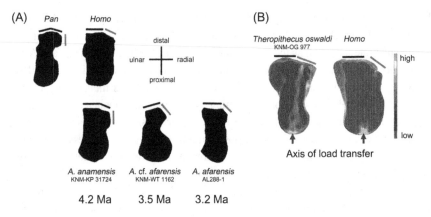

Figure 12.1
Coronal sections through capitates showing the angulation of the MCII and MCIII facets. In (A) a modern chimpanzee and human capitate are contrasted (top row), and the changes from *A. anamensis* to *A. afarensis* are outlined (bottom); all capitates are scaled to the same size. The MCIII orientation is indicated by a black bar, and the MCII facet is shown by a red bar (modified from Leakey et al. 1998). In (B) the results of finite element analyses are shown. The stress flow from the capitate head towards the distal metacarpal facets is illustrated for a fossil *Theropithecus oswaldi* from Olorgesailie (ca. 0.9 Ma) and an extant *Homo sapiens*. For analyses, all finite element models were scaled to the same size, the distal joint surfaces were fixed and a theoretical load of 100N was applied to the capitate head (arrow) in proximo-distal direction. Note the radially directed axis of stress flow in both taxa due to the combined effects of external morphology and trabecular arrangement (modified from Macho et al. 2010).

the radial aspect of the hand through the carpus (figure 12.1B). *Australopithecus anamensis* from Kanapoi at 4.2 Ma (Leakey et al. 1998) still has a radially oriented MCII facet, similar to the condition seen in chimpanzees, whereas later australopiths consistently exhibit distally oriented facets. It seems no coincidence that this change in carpus morphology roughly coincided with a shift towards consumption of C_4 foods (Cerling et al. 2013a), that is, most probably USOs. From a broader evolutionary and ecological, as well as comparative perspective it is thus parsimonious to consider the hominin hand first and foremost to be an adaptation to better extract and process novel food sources (Almécija et al. 2015). If so, the hominin hand would be exapted for tool use, rather than being selected for it. Such a proposition is not unreasonable; the similarities in capitate joint figuration and force transmission between theropiths and modern humans (and hominins; figure 12.1B) are striking, while hominin teeth are indistinguishable from those of theropiths in stable isotope composition (Cerling et al. 2013b). Both taxa were ecological generalists and were likely to have exploited the same resources. Initially, they may have done so in a similar manner. Furthermore, by virtue of being terrestrial primates, which share a plesiomorphic pattern of hand innervation (Hashimoto et al. 2013), greater manual dexterity and larger brains would have been selected for in both lineages (Heldstab et al. 2016), and mediated by ecological demands (Koops et al. 2014). Tool use in a cognitively advanced primate would conceivably be the logical progression of this trend, rather than its cause.

All hominoids use tools (e.g., Kinani and Zimmermann 2015, McGrew 1992). These are mostly made of perishable materials, such as sticks and leaves. Hominins almost certainly used such tools also, although this can never be verified, as such materials do not fossilize. Importantly, these perishable tools do not require complex manipulation for their modification. In fact, the stiff, specialized wrist of hominoids is poorly equipped to carry out complex manipulatory tasks. In the great apes, the wrist is designed to provide stability to the hand during knuckle walking (Dainton and Macho 1999, Kivell and Schmitt 2009), whilst the elongated curved fingers, especially in chimpanzees, are adaptations to arboreal behaviors. This anatomical setup limits the range of movements possible and, consequently, constrains the complex manipulation necessary to produce stone tools (or extract and process USOs). It further follows that differences in behavior are probably constrained by anatomy, rather than cognition, and triggered by the ecological needs imposed by the animal's respective habitat, that is, closed versus open. Unsurprisingly therefore, there is unequivocal evidence for tool use in the hominin fossil record before significant encephalization had taken place; brain sizes in early australopithecines fall within the range of gorillas (Macho 2015b). The first stone tools date to about 3.3 Ma (Harmand et al. 2015) and cut-marks on animal bones at 3.39 Ma (McPherron et al. 2010) indicate an even earlier use of stone tools in the middle Pliocene. Thus, hominin material culture appears to have a long history. Interpretation of this observation needs to be nuanced though. Even at comparable cognitive abilities between chimpanzees and early hominins, the material manifestations of the species' abilities would have been different. This is because of species-specific anatomical constraints, on the one hand, and differences in ecological setting, on the other. To conflate the prerequisites for tool use/manufacture, that is, cognition, morphology, and need (e.g., the food sources on which they depend/ed), will not yield the insights into early hominin behavior that we seek.

This brief synopsis illustrates that the African ape and hominin lineages split in rapid succession of each other, and that genetic relatedness is a poor predictor of phenotypic outcomes. Neither conclusion is new, but the implications for paleoanthropological studies are rarely explored. Mostly, morphology is appraised within the framework of cladistics, that is, shared derived and primitive characters, and homoplasies are considered a nuisance rather than an opportunity to better understand the drivers of convergent morphologies. Yet, it is the interplay between morphological evolution, environment, behavior, and genes that has much to offer for an understanding of evolutionary processes and its ultimate drivers, as outlined for the hominin hand. Disentangling the various influences on evolutionary pathways is difficult and, for early hominins, is hampered by the lack of ancient DNA. Alternatively, a greater appreciation of developmental plasticity and reaction norms in the great apes in response to environmental conditions promises to provide invaluable insights.

The Influence of Ecology on Hominin Evolution

Morphologies are brought about by developmental processes, which, in turn, respond to and create novel environmental conditions (Laland et al. 2015, Müller 2007). In other words, phenotypes are the outcomes of the dynamic, intricate interplay between genes and their surroundings. Recent studies on gorillas, in particular, inform hominin evolution for two reasons: first, when analyzing the binding in human and gorilla of CTCF, a protein essential to vertebrate development, Ong and Corces (2014) and Scally et al. (2012) found approximately 70 percent of gorilla CTCF binding regions to be shared with humans. It is tempting to speculate that this could, in some instances, translate to similarities in reaction norms between the two species, that is, their ability to respond phenotypically to changed environmental conditions without changes in genotype. Second, gorillas exhibit considerable diversity in habitat exploitation as well as in substrate use and are therefore more similar to hominins than are chimpanzees. These facts are worth assessing against propositions that the lineage *Homo* is unique in its ability to plastically alter developmental processes and life histories (Kuzawa and Bragg 2012), a trait, which they assumed we acquired in response to environmental changes and fluctuations during our evolutionary history.

Hominoids evolved in forested environments. Of the African apes, chimpanzees largely retained their preference for closed habitats and they continue to be specialized soft fruit consumers, whereas gorillas are more diverse both with regard to substrate use, that is, arboreality vs. terrestriality, and dietary ecology. Hominins, in contrast, evolved in more open and mosaic habitats, became committed terrestrial bipeds, and broadened their dietary niche beyond what is seen in the great apes. This diversification occurred against the backdrop of substantial abiotic changes and fluctuations during the late Miocene to Pliocene. In Africa, global climatic changes (deMenocal 1995) were amplified by low-latitudinal orbital forcing (Levin 2015) and tectonic uplift (Sepulchre et al. 2006), and were modulated by local basin dynamics (Trauth et al. 2007, 2009, 2010). The combination of these factors influenced local moisture regimes and, ultimately, vegetation structures (Bonnefille 2010), even at small, localized scales (Bonnefille et al. 2004, Magill et al. 2013). The changes in water levels and vegetation cover led to the opening and closing of corridors (Trauth et al. 2010), with consequences for primate population dynamics and vicariance (e.g., Macho 2014b, 2015c). Therefore, and whatever the exact mechanisms may have been, most paleoanthropologists concede that climate variability shaped hominin evolution (i.e., extinction, speciation, and morphology) and selected for hominin developmental and behavioral flexibility, including the evolution of material culture (Potts and Faith 2015) and life history (Kuzawa and Bragg 2012); human life histories uniquely combine high reproductive rates with slow growth rates (Macho 2017). To better appraise the uniqueness (or otherwise) of modern human plasticity and life history, a closer look at *Gorilla* diversification and (paleo)biology is informative.

Genetic evidence indicates that the ancestral western (*Gorilla gorilla*) and eastern (*Gorilla beringei*) gorillas split some 0.9–1.75 Ma ago (Roy et al. 2014, Scally et al. 2012, Thalmann et al. 2011) with gene flow persisting until 200,000 years ago (Roy et al. 2014). The split of the two eastern gorillas, *G. beringei graueri* and *G. beringei beringei*, occurred much more recently and dates back some 10,000 years, with gene flow persisting until at least 5,000 years ago (Roy et al. 2014). This time frame for the split of gorillas is shorter than the average error margins assigned to individual early hominin fossil finds (Behrensmeyer and Reed 2013); time averaging of Plio-Pleistocene fossils within assemblages is often in the order of 100 ka or more. This uncertainty in dating obscures information about when during the precessional cycle the fossils were deposited (Hopley and Maslin 2010). This is problematic, as moisture regimes during the precessional cycle (i.e., at a periodicity of approximately 21 ka) can be so dramatic as to alter the local vegetation structure from closed forests to open savannas (Magill et al. 2013). This makes it difficult, if not impossible, to fully appraise the influence of climate on early hominin evolution and to assess whether differences in morphology between stratigraphic layers, or sites, constitute plastic responses to abiotic conditions, or point towards species differences. Such concerns are clearly warranted when gorillas are used as a yardstick.

Eastern gorilla subspecies/populations became reproductively isolated due to climate and habitat change; the once continuous population now occupies different vegetation zones, which has affected the species' biology. The habitats occupied by *G. b. graueri* are at lower altitudes, while *G. b. beringei* inhabit regions at higher elevations, where shrubs and herbaceous vegetation dominate (Tocheri et al. 2016). Mountain gorillas subsist almost entirely on herbaceous matter, with fruits accounting for less than five percent of the gorilla diet (Watts 1984). Because of their reliance on vegetative matter, which is easily available, life histories of mountain gorillas are accelerated and their brains are small in relation to their bodies (McFarlin et al. 2013). This contrasts markedly with western lowland gorillas, who feed predominantly on ripe fruits like the sympatric chimpanzees (Rogers et al. 2004); these foods are clumped and seasonal. In order to mitigate the risks of starvation, the life history stages of *G. gorilla* are significantly protracted, such that infancy and the juvenile periods are nearly doubled compared to those of *G. beringei* (Breuer et al. 2009). Brain growth seemingly continues until well after infants have been weaned (Macho and Lee-Thorp 2014). The effects of this prolonged developmental period on cognitive abilities have yet to be explored, although it may be no coincidence that tool use in gorilla was first reported for lowland gorillas (Breuer et al. 2005) rather than mountain gorillas, even though the latter have been subject of investigation for decades. Perhaps even more directly relevant for an interpretation of hominin fossil record are the difference in anatomy found between gorilla (sub)species, particularly those that relate to substrate use, for example, the foot (although there are also marked differences in the hand).

Adolph Schultz (1926) was the first to document the remarkable diversity in foot morphology among the African apes and noticed the similarities between modern humans and the terrestrial mountain gorillas (figure 12.2, top drawings). Recent sophisticated analyses of individual tarsals confirmed this diversity in gorilla morphology: uni- and multivariate

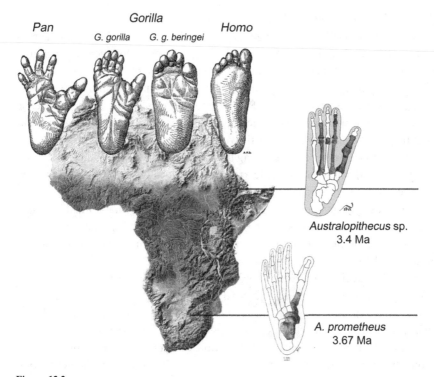

Figure 12.2
Schultz (1927) noted differences in foot morphology between *Pan*, *Gorilla*, and *Homo* (top line drawing). The more arboreal lowland gorillas appear to resemble chimpanzees, whereas the terrestrial mountain gorilla is more similar to modern humans in overall morphology. Gene flow between the gorilla (sub)species persisted until some 200ka. A similar time span separates *Australopithecus* sp. from Ethiopia (Haile-Selassie et al. 2012) from *A. prometheus* from South Africa (Clarke and Tobias 1995, Granger et al. 2015); line drawings on the right were modified from their original publications and mirrored, where necessary, for easy comparison. The fossil taxa are from widely different geographic, climate, and ecological zones.

analyses yielded complete separation not only between (sub)species (Tocheri et al. 2011) but also between populations of eastern lowland gorillas (Dunn et al. 2014). *Gorilla* ability to respond morphologically to different habitats within such a short evolutionary time-frame, that is, in less than 200 ka and 10 ka respectively, is remarkable and raises questions about our ability to appropriately interpret the hominin fossil record (Figure 12.2). Inter-pretations are compounded by time averaging and biogeography alike. Given the distinct habitat zones occupied by early hominins, differences in behavior and morphology are expected. What remains unclear however is the extent to which such morphological dif-ferences reflect species differences. To resolve these issues, a better understanding of geographic barriers and, hence, restricted gene flow (i.e., refugia) will ultimately inform this debate (e.g., Waters et al. 2013). This has yet to be accomplished in paleoanthropol-ogy, although some attempts have already been made, that is, for *A. bahrelghazali* (Macho

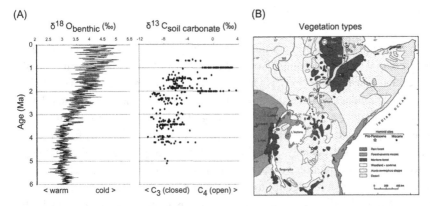

Figure 12.3
(A) Benthic foraminifer oxygen (left) and pedogenic (right) isotopes reveal that the climate over the last 6 myr became generally cooler (left) and more open (right); the pedogenic isotope data for East African hominin sites were selected from a compilation provided by Lister (2013). In (B) the heterogeneity of East African vegetation types is shown (modified from Bonnefille 2010). Both global and local factors will have affected the habitat in which hominins evolved and to which they will have responded both morphologically and behaviorally.

2015c). For Plio-Pleistocene hominins, this will probably be a little less straightforward, however: Hominins are eurybiomic, exhibit great developmental plasticity, and are a young clade that diverged only recently. High levels of reticulation are therefore possible, even likely, that is, intercrossing between lineages with recombination between genes, and hybridization. This situation has parallels in *Papio,* where morphologically and behaviorally different allotaxa are associated with distinct ecotones, yet the possibility of gene transfer persists (Jolly 1993, 2001). This blurs species boundaries, particularly when interpreting extinct taxa. Hence, and using *Gorilla* as a yardstick, it will be difficult to determine whether differences in hominin morphology, for example, the foot, reflect differences in ecology or phylogeny (Figure 12.2, right line drawings; Clarke and Tobias 1995, Haile-Selassie et al. 2012, Lovejoy et al. 2009b). By the same token, there is no foolproof way to assess whether similarities in morphology between fossils from vastly different regions/ecotones, for examle from East and South Africa, are due to homoplasy or shared phylogeny. Either way, the study of morphology without due regard of the environment, (bio)geography and vicariance, (paleo)climatology, and consideration of the effects of developmental plasticity is unlikely to resolve these issues (figure 12.3). It is within this wider context that the study of gorilla evolution, diversity, and developmental plasticity is highly relevant for the paleoanthropologist, while baboons remain good ecological models. Of course, once hominins advanced cognitively, began to actively alter their surroundings through tool-assisted behaviors and culture, and engaged in long-distance travel, many of the feedback mechanisms between morphology, environment, and development would have changed. The interplay between various factors will then have become more complex still.

Conclusions

Hominins are unique in many aspects of their morphology and behavior, and choosing an appropriate referential model for comparison and as proxy for the LCA has been difficult. Recent advances and research outcomes in genetics, developmental biology, and field studies indicate that no such model exists. Furthermore, inferences solely based on genetic relatedness, or on morphology, are likely to be misleading. They do not inform evolutionary processes. Instead, paleoanthropologists must find ways to integrate insights from diverse sources, such as morphology, ecology, (developmental) biology, geology, and climatology in order to decipher macroevolutionary patterns. Although intrinsic factors, such as genes, will influence the dynamics of lineage splitting, extinction, and adaptation, these processes will ultimately be modulated by extrinsic factors, such as climate, resource availability, intrinsic and extrinsic mortality rates, and biogeography. To unravel these interactions is the challenge faced by paleoanthropologists. This requires a multi- and interdisciplinary research protocol and a willingness to draw on different models. In many instances, gorillas (and baboons) may be better proxies for early hominin paleobiology than chimpanzees (save for the large size of gorillas, which imposes constraints on the animals' biology). Far from being generalized, chimpanzees are derived in many aspects of their morphology and habitat use. Their utility as a model species for the LCA of the panin-hominin clade is called into question.

Acknowledgments

I thank Jeffrey Schwartz for inviting me to the workshop entitled "*Is paleoanthropology an evolutionary science? Or, Are analyses of human evolution biological?*" which formed the basis of this contribution, and for overseeing the editing of this book. Gerd Müller and the Konrad Lorenz Institute (KLI) are thanked for hosting the event, whilst Isabella Sarto-Jackson and Eva Lackner superbly organized the workshop.

References

Alberts, S. C., J. A. Hollister-Smith, R. S. Mututua, S. N. Sayialel, P. M. Muruthi, J. K. Warutere, and J. Altmann. 2005. "Seasonality and Long-Term Change in a Savanna Environment." In *Seasonality in Primates,* edited by D. K. Brockmann and C. P. van Schaik, 157–195. Cambridge: Cambridge University Press.

Alberts, S. C., and J. Altmann. 2012. "The Amboseli Baboon Research Project: 40 Years of Continuity and Change." In *Long-Term Field Studies of Primates,* edited by P. M. Kappeler and D. P. Watts, 261–287. Berlin: Springer Verlag.

Altmann, S. A. 1998. *Foraging for Survival.* Chicago: University of Chicago Press.

Almécija, S., and D. M. Alba. 2014. "On Manual Proportions and Pad-to-Pad Precision Grasping in *Australopithecus Afarensis*." *Journal of Human Evolution* 73:88–92.

Almécija, S., J. B. Smears, and W. L. Jungers. 2015. "The Evolution of Human and Ape Hand Proportions." *Nature Communications* 6:7717.

Bakewell, M. A., P. Shi, and J. Zhang. 2007. "More Genes Underwent Positive Selection in Chimpanzee Evolution than in Human Evolution." *Proceedings of the National Academy of Science USA* 104:7489–7494.

Behrensmeyer, A. K., and K. E. Reed. 2013. "Reconstructing the Habitats of *Australopithecus*: Paleoenvironments, Site Taphonomy, and Faunas." In *The Paleobiology of* Australopithecus, edited by K. Reed, J. G. Fleagle, and R. E. Leakey, 41–60. Dordrecht: Springer Verlag.

Bonnefille, R. 2010. "Cenozoic Vegetation, Climate Changes and Hominid Evolution in Tropical Africa." *Global and Planetary Change* 72:390–411.

Bonnefille, R., R. Potts, F. Chalie, D. Jolly, and O. Peyron. 2004. "High-Resolution Vegetation and Climate Change Associated with Pliocene *Australopithecus Afarensis*." *Proceedings of the National Academy of Science USA* 101:12125–12129.

Breuer, T., M. Breuer-Ndoundou Hockemba, C. Olejniczak, R. J. Parnell, and E. J. Stokes. 2009. "Physical Maturation, Life-History Classes and Age Estimates of Free-Ranging Western Gorillas—Insights from Mbeli Bai, Republic of Congo." *American Journal of Primatology* 71:106–119.

Breuer, T., M. Ndoundou-Hockemba, and V. Fishlock. 2005. "First Observation of Tool Use in Wild Gorillas." *PLoS ONE* 3:e380.

Brunet, M., F. Guy, D. Pilbeam, H. T. Mackaye, A. Likius, D. Ahounta, A. Beauvillain, C. Blondel, H. Bocherens, J. R. Boisserie, L. de Bonis, Y. Coppens, J. Dejax, C. Denys, P. Duringer, V. Eisenmann, G. Fanone, P. Fronty, D. Geraads, T. Lehmann, F. Lihoreau, A. Louchart, A. Mahamat, G. Merceron, G. Mouchelin, O. Otero, P. P. Campomanes, M. Ponce de Leon, J. C. Rage, M. Sapanet, M. Schuster, J. Sudre, P. Tassy, X. Valentin, P. Vignaud, L. Viriot, A. Zazzo, and C. Zollikofer. 2002. "A New Hominid from the Upper Miocene of Chad, Central Africa." *Nature* 418:145–151.

Cerling, T. E., K. L. Chritz, N. G. Jablonski, M. G. Leakey, and F. K. Manthi. 2013b. "Diet of *Theropithecus* from 4 to 1 Ma in Kenya." *Proceedings of the National Academy of Science USA* 110:10507–10512.

Cerling, T. E., F. K. Manthi, E. N. Mbua, L. N. Leakey, M. G. Leakey, R. E. Leakey, F. H. Brown, F. E. Grine, J. A. Hart, P. Kaleme, H. Roche, K. T. Uno, and B. A. Wood. 2013a. "Stable Isotope-Based Diet Reconstructions of Turkana Basin Hominins." *Proceedings of the National Academy of Science USA* 110:10501–10506.

Clarke, R. J., and P. V. Tobias. 1995. "Sterkfontein Member 2 Foot Bones of the Oldest South African Hominid." *Science* 269:521–524.

Crompton, P. H. 2016. "The Hominins: A Very Conservative Tribe? Last Common Ancestors, Plasticity and Ecomorphology in Hominidae. Or, What's in a Name?" *Journal of Anatomy* 228:686–699.

Dainton, M., and G. A. Macho. 1999. "Did Knuckle Walking Evolve Twice?" *Journal of Human Evolution* 36:171–194.

deMenocal, P. B. 1995. "Plio-Pleistocene African Climate." *Science* 270:53–59.

Dunn, R. H., M. W. Tocheri, C. M. Orr, and W. L. Jungers. 2014. "Ecological Divergence and Talar Morphology in Gorillas." *American Journal of Physical Anthropology* 153:526–541.

Elton, S. 2006. "Forty Years on and Still Going Strong: The Use of Hominin-Cercopithecid Comparisons in Palaeoanthropology." *Journal of the Royal Anthropological Institute* 12:19–38.

Gordon, D., J. Huddleston, M. J. P. Chaisson, C. M. Hill, Z. N. Kronenberg, K. M. Munson, M. Malig, A. Raja, O. Fiddes, L. W. Hillier, C. Dunn, C. Baker, J. Armstrong, M. Dickhans, B. Paten, J. Shendure, R. K. Wilson, D. Haussler, C. S. Chin, and E. E. Eichler. 2016. "Long-Read Sequence Assembly of the Gorilla Genome." *Nature* 352:aae0344.

Granger, D. E., R. J. Gibbon, K. Kuman, R. J. Clarke, L. Bruxelles, and M. W. Caffee. 2015. "New Cosmogenic Burial Ages for Sterkfontein Member 2 *Australopithecus* and Member 5 Oldowan." *Nature* 522:85–88.

Grueter, C. C., B. Chapais, and D. Zinner. 2012. "Evolution of Multilevel Social Systems in Nonhuman Primates and Humans." *International Journal of Primatology* 33:1002–1037.

Haile-Selassie, Y., B. M. Latimer, M. Alene, A. L. Deino, L. Gilbert, S. M. Melillo, B. Z. Saylor, G. R. Scott, and C. O. Lovejoy. 2010. "An early *Australopithecus Afarensis* Postcranium from Woranso-Mille, Ethiopia." *Proceedings of the National Academy of Science USA* 107:12121–12126.

Haile-Selassie, Y., B. Z. Saylor, A. Deino, N. E. Levin, M. Alene, and B. M. Latimer. 2012. "A New Hominin Foot from Ethiopia Shows Multiple Pliocene Bipedal Adaptations." *Nature* 483:565–570.

Harmand, S., J. E. Lewis, C. S. Feibel, C. J. Lepre, S. Prat, A. Lenoble, X. Boës, R. L. Quinn, M. Brenet, A. Arroyo, N. Taylor, S. Clément, G. Daver, J. P. Brugal, L. Leakey, R. A. Mortlock, J. D. Wright, S. Lokorodi, C. Kirwa, D. V. Kent, and H. Roche. 2015. "3.3-Million-Year-Old Stone Tools from Lomekwi 3, West Turkana, Kenya." *Nature* 521:310–315.

Hashimoto, T., K. Ueno, A. Ogawa, T. Asamizuya, C. Suzuki, K. Cheng, M. Tanaka, M. Taoka, Y. Iwamura, G. Suwa, and A, Iriki. 2013. "Hand before Foot? Cortical Somatotopy Suggests Manual Dexterity Is Primitive and Evolved Independently of Bipedalism." *Philosophical Transactions of the Royal Society B* 368:20120417.

Heldstab, S. A., Z. K. Kosonen, S. E. Koski, J. M. Burkart, C. P. van Schaik, and K. Isler. 2016. "Manipulation Complexity in Primates Coevolved with Brain Size and Terrestriality." *Nature Scientific Reports* 6:24528.

Hopley, P. J., and M. A. Maslin. 2010. "Climate-Averaging of Terrestrial Faunas: An Example from the Plio-Pleistocene of South Africa." *Palaeobiology* 36:32–50.

Hunt, K. D. 2016. "Why Are There Apes? Evidence for the Co-evolution of Ape and Monkey Ecomorphology." *Journal of Anatomy* 228:630–685.

Jolly, C. J. 1970. "The Seed-Eaters: A New Model of Hominid Differentiation Based on a Baboon Analogy." *Man* 5:5–26.

Jolly, C. J. 1993. "Species, Subspecies, and Baboon Systematics." In *Species, Species Concepts, and Primate Evolution*, edited by W. H. Kimbel and L. B. Martin, 67–107. New York: Plenum Press.

Jolly, C. J., 2001. "A Proper Study for Mankind: Analogies from the Papionin Monkeys and Their Implications for Human Evolution." *Yearbook of Physical Anthropology* 44:177–204.

Katoh, S., Y. Beyene, T. Itaya, H. Hyodo, M. Hyodo, K. Yagi, C. Gouzu, G. WoldeGabriel, W. K. Hart, S. H. Ambrose, H. Nakaya, R. L. Bernor, J. R. Boisserie, F. Bibi, H. Saegusa, T. Sasaki, K. Sano, B. Asfaw, and G. Suwa. 2016. "New Geological and Palaeontological Age Constraint for the Gorilla-Human Lineage Split." *Nature* 530:215–220.

Kinani, J. F., and D. Zimmermann. 2015. "Tool Use for Food Acquisition in a Wild Mountain Gorilla (*Gorilla Beringei Beringei*)." *American Journal of Primatology* 77:353–357.

Kivell, T. L., and D. Schmitt. 2009. "Independent Evolution of Knuckle-Walking in African Apes Shows that Humans Did Not Evolve from a Knuckle-Walking Ancestor." *Proceedings of the National Academy of Science USA* 106:14241–14246.

Koops, K., E. Visalberghi, and C. P. van Schaik. 2014. "The Ecology of Primate Material Culture." *Biology Letters* 10:20140508.

Kuzawa, C. W., and J. M. Bragg. 2012. "Plasticity in Human Life History Strategy: Implications for Contemporary Human Variation and the Evolution of the Genus *Homo*." *Current Anthropology* 53:S369–S382.

Laland, K. N., T. Uller, M. W. Feldman, K. Sterelny, G. B. Müller, A. Moczek, E. Jablonka, and J. Odling-Smee. 2015. "The Extended Evolutionary Synthesis: Its Structure, Assumptions and Predictions." *Proceedings of the Royal Society B* 282:20151019.

Langergraber, K. E., K. Prüfer, C. Rowney, C. Boesch, C. Crockford, K. Fawcett, E. Inoue, M. Inoue-Muruyama, J. C. Mitani, M. N. Muller, M. M. Robbins, G. Schubert, T. S. Stoinski, B. Viola, D. Watts, R. M. Wittig, R. W. Wrangham, K. Zuberbühler, S. Pääbo, and L. Vigilant. 2012. "Generation Times in Wild Chimpanzees and Gorillas Suggest Earlier Divergence Times in Great Ape and Human Evolution." *Proceedings of the National Academy of Science USA* 109:15716–15721.

Leakey, M. G., C. S. Feibel, I. McDougall, C. Ward, and A. Walker. 1998. "New Specimens and Confirmation of an Early Age for *Australopithecus Anamensis.*" *Nature* 393:62–66.

Levin, N. E. 2015. "Environment and Climate of Early Human Evolution." *Annual Review of Earth and Planetary Sciences* 43:405–429.

Lister, A. M. 2013. "The Role of Behaviour in Adaptive Morphological Evolution of African Proboscideans." *Nature* 500:331–334.

Lovejoy, C. O., B. Latimer, G. Suwa, B. Asfaw, and T. D. White. 2009a. "The Foot of *Ardipithecus Ramidus.*" *Science* 326:72–72e8.

Lovejoy, C. O., G. Suwa, L. Spurlock, B. Asfaw, and T. D. White. 2009b. "The Pelvis and Femur of *Ardipithecus Ramidus*: The Emergence of Upright Walking." *Science* 326:71–71e6.

Macho, G. A. 2014a. "Baboon Feeding Ecology Informs the Dietary Niche of *Paranthropus boisei.*" *PLoS ONE* 9:e84942.

Macho, G. A. 2014b. "An Ecological and Behavioural Approach to Hominin Evolution during the Pliocene." *Quaternary Science Reviews* 96:23–31.

Macho, G. A. 2015a. "Can Extant Primates Serve as Models to Determine the Dietary Ecology of Hominins? The Case of Paranthropines." *Trends in Biological Anthropology* 1:1–10.

Macho, G. A. 2015b. "Causes, Mechanisms and Consequences of Early Hominin Encephalisation: A Hypothesis." *Human Evolution* 30:69–83.

Macho, G. A. 2015c. "Pliocene Hominin Biogeography and Ecology." *Journal of Human Evolution* 87:78–86.

Macho, G. A. 2017. "From Rainforests to Savannas and Back: The Impact of Abiotic Factors on Non-human Primate and Hominin Life Histories." *Quaternary International,* 448:5–13.

Macho, G. A., and J. A. Lee-Thorp. 2014. "Niche Partitioning in Sympatric *Gorilla* and *Pan* from Cameroon: Implications for Life History Strategies and for Reconstructing the Evolution of Hominin Life Histories." *PLoS ONE* 9:e102794.

Macho, G. A., I. R. Spears, M. G. Leakey, D. J. McColl, Y. Jiang, R. Abel, M. Nakatsukasa, Y. Kunimatsu. 2010. "An Exploratory Study on the Combined Effects of External and Internal Morphology on Load Dissipation in Primate Capitates: Its Potential for an Understanding of the Positional and Locomotor Repertoire of Early Hominins." *Folia Primatologica* 81:292–304.

Magill, C. R., G. M. Ashley, and K. H. Freeman. 2013. "Ecosystem Variability and Early Human Habitats in Eastern Africa." *Proceedings of the National Academy of Science USA* 110:1167–1174.

Marzke, M. W., and R. F. Marzke. 2000. "Evolution of the Human Hand: Approaches to Acquiring, Analysing and Interpreting the Anatomical Evidence." *Journal of Anatomy* 197:121–140.

McFarlin, S. C., S. K. Barks, M. W. Tocheri, J. S. Massey, A. B. Eriksen, K. A. Fawcett, T. S. Stoinski, P. R. Hof, T. G. Bromage, A. Mudakikwa, M. R. Cranfield, and C. C. Sherwood. 2013. "Early Brain Growth Cessation in Wild Virunga Mountain Gorillas (*Gorilla Beringei Beringei*)." *American Journal of Primatology* 75:450–463.

McGrew, W. C. 1992. *Chimpanzee Material Culture. Implications for Human Evolution.* Cambridge: Cambridge University Press.

McPherron, S. P., Z. Alemseged, C. W. Marean, J. G. Wynn, D. Reed, D. Geraads, R. Bobe, and H. A. Béarat. 2010. "Evidence for Stone-Tool-Assisted Consumption of Animal Tissues before 3.39 Million Years Ago at Dikika, Ethiopia." *Nature* 466:857–860.

Müller, G. B. 2007. "Evo-Devo: Extending the Evolutionary Synthesis." *Nature Reviews Genetics* 8:943–949.

Napier, J. 1993. *Hands*. Revised Edition. Princeton: Princeton University Press.

Ong, C. T., and V. G. Corces. 2014. "CTCF: An Architectural Protein Bridging Genome Topology and Function." *Nature Reviews Genetics* 15:235–246.

Potts, R., and J. T. Faith. 2015. "Alternating High and Low Climate Variability: The Context of Natural Selection and Speciation in Plio-Pleistocene Hominin Evolution." *Journal of Human Evolution* 87:5–20.

Rogers, M. E., K. Abernethy, M. Bermejo, C. Cipolletta, D. Doran, K. McFarland, T. Nishihara, M. Remis, and C. E. G. Tutin. 2004. "Western Gorilla Diet: A Synthesis from Six Sites." *American Journal of Primatology* 64:173–192.

Roy, J., M. Arandjelovic, B. J. Bradley, K. Guschanski, C. R. Stephens, D. Bucknell, H. Cirhuza, C. Kusamba, J.C. Kyungu, V. Smith, M. M. Robbins, and L. Vigilant. 2014. "Recent Divergences and Size Decreases of Eastern Gorilla Populations." *Biology Letters* 10:20140811.

Scally, A., and R. Durbin. 2012. "Revising the Human Mutation Rate: Implications for Understanding Human Evolution." *Nature Reviews Genetics* 13:745–753.

Scally, A., J. Y. Dutheil, L. W. Hillier, G. E. Jordan, I. Goodhead, J. Herrero, A. Hobolth, T. Lappalainen, T. Mailund, T. Marques-Bonet, S. McCarthy, S. H. Montgomery, P. C. Schwalie, A. Tang, M. C. Ward, Y. Xue, B. Yngvadottir, C. Alkan, L. N. Andersen, Q. Ayub, E. V. Ball, K. Beal, B. J. Bradley, Y. Chen, C. M. Clee, S. Fitzgerald, T. A. Graves, Y. Gu, P. Heath, A. Heger, E. Karakoc, A. Kolb-Kokocinski, G. K. Laird, G. Lunter, S. Meader, M. Mort, J. C. Mullikin, K. Munch, T. D. O'Connor, A. D. Phillips, J. Prado-Marinez, A. S. Rogers, S. Sajjadian, D. Schmidt, K. Shaw, J. T. Simpson, P. D. Stenson, D. J. Turner, L. Vigilant, A. J. Vilella, W. Whitener, B. Zhu, D. N. Cooper, P. de Jong, E. T. Dermitzakis, E. E. Eichler, P. Flicek, N. Goldman, N. I. Mundy, Z. Ning, D. T. Odom, C. P. Ponting, M. A. Quail, O. A. Ryder, S. M. Searle, W. C. Warren, R. K. Wilson, M. H. Schierup, J. Rogers, C. Tyler-Smith, and R. Durbin. 2012. "Insights into Hominid Evolution from the Gorilla Genome Sequence." *Nature* 483:169–175.

Schultz, A. H. 1926. "Fetal Growth of Man and Other Primates." *Quarterly Review of Biology* 1:465–521.

Schultz, A. H. 1927. "Studies of the Growth of Gorilla and Other Higher Primates with Special Reference to a Fetus or Gorilla, Preserved in the Carnegie Museum." *Memoirs Carnegie Museum* 11:1–86.

Senut, B., M. Pickford, D. Gommery, P. Mein, K. Cheboi, and Y. Coppens Y. 2001. "First Hominid from the Miocene (Lukeino Formation, Kenya)." *Comptes rendus de l'Académie des sciences* 332:137–144.

Sepulchre, P., G. Ramstein, F. Fluteau, M. Schuster, J.J. Tiercelin, and M. Brunet M. 2006. "Tectonic Uplift and Eastern Africa Aridification." *Science* 313:1419–1423.

Suwa, G., R.Y. Kono, S. Katho, B. Asfaw, and Y. Beyene. 2007. "A New Species of Great Ape from the Late Miocene Epoch in Ethiopia." *Nature* 448:921–924.

Swedell, L. and T. Plummer. 2012. "A Papionin Multilevel Society as a Model for Hominin Social Evolution." *International Journal of Primatology* 33:1165–1193.

Thalmann, O., D. Wegmann, M. Spitzner, M. Arandjelovic, K. Guschanski, C. Leuenberger, R.A. Bergl, and L. Vigilant. 2011. "Historical Sampling Reveals Dramatic Demographic Changes in Western Gorilla Populations." *BMC Evolutionary Biology* 11:85.

Teaford, M.F., and P.S. Ungar. 2000. "Diet and the Evolution of the Earliest Human Ancestors." *Proceedings of the National Academy of Science USA* 97:13506–13511.

Tocheri, M.W., A. Razdan, R.C. Williams, and M.W. Marzke. 2005. "A 3D Quantitative Comparison of Trapezium and Trapezoid Relative Articular and Nonarticular Surface Areas in Modern Humans and Great Apes." *Journal of Human Evolution* 49:570–586.

Tocheri, M.W., C.R. Solhan, C.M. Orr, J. Femiani, B. Frohlich,C.P. Groves, W.E. Harcourt-Smith, B.G. Richmond, B. Shoelson, and W.L. Jungers. 2011. "Ecological Divergence and Medial Cuneiform Morphology in Gorillas." *Journal of Human Evolution* 60:171–184.

Tocheri, M.W., R. Dommain, S.C. McFarlin, S.E. Burnett, D.T. Case, C.M. Orr, N.T. Roach, B. Villmoare, A.B. Eriksen, D.C. Kalthoff, S. Senck, Z. Assefa, C.P. Groves, and W.L. Jungers. 2016. "The Evolutionary Origin and Population History of the Grauer Gorilla." *Yearbook of Physical Anthropology* 159:S4–S18.

Trauth, M. H., J. C. Larrasoaña, and M. Mudelsee. 2009. "Trends, Rhythms and Events in Plio-Pleistocene African Climate." *Quaternary Science Reviews* 28:399–411.

Trauth, M. H., M. A. Maslin, A. L. Deino, A. Junginger, M. Lesoloyia, E. O. Odada, D. O. Olago, L. A. Olaka, M. R. Strecker, and R. Tiedemann. 2010. "Human Evolution in a Variable Environment: The Amplifier Lakes of Eastern Africa." *Quaternary Science Reviews* 29:2981–2988.

Trauth, M. H., M. A. Maslin, A. L. Deino, M. R. Strecker, A. G. N. Bergner, and M. Duhnforth. 2007. "High- and Low-Latitude Forcing of Plio-Pleistocene East African Climate and Human Evolution." *Journal of Human Evolution* 53:475–486.

Ward, C. V., M. G. Leakey, B. Brown, F. Brown, J. Harris, and A. Walker.1999. "South Turkwel: A New Pliocene Hominid Site in Kenya." *Journal of Human Evolution* 36:69–95.

Ward, C. V., M. G. Leakey, and A. Walker. 2001. "Morphology of *Australopithecus Anamensis* from Kanapoi and Allia Bay, Kenya." *Journal of Human Evolution* 41:255–368.

Waters, J. M., C. I. Fraser, and G. M. Hewitt. 2013. "Founder Takes All: Density-Dependent Processes Structure Biodiversity." *Trends in Ecology and Evolution* 28:78–85.

Watts, D. P. 1984. "Composition and Variability of Mountain Gorilla Diets in the Central Virungas." *American Journal of Primatology* 7:323–56.

White, T. D., B. Asfaw, Y. Beyene, Y. Haile-Selassie, O. Lovejoy, G. Suwa, and G. WoldeGabriel. 2009. "*Ardipithecus Ramidus* and the Paleobiology of Early Hominids." *Science* 326:64–86.

White, T. D., C. O. Lovejoy, B. Asfaw, J. P. Carlson, and G. Suwa. 2015. "Neither Chimpanzee nor Human, *Ardipithecus* Reveals the Surprising Ancestry of Both." *Proceedings of the National Academy of Science USA* 112:4877–4884.

Whiten, A., R. W. Byrne, and S. P. Henzi. 1987. "The Behavioural Ecology of Mountain Baboons." *International Journal of Primatology* 8:367–388.

Wood, B., and T. Harrison. 2011. "The Evolutionary Context of the First Hominins." *Nature* 470:348–352.

13 Archeological Sites from 2.6–2.0 Ma: Toward a Deeper Understanding of the Early Oldowan

Thomas W. Plummer and Emma M. Finestone

Introduction

By 2.6 million years ago (Ma), hominins in East Africa were striking stones held in their hands to knock off sharp shards of rock, forming concentrations of artifacts and flaking debris dense enough to be readily visible in the geological record. This early technology is referred to as the Oldowan Industry within the Oldowan Industrial Complex (type site Olduvai Gorge; Leakey 1971). The term Oldowan is applied to this ancient technology, to the artifacts produced by these technological practices, and to the sites where these artifacts are found. Here we also use this term in referring to the hominin producers of artifacts (i.e., Oldowan hominins), as more than one taxon is likely responsible for Oldowan artifact production.

While it is clear that the hominins forming Oldowan sites had a good sense of the geometry and force necessary to successfully produce flakes, and often selectively utilized stone raw material based on its physical property, many questions remain about the uses the artifacts were put to, and the adaptive significance of early Oldowan tools. Here we review the record of the earliest Oldowan sites, known from roughly 2.6 to 2.0 million years ago. During this time interval there is a shift from frequently low-density artifact scatters in a narrow set of depositional contexts to more dense concentrations of archeological material, including abundant fossils in a broader array of habitat settings. The suite of behaviors leading to Oldowan site formation, and their place within the daily and seasonal rhythms of hominin life, are very poorly understood. We discuss the conclusions that can reasonably be drawn about the behavior and ecology of the hominins forming the earliest Oldowan sites from the data at hand, and then consider the research needed to obtain a richer understanding of early Oldowan technology.

The Earliest Lithic Technologies: Timing and General Characteristics

Humans are unique among primates in the extent to which technology has figured in niche construction. Instead of filling a narrow ecological role, humans occupy an exceptionally broad and flexible niche, capable of altering behavioral strategies in response to changes in the environment and exploiting a variety of diverse resources for tools and material culture. These technologies can accumulate over time, altering landscapes and shaping the selective pressures within them. From an evolutionary perspective, there is little doubt that cumulative technology, niche construction, and behavioral flexibility have co-directed human evolution (Laland et al. 2010). The development and transmission of technology in nonhuman primates, as well as the appearance of subsistence-related technology in the archeological record, are therefore of inherent interest. Because of preservation bias against non-lithic raw materials, the oldest technologies are recognized exclusively from hominin modification of stones.

The oldest evidence for stone tool manufacture comes from the 3.3 Ma site of Lomekwi 3, on the western shore of Lake Turkana, Kenya (Harmand et al. 2015, Hovers 2015). The context and behaviors associated with this early industry, named the Lomekwian, are still under investigation, but it is clear that flake production was generally carried out by hitting a stone against stationary anvils on the ground. How well established stone tool use was in the late Pliocene, what tasks the tools were used for, and the identity of the hominin (living humans and related fossil species) taxon or taxa that produced the earliest stone artifacts all require further investigation. One possible use of the tools was for animal butchery. While stone tool–modified bones have yet to be found associated with Lomekwian artifacts, linear marks on roughly coeval (3.39 Ma) surface fossils from Dikika, Ethiopia, may have been produced by hominins cutting meat off of bones using stone tools (McPherron et al. 2010, 2011, though see Dominguez-Rodrigo et al. 2010, 2012 for a contrary interpretation).

Beginning by about 2.6 Ma, artifacts referred to the Oldowan Industry appear in the geological record of East Africa (Leakey 1971, Plummer 2004, Schick and Toth 2006, Rogers and Semaw 2009). In contrast to the Lomekwian, Oldowan artifact production was frequently carried out with stones held in both hands, using a method termed hard-hammer percussion. One stone, often a rounded "hammerstone," was used to hit another stone held in a fixed position in the opposite hand (a "core" or "flaked piece") in order to knock off "flakes," or "detached pieces." Hominins at most Oldowan sites were selective in the raw materials they transported and used to make tools, suggesting at least a nominal appreciation for the material properties of different stone lithologies. Animal carcasses were also transported by at least 2 Ma. It is the transport of materials, both stones and foodstuffs, that is one of the novelties of Oldowan hominin behavior (Potts 1991, Plummer 2004).

There is no clear last appearance date for the Oldowan, as the basic practices used to produce Oldowan tools continue on through space and time in the Early Stone Age, often alongside assemblages containing type fossils for other archeological industries. Here we are restricting the term "Oldowan" to sites that precede the appearance of the large bifacial tools of the Acheulean Industrial Complex by 1.7 Ma in East Africa (Lepre et al. 2011, Beyene et al. 2013).

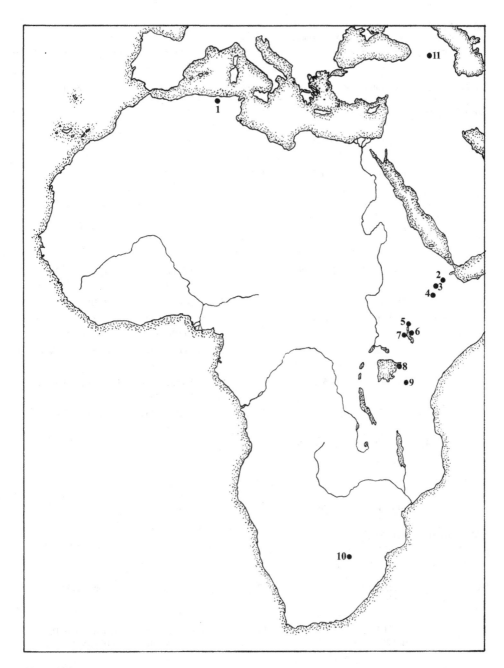

Figure 13.1
Location of sites mentioned in text. Key to sites: 1=Ain Hanech, 2=Hadar, 3=Gona, 4=Middle Awash, 5=Omo, 6=FwJj20, 7=West Turkana, 8=Kanjera South, 9=Olduvai Gorge, 10=Sterkfontein, 11=Dmanisi. After Plummer (2004), figure 1.

Table 13.1
Major Oldowan Occurrences from 2.6–2.0 Ma.

Locality	Excavation	Age (my)	Excavation Size (m²)	Number of Excavated Artifacts	Raw Material
Gona, Ethiopia	Kada Gona 2-3-4	2.6–2.5	NR	21	B, T
	WG-1	2.5–2.3	10	19	B, T
	EG-10	2.6–2.55	13	667	T > 70%
	EG-12	2.6–2.55	9	444	T>70%, R, B
	EG-13	2.6–2.55	2	Surface: 152 In situ: 27	T, R, La, AL, B, QL, VV
	EG-24	2.6–2.55	Not described	n/a	T, R, B?
	DAN-1	2.6–2.5	2.8	Surface: 67 In situ: 45	T, R, AL, La, VV, QL, B
	DAN-2d	2.58–2.27	10	Surface: 24 In situ: 36	T, R, AL, La, VV, QL, B
	OGS-6a	2.6–2.55	4	Surface: 48 In situ: 52	T, AL, R, B, La, VV, QL
	OGS-7	2.6–2.55	2.6	Surface: 65 In situ: 188	R, AL, La, QL, VV, T, B
Hadar Ethiopia	AL 666	2.36–2.33	20	224	R, Q, B, T, C
	AL894	2.36	21.5	4749	R, B, T
Middle Awash, Ethiopia	Hata Mbr, Bouri Fm	2.5	NR	0	n/a
West Turkana, Kenya	Lokalalei 1 (GaJh5)	2.34	67	417	Predominantly Lava
	Lokalalei 2C (LA2C)	2.34	17	492 surface, 2122 in situ	Predominantly B, P (10 types)
Omo Shungura Fm, Mbr F, Ethiopia	FtJi 1	2.34–2.3 Ma	18	367	Q, C, L

Excavated Vertebrate Fossils	Geomorphological and Paleoenvironmental settings	Representative References
No	Streambank or adjacent floodplain	Roche 1996; Roche and Tiercelin 1980
5	Streambank or adjacent floodplain with seasonal flooding	Harris, 1983; Harris and Capaldo 1993
No	Floodplains close to stream channel margins; artifacts occurred in 2 discrete 10-cm-thick levels separated by 40 cm of sterile sediment, within a clay	Semaw et al. 1997; Semaw 2000; Stout et al. 2010
No	Floodplains close to stream channel margins; artifacts occurred in single 40-cm-thick layer in a blocky clay	Semaw et al. 1997; Semaw 2000; Stout et al. 2010
27, 1 cutmarked surface rib	Proximal floodplain of paleo-Awash	Stout et al. 2005; Rogers and Semaw, 2009
No	Floodplain of paleo-Awash	Quade et al. 2004
No	Proximal floodplain of paleo-Awash	Stout et al. 2005; Rogers and Semaw, 2009
>100, 5 cutmarked surface specimens	Proximal floodplain of paleo-Awash	Stout et al. 2005; Rogers and Semaw, 2009
>50, 1 cutmarked surface specimen	Floodplain of paleo-Awash	Semaw et al. 2003; Dominguez-Rodrigo et al. 2005
fossils present	Paleo-Awash channel bank; artifacts from <10 cm thick layer	Semaw et al. 2003; Stout et al. 2010
3	Proximal floodplain, paleo-Awash River. Main level 30–35 cm thick. Predominantly open, with wetlands and bushed or wooded grasslands and with trees close to a water source	Kimbel et al. 1996; Goldman-Neuman and Hovers 2012; Hovers, 2009, 2012
330	Proximal floodplain, paleo-Awash River. Densest concentration in 30-cm layer.	Goldman-Neuman and Hovers 2012; Hovers, 2009, 2012
Approx. 6 surface and in situ fossils with possible stone tool damage	Broad, grassy, featureless margin of a shallow freshwater lake	de Heinzelin et al. 1999
>3415, some with possible stone tool damage	Near intersection of ephemeral basin-margin streams and meandering, axial, ancestral Omo river.	Kibunjia 1994; Harmand 2009a,b
239, surface and in situ	Near intersection of ephemeral basin-margin streams and meandering, axial, ancestral Omo river. Open environment on alluvial plain, with patches of bushes or forest along ephemeral river	Roche et al. 1999; Delagnes and Roche 2005; Harmand 2009a,b
Present, derived context	Deposited in braided stream system	Howell et al. 1987; Merrick and Merrick 1976

(continued)

Table 13.1 (*continued*)

Locality	Excavation	Age (my)	Excavation Size (m²)	Number of Excavated Artifacts	Raw Material
	FtJi 2	2.34–2.3 Ma	22	224	Q
	FtJi 5	2.34–2.3 Ma	8	24	Q
	Omo 57	2.34–2.3 Ma	NR	34	Q, C, L
	Omo 123	2.34–2.3 Ma	NR	730	Q, C, L
Kanjera Fm (S), Kenya	Excavation 1	ca. 2.0 Ma	175	>3100 (under analysis)	Still under analysis, but includes A, B, C, I, Li, J, M, N, P, Q, Qt, R, S
Il Dura region, Koobi Fora Fm	FwJj20	1.95 Ma	130	2,633	B, C, J, Q

Artifact lithology abbreviations as follows: A, andesite; AL, aphanitic lava; B, basalt; C, chert; F, felsite; Fen, fenites; I, ijolite; J, jasper; L, lava unspecified; La, latite; Li, limestone; M, microgranite; N, nephelinite; P, phonolite; Q, quartz; QL, quartz latite; Qt, quartzite; R, rhyolite; S, sandstone; T, trachyte; VV, vitreous volcanic.
Table after Potts (1991, p. 156–157), Plummer (2004, p. 120–121), and Rogers and Semaw (2009, p. 157–160).

Oldowan occurrences are best known from East Africa, and sites in the 2.6–2.0 Ma range are found exclusively in this region (table 13.1, figure 13.1). Important localities from this time interval include Gona, Hadar, and the lower Omo River Valley in Ethiopia, and West Turkana, Koobi Fora, and Kanjera South in Kenya. Whether the inception and earliest usage of Oldowan tools were restricted to East Africa, or whether behaviors forming Oldowan sites were more broadly distributed across Africa, is at this point unclear. By 2.0–1.8 Ma sites attributed to the Oldowan are found in North (e.g., Ain Hanech, Algeria), South (e.g., Sterkfontein, South Africa), and East (e.g., Olduvai Gorge, Tanzania) Africa (Plummer 2004). The recovery of a comparable technology to the Oldowan at Dmanisi, Georgia, at over 1.8 Ma suggests that the earliest travelers out of Africa brought the Oldowan tool kit with them (Gabunia et al. 2001).

The Oldowan archeological record appears and spreads in the context of global and regional environmental shifts and the appearance and turnover of multiple hominin taxa. During the late Pliocene, global cooling and drying, increased climatic variability, and regional tectonic uplift changed African environments on continental to local scales (Potts

Excavated Vertebrate Fossils	Geomorphological and Paleoenvironmental settings	Representative References
0	Overbank deposits of a meandering stream system near interface between gallery forest lining the proto-Omo channel and grasslands that lay beyond the forest	Howell et al. 1987; Merrick and Merrick 1976
Present, derived context	Deposited in braided stream system	Howell et al. 1987; Merrick and Merrick 1976
Present	Deposited in braided stream system	Chavaillon 1976; Howell et al. 1987; de la Torre 2004
0	Overbank deposits of a meandering stream system near interface between gallery forest lining the proto-Omo channel and grasslands that lay beyond the forest	Chavaillon 1976; Howell et al. 1987; de la Torre 2004
>3600 (under analysis), surface and in situ bones cutmarked and percussed; persistent carnivory	Three-meter-thick sequence of sandy silts, pebbly sands, and diffuse conglomerate lenses. Sites with alluvial and fluvial deposition, wooded grassland to open grassland	Plummer et al. 1999; Plummer 2004; Plummer 2009a,b; Braun et al. 2008a,b; Braun et al. 2009a,b; Ferraro et al. 2013; Plummer and Bishop 2016
3648, surface and in situ bones some cutmarked and percussed, includes aquatic fauna	Incipient soil formed on deltaic floodplain, archaeological horizon 15 cm thick; formed in riparian gallery forest or woodland	Braun et al. 2010; Bamford 2011; Archer et al. 2014

1998, 2012, deMenocal 2004, Kingston 2007, Anton et al. 2014, Cerling et al. 2015). The relatively wooded habitats characteristic of the early and middle Pliocene were replaced by more complex mosaic landscapes incorporating larger amounts of C_4 grasses adapted to warmer and drier climates (Bobe and Behrensmeyer 2004, Potts 2007, Cerling et al. 2015). This rapid remodeling of landscapes, coupled with an increase in habitat heterogeneity and broadening of habitat spectrums, introduced new selective pressures that undoubtedly influenced hominin evolution. Indeed, these environmental changes are linked to key adaptive shifts in the hominin biological and behavioral record such as changes in body size, mobility, and diet (Plummer 2004, Potts 2007, Pontzer 2012).

Research over the last several decades has provided a great deal of data, and some contradictory interpretation, regarding Oldowan hominin behavior and the adaptive significance of the first stone tools. It is clear that Oldowan hominins had a sophisticated sense of fracture mechanics from the first appearance of the industry at ca. 2.6 million years ago at Gona, Ethiopia (Semaw et al. 2003, Stout et al. 2005, 2010, Toth et al. 2006,

Rogers and Semaw 2009). However, it has been argued by some that there were important differences in the behavior of hominins forming the early sites (ca. 2.6–2.3 Ma) relative to the later ones (ca. 2.0–1.7 Ma) attributed to this industrial complex (Harmand 2009a, b, Goldman-Neuman and Hovers 2012, Potts 2012), with the gap in these time ranges reflecting a dearth of sites from ca. 2.3–2.0 Ma. Here we review the data for the earliest Oldowan sites, and compare these data to those from two approximately 2.0 Ma sites in Kenya. This comparison will serve to highlight what may be behavioral differences reflected in site formation in the early versus late Oldowan.

Overview of Sites from 2.6–2.3 Ma

The sites in the earlier time interval are more geographically restricted, appearing first in Gona, Ethiopia, at 2.6–2.5 Ma, and then between 2.36–2.32 Ma at Hadar and Omo, Ethiopia, and Lokalalei, Kenya (table 1 and references therein). Fossils with possible stone tool damage in the Bouri Formation, Middle Awash, Ethiopia, may also evince hominin activity at 2.5 Ma (de Heinzelin et al. 1999). These localities are considered in turn below.

Gona, Ethiopia

The Busidima Formation in the Gona Research Project study area has yielded the largest number of early Oldowan sites, dating 2.6–2.5 Ma (table 1; Semaw et al. 2003, Quade et al. 2004, Rogers and Semaw 2009). The excavations are all small (<13 m²), partly reflecting the difficulty excavating on the steep slopes in the study area, but several have yielded samples of hundreds of in situ artifacts. Fossils are infrequently present and often not well preserved in the archeological levels, but cutmarks on a small number of surface collected fossils have been documented (Dominguez-Rodrigo et al. 2005). The archeological occurrences are all found in the middle to upper part of fining upward sequences deposited by the paleo-Awash River. The sites were deposited on the banks or point bars of the river, most likely associated with gallery forest lining the channel, or up to several hundred meters lateral to the channel in the edaphic grasslands beyond the forest (Levin et al. 2004, Quade et al. 2004). Hominins would have had access to water, cobbles for artifact manufacture, woodland resources from the gallery forest, as well as food from the woodland/grassland ecotone and the edaphic grasslands themselves.

Collectively, three localities (EG-10, EG-12, and OGS-7) provide over one thousand excavated artifacts in minimally disturbed sediments (Semaw 1997, 2000, 2006, Stout et al. 2005, Stout et al. 2010). Despite their antiquity, cores were systematically reduced, and there are numerous large and well-formed invasive flakes produced through free-hand, hard-hammer percussion. The ability of hominins to detach these large flakes is indicative of effective sensorimotor control and the ability to utilize advantageous platform angles (Toth et al. 2006, Stout and Chaminade 2007, Stout et al. 2008). It also allowed for a higher degree of core reduction without premature exhaustion of usable platform angles (Stout et al. 2010).

Unifacial cores (75–78%) and cortical platform flakes (80%) dominate the East Gona (EG-10 and EG-12) assemblage (figure 13.2; Semaw 2000, 2006, Schick and Toth 2006, Toth et al. 2006, Stout et al. 2010). In contrast, the artifact assemblage from OGS-7 (Ounda Gona South) is characterized by a high proportion of bifacial cores (57%) and multifacial cores (29%), and the presence of noncortical flake platforms (59%; Stout et al. 2010), indicating that the Gona hominins were not limited to unifacial core reduction.

At East Gona hominins were transporting partially flaked cores to flake on site, and removing large flakes for use elsewhere (Toth et al. 2006). The assemblage is minimally disturbed and indicative of high-intensity flaking predominantly late in the reduction sequence (Semaw 1997, 2000, 2006, Toth et al. 2006, Stout et al. 2010). The OGS-7 assemblage, on the other hand, contains both early and late reduction stages (Stout et al. 2010) and more closely resembles unbiased flake populations produced in knapping experiments (Toth 1987, Stout et al. 2010).

There was a strong preference for felsic volcanic rock in the Gona artifact assemblages (Semaw 2000, Semaw et al. 2003, Stout et al. 2005, Stout et al. 2010). East Gona lithics are dominated by trachyte (Semaw 2000, Stout et al. 2005, Stout et al. 2010), while Ounda Gona South artifacts were produced primarily with aphanitic and vitreous volcanic raw materials (Stout et al. 2005, Stout et al. 2010). At both East and South Gona, felsic volcanic rock is represented in much higher proportions in the artifact samples than in the local channel gravels used as raw material sources, suggesting that the earliest known Oldowan hominins preferentially selected and transported high-quality, fine-grained volcanic material.

Middle Awash, Ethiopia

The 2.5 Ma Hata Member of the Bouri Formation may preserve evidence of Oldowan hominin processing of large mammal remains (de Heinzelin et al. 1999). Unlike the fluvial deposits that most early Oldowan sites are associated with, the sediments of the Hata Member were deposited at the margin of a shallow, freshwater lake. The paleoenvironmental context differed from the other sites in the 2.6–2.3 Ma interval as well, as the lake margin zone was thought to consist of broad grassy plains, lacking the woodland habitat inferred elsewhere. Furthermore, the lake margin was devoid of conglomerates or volcanic outcrops that could have served as local sources of lithic raw material.

No in situ artifacts were found. Surface artifacts were apparently present but rare, and were not described. Approximately six isolated faunal elements with possible stone tool–cut and/or percussion marks were found in the same stratigraphic horizon across 2 km of outcrop. Three of these (a bovid mandible, a bovid tibia, and an equid femur) were excavated from an in situ context at two different sites.

Omo Shungura Formation, Ethiopia

Five archeological sites have been described from the Shungura Formation in the lower Omo River Valley, Ethiopia, deposited in 2.34–2.3 Ma sediments associated with the paleo-Omo River system (Chavaillon 1976, Merrick 1976, Howell et al. 1987, de la Torre 2004). Three of these sites (FtJi 1, FtJi 5, Omo 57) were found in secondary context in braided stream deposits, while the other two (FtJi 2 and Omo 123) were found in overbank deposits of a meandering stream system, thought to have been near the interface between woodlands lining the proto-Omo channel and grasslands that lay beyond the zone of riparian woodland.

Hominin knapping at the Omo sites was focused on the production of small flakes and waste from the reduction of quartz pebbles through freehand and occasionally bipolar percussion (Chavaillon 1976, Merrick 1976, de la Torre 2004). More than 90 percent of artifacts were made of quartz. This was originally thought to reflect the predominance of quartz in the clast populations of the local conglomerates. Recent lithic surveys in proximity to the archeological sites have shown that quartz makes up only 20–40 percent of the clast population in the pebble conglomerates (Delagnes et al. 2011). Thus, the high frequency of quartz in artifact assemblages reflects a strong hominin preference for this material.

Small initial clast size and the brittle nature of quartz imposed constraints on knappers so that Omo cores were less productive than many of the larger cores found at other Oldowan sites (figure 13.2). Hominins chose pebbles with natural platforms and discarded them, often after only removing a few flakes, when suitable angles for knapping were lost (de la Torre 2004). Rather than focusing on the apparent simplicity of the production sequence, de la Torre (2004) emphasized the technical mastery displayed by the Omo hominins, who efficiently produced usable flakes from a difficult material. Taking advantage of natural planes and angles of quartz pebbles, they worked small cores but produced flakes with few accidents that were nearly identical to the length of the core, with thin cross-sections and good cutting edges.

Hadar, Ethiopia

Two in situ Oldowan occurrences were found in the Makamitalu Basin of the Hadar Research project area (Hovers 2009, Neuman and Hovers 2012). The sites, A.L. 666 and A.L. 894, are in the Busidima Formation below the BKT-3 tephra dated to 2.36 Ma. Both sites were stratigraphically overlying a massive conglomerate from the paleo-Awash River, and were situated within crevasse-splay deposits in the proximal flood plain of the river. This would have situated them in a well-watered setting, possibly near riparian woodlands as has been reconstructed for the Gona sites.

A.L. 666 was a 20 m^2 excavation in sediments nearby the discovery site of early *Homo* mandible A.L. 666-1, and yielded a small number of surface and in situ artifacts (Kimbel et al. 1996). A.L. 894 was a 21.5 m^2 excavation approximately 170 m away from A.L. 666, and stratigraphically a bit lower (and therefore slightly older) than the latter. Both sites yielded

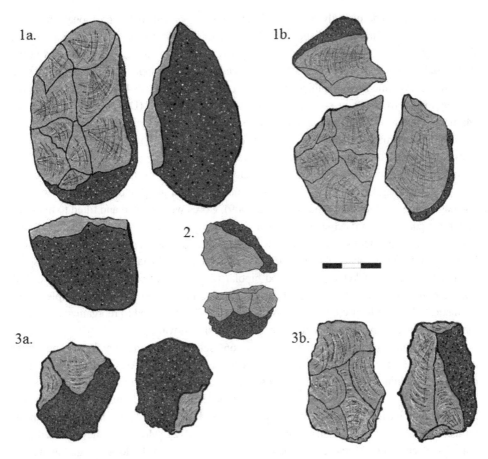

Figure 13.2
Oldowan artifacts from three East African localities. 1. Cores from Gona, Ethiopia, illustrating simple unifacial reduction from East Gona (1a) and an irregular multifacial core from Gona South (1b) (after Stout et al. 2010, figures 7 and 8). 2. A small quartz core from Omo, Ethiopia, where a removal on the horizontal surface served as a striking platform for three unidirectional flakes on the transversal surface (after de la Torre, 2004, figure 4). 3. Cores from Kanjera South, Kenya, illustrating the reduction of local (3a) versus nonlocal (3b) raw materials. 3a is a core made from locally available fenetized andesite and 3b is a flake used as a core made from Nyanzian rhyolite transported over 10 km (after Braun and Plummer, 2013, figure 2). Scale equals 3 centimeters.

in situ artifacts predominantly in a 30–35 cm thick layer, with postdepositional processes playing a role in their vertical dispersion. Refits were found in both assemblages, suggesting that site integrity was good, with the much larger A.L. 894 assemblage being the most informative (table 1). This is interpreted as being a location where knapping took place, and where there was a dynamic flow of lithics into and out of the site. A.L. 894 is also notable in preserving fossils (n=330) in spatial association with stone tools (Dominguez-Rodrigo and Martinez-Navarro 2012). Taphonomic analysis of the faunal assemblage found no evidence of stone tool damage on the bones, suggesting that there was no behavioral association between the stone tools and fauna from the site.

Similar to reduction strategies observed at Gona and (as will be discussed later) Lokalalei 2C, the assemblages from A.L. 894 and A.L. 666 exhibit a good understanding of flaking mechanics and conchoidal fracture (Hovers 2009). Cores were rotated and tilted to manipulate the impact angle and achieve large invasive flake removals in order to efficiently reduce cores and rectify mistakes. Most flake scars are unipolar. However, when mistakes occurred, knappers sometimes rotated the surface of cores, so that accidental removals occur disproportionally often on flakes that contained bipolar (23.54%) and centripetal (7.55%) dorsal scar patterns (Hovers 2009). Thus, while reduction strategies were generally simple, more sophisticated methods of core rotation were occasionally employed to reorganize cores and maximize production, especially following accidental removals.

Although hominins at both A.L. 894 and A.L. 666 selected from the same local conglomerates, there were differences in the degree of selectivity practiced. At A.L. 894, the slightly older site, hominins selected against basalts and against rocks with impurities in their matrix. In contrast, hominins at A. L. 666 applied stricter selection criteria, with the artifact assemblage containing a higher frequency of homogenous and fine-grained raw materials than those utilized in A.L. 894 (Goldman-Neuman and Hovers 2012). Moreover, they utilized raw materials such as quartz and chert that were either very rare in the local conglomerates or transported to the site from farther afield. This may evidence a trend towards higher selectivity and more complex transport patterns through time at Hadar.

West Turkana, Kenya

The two oldest Oldowan sites found thus far in Kenya are the Lokalalei 1 and 2C sites from the Nachukui Formation on the western side of Lake Turkana (Roche et al. 1999, Delagnes and Roche 2005, Harmand 2009a, b). The two sites are one kilometer apart, and are both dated to 2.34 Ma. At the time the sites were formed the proto-Omo River flowed through the Turkana basin. The sites were deposited in an alluvial plain across which small eastward-flowing streams flowed to join the main axial system of the river. These streams formed debris-flow conglomerates between 10 and a few hundred meters from where the archeological sites formed, providing a ready source of raw material for artifact production. Thus, like the other sites reviewed thus far, the cost of search, acquisition, and transport of lithic raw material was relatively low. Stable isotopic analysis of pedogenic carbonates associated with the sites suggests that they were situated near or within riparian woodlands that would have provided hominins with shade, refuge from predators, plant food resources, and potable water (Quinn et al. 2013).

The 60 m² Lokalalei 1 excavation yielded 392 artifacts, including flakes, cores, hammerstones, and unmodified split cobbles (Kibunjia 1994, Harmand 2009a). Notably, the site also yielded a fossil assemblage (n=3,000 fragments). Preliminary faunal analysis suggests the assemblage is predominantly composed of small bovids, which may preserve traces of hominin and carnivore damage (Kibunjia 1994, Brugal et al. 2003). The 17 m² Lokalalei 2C excavation yielded 2624 artifacts including cores, whole or broken flakes, a few possibly retouched pieces, hammerstones, and unmodified split cobbles. The small faunal sample found spatially associated with the artifacts did not provide any substantive taphonomic information. Artifacts from Lokalalei 1 and 2C are primarily composed of locally available lavas, with phonolites in particular being well-represented (Harmand 2009a). The parallel mineral orientation of the medium-grained phonolite facilitated fracture alone its foliation plane, allowing for greater fracture predictability during knapping.

Though these sites were close geographically and temporally, there may have been differences in the technological competency of the hominins producing their lithic assemblages. At Lokalalei 1, cobbles seemed to have been opportunistically gathered, with little attention paid to shape (Harmand 2009a). The low frequency of natural platforms on the globular cobbles used resulted in low productivity reduction sequences. Moreover, the Lokalalei 1 hominins may have been less dexterous, as the artifacts there exhibit numerous knapping accidents, flakes tend to be smaller, and the frequency of impact damage from blows not yielding a flake is high (Kibunjia 1994, Harmand 2009a).

In contrast, the angular blocks providing natural platforms preferred at Lokalalei 2C promoted far longer reduction sequences. The assemblage contains a remarkably high number of refits between flakes and cores, providing exceptional insight into hominin technological practices. Lokalalei 2C cores exhibit unidirectional and multidirectional removals, and hominins adeptly maintained adequate flaking angles and core volumetric structure during knapping (Roche et al. 1999, Delagnes and Roche 2005).

Overview of Sites ca. 2.0 Ma

Kanjera South, Kenya

The Oldowan occurrences at Kanjera South were first recognized in a small (~0.5 km²) amphitheater in 1995 (figure 13.1). Abundant artifacts and associated fauna are found within the Southern Member of the Kanjera Formation, which is approximately 12 m thick and comprised of six beds; from oldest to youngest, KS-1 to KS-6 (Behrensmeyer et al. 1995, Ditchfield et al. 1999, Plummer et al. 1999). Oldowan artifacts and associated fauna are found in a 3-m-thick sequence in the fine sands and silts of upper KS-1 through KS-3. These were deposited in a generally low-energy fluvial setting by sheetwash and ephemerally active, shallow rivulets flowing to a depositional low to the north. Several discontinuous, thin conglomerate levels provide evidence of brief episodes of rapid water flow. Biostratigraphy and magnetostratigraphy constrain the age of the archeological levels to 1.95 Ma to 2.3 Ma. Given the relatively rapid deposition of the sediments, the archeological levels are thought to date to approximately 2 Ma (Plummer et al. 2009a). In 1997, 2000, and 2001, an 169 m² excavation (Excavation 1) yielded >2,900 Oldowan artifacts and 3,600 fossils, including bones with stone tool damage. Most of this material was recovered from KS-2, the middle of the three archeological levels. Taphonomic and zooarchaeological analyses indicate that, with the exception of the conglomerate levels, hominins were the primary agent of site formation in all three beds (Plummer et al. 1999, Plummer 2004, Ferraro 2007, Ferraro et al. 2013). Estimated rates of sedimentation and pedogenesis suggest that archeological materials accumulated relatively rapidly over a period of decades to centuries per bed.

Analyses so far indicate: (1) that the Kanjera South hominins had early access to small (size 1 and 2) bovids, probably by hunting, but were likely having mixed access to size 3 and larger taxa (Ferraro et al. 2013); (2) They situated their activities in a grassland-dominated ecosystem as indicated by the inferred habitat preferences of the recovered fauna and stable isotopic chemistry of paleosol carbonates and ungulate teeth (Plummer et al. 2009a, b); (3) They habitually transported stone materials (30% of the lithic assemblage) selectively collected from conglomerates, over longer distances (>10 km) than typical for the Oldowan (Braun et al. 2008a), reflecting their preference for hard, easily flaked materials unavailable on the northern half of the Homa Peninsula (Braun et al. 2009a, b, Braun and Plummer 2013); (4) They deployed different technological strategies to more intensively utilize these hard, non-local raw materials, including exploiting multiple core surfaces, removing old platforms to develop new exploitation surfaces (platform rejuvenation flakes), maintaining convex surfaces to allow longer debitage sequences, producing flakes that removed less core volume (flakes with higher edge to mass ratios), and retouching flakes (figure 13.2); local raw materials show far less complex reduction sequences, and no retouch (Braun et al. 2009a, b, Braun and Plummer 2013); and (5) They used these "exotic" raw materials for a wide variety of tasks, which, based on use-wear, included woodworking, processing of herbaceous plant tissues, cutting animal tissue, scraping bone, and peeling/cutting underground storage organs (USOs; Lemorini et al. 2014).

FwJj20 in Il Dura, Kenya

The Il Dura region of the Turkana Basin has not been as well studied as the other parts of East Turkana. However, recent field work (Braun et al. 2010) and extensive geological mapping (Gathogo and Brown 2006) highlight this region as a focus of hominin activity during the Upper Burgi Member. Excavations at FwJj20 in Il Dura provide data complementary to those from the roughly coeval Kanjera South. The archeological horizon was deposited lateral to a channel bar associated with a major tributary river system draining the Surgei Highlands to the east. Excavation and surface collection at a single stratigraphic horizon and associated tephrostratigraphic and paleomagnetic data confirm an age of 1.95 Ma for the archeological occurrence. Excavation has yielded a similarly sized lithic and faunal assemblage to Kanjera South (2,633 artifacts; 3,648 fossils). The FwJj20 sample derives from a single, 6–15 cm deposit of clay and silt that was rapidly buried by a sand unit. While small (size 1 and 2) bovids are the most common animals found at Kanjera, medium (size 3) mammals and aquatic species (Chelonia, *Crocodylus*; *Hipopotamus*) predominate at FwJj20. Multiple specimens preserve clear evidence of hominin butchery (McCoy 2009, Braun et al. 2010, Archer et al. 2014). Paleoenvironmental indicators, including fossil wood, bovid tribal representation, and pedogenic carbonate stable isotopic evidence (Quinn et al. 2013) demonstrate that hominin activities occurred in well-watered habitat possibly near a riverine forest. This appears to represent the opposite end of the East African habitat spectrum from Kanjera's open to wooded grassland. While core forms are almost exclusively made on locally available basalt cobbles, the debitage portion of the assemblage is made on a variety of raw materials including some (e.g., chert, jasper, quartz) not common in nearby channel conglomerates. This seems to be an indication of variable artifact discard patterns mediated by the physical properties and availability of different raw materials.

Who made the Oldowan tools?

Tool use and a relatively large cranial capacity are frequently associated with the definition of the genus *Homo* (e.g., Leakey et al. 1964). However, tool use is now recognized in a number of mammals and birds (Sanz et al., 2013), and so cannot be used as a defining trait of the human genus. Moreover, tool use may have been widespread among hominins (Panger et al. 2002, Plummer 2004), and the newly described 3.3 Ma artifacts at Lomekwi (Harmand et al. 2015), if confirmed by further research, would likely reflect stone tool usage by a hominin genus other than *Homo*. Given the time span the Oldowan is known from and the complexity of the Plio-Pleistocene hominin fossil record, it seems possible that multiple hominin species, perhaps even multiple hominin genera, made Oldowan tools. The oldest fossil attributed to the genus *Homo* is a 2.8 Ma mandible from Ledi-Geraru in Ethiopia (Villmoare et al. 2015, but see Schwartz and Tattersall 2015 for complications in defining the genus *Homo*). The genus *Paranthropus* is only slightly younger, with a first appearance at 2.66 Ma from the Upper Ndolanya Beds, Laetoli, Tanzania (Harrison 2011). The youngest East African gracile australopithecin, *Australopithecus garhi*, is found in the same sediments as some of the oldest possible evidence of butchery with stone artifacts (2.5 Ma from the Bouri Formation, Middle Awash, Ethiopia; Asfaw et al. 1999, de Heinzelin et al. 1999). This has led some to speculate that *A. garhi* was the ancestor of *Homo* and the first stone tool-user (Asfaw et al. 1999, de Heinzelin et al. 1999, Semaw et al. 2003), though the Ledi-Geraru specimen makes the former point unlikely and the latter point cannot be confirmed at this time. The geographical and temporal distribution of *Paranthropus* largely overlaps that of Oldowan archeological occurrences in East and South Africa, and the argument that *Paranthropus* made Oldowan tools has been made (e.g., Susman 1988, 1991) but again is difficult to test.

While the identity of the hominin(s) forming Oldowan archeological sites older than 2 Ma is unclear, the larger-bodied *Homo* present in Africa by 1.8 Ma (*H. erectus*; *H. ergaster* to some; Anton 2003) likely used the Oldowan prior to the development of the Acheulean or Karari Industries (Isaac 1997). Additional cranial material may push the first appearance of this taxon back to 1.88–1.9 Ma (Feibel et al. 1989, Wood 1992, Anton 2003). Isolated post-cranial elements from East Turkana, Kenya (e.g., femora KNM-ER 1472 and 1481a at 1.89 Ma, innominate KNM-ER 3228 at 1.95 Ma) provide evidence of large-bodied *Homo* at nearly 2 Ma (Rose 1984, Anton 2003). An improved fossil record allowing the *Homo* lineage to be traced back in time to the ancestor of early African *H. erectus* may one day identify with some certainty an Oldowan-using taxon older than 2 Ma.

Interpreting the Earliest Archeological Sites

It is clear that Oldowan hominins had a sophisticated sense of fracture mechanics from the first appearance of the industry at ca. 2.6 million years ago at Gona, Ethiopia (Semaw et al. 2003, Stout et al. 2005, 2010, Toth et al. 2006, Rogers and Semaw 2009). However, there is variation in flaking strategies and raw material utilization within the small sample of sites excavated from 2.6–2.3 Ma, and between those sites and the sites known after about 2.0 Ma. How this variation is interpreted differs between research groups. Some researchers (e.g., Semaw 2000, Semaw et al. 2003, Rogers and Semaw 2009, Stout et al. 2010) feel that the earliest makers of Oldowan tools at 2.6 Ma were as proficient in their reduction strategies as those making tools later in time, and believe that the technological variation seen between sites can be explained by factors such as differences in the quality, size, and shape of the clasts available for artifact manufacture.

Others (e.g., Delagnes and Roche 2005, Harmand 2009a, b, Hovers 2009, Goldman-Neuman and Hovers 2012, Potts 2012) believe that there is more technological variability in the earlier set of sites, and/or that site assemblages in the 2.6–2.3 Ma time bracket resulted from a somewhat different suite of behaviors from those forming sites at approximately 2.0–1.7 Ma. The most common arguments for how the "earlier" Oldowan sites differ from sites appearing after ca. 2 Ma are as follows:

The earlier Oldowan sites are more geographically restricted

The Oldowan appears in Gona, Ethiopia, nearly 2.6 Ma, and after a hiatus of several hundred thousand years appears at 2.36–2.32 Ma at Hadar and Omo, Ethiopia, and West Turkana, Kenya. Whether the spread of the Oldowan reflects cultural diffusion or independent invention of flake-based technologies is unclear (Plummer 2004, Hovers 2012), but the lengthy dispersal time of the Oldowan as currently documented does not suggest lithic technology was rapidly adopted across hominin populations. This is a clear difference with the more widespread distribution of later Oldowan sites. Starting at ca. 2.0 Ma sites are more geographically widespread, appearing for the first time in new regions of East Africa (e.g., the Homa Peninsula, Kenya), North Africa (e.g., Ain Hanech, Algeria) as well as in South Africa (e.g., Sterkfontein, Swartkrans, and Malapa), and across a wider variety of depositional settings and habitats (Potts 1998, 2012, Plummer 2004, Plummer et al. 2009a, Braun et al. 2010, Magill et al. 2013).

Artifacts and fossils at the oldest sites are confined to narrow stratigraphic intervals

Most sites have a single restricted vertical distribution suggesting a limited span for onsite hominin activities, with the Gona site of EG-10 being one of the only instances of a stratigraphically stacked set of two separate archeological levels (Semaw 2000). This suggests that hominin use and discard of stone artifacts at particular spots on the landscape was less iterative than the later Oldowan (Potts 2012). By 2.0 Ma, hominin activities appear more persistent at specific locales, for example occurring through consecutive layers in a 3-meter sequence of sediments at Kanjera South, Kenya. The locality was apparently attractive to hominins for decades, if not centuries, and hominins repeatedly returned and discarded artifacts and bones there (Plummer et al. 1999).

The early Oldowan sites occur within a narrow range of depositional settings and habitats

All of the sites save Bouri, Ethiopia, were formed near a large stream or river, where overbank deposits sealed the archeological levels (table 1). These sites were most likely formed in riparian woodlands, or near riparian woodland/grassland ecotones (Plummer 2004, Quade et al. 2004, Rogers and Semaw 2009). The Bouri fossils were deposited in an open, lake margin context, but the sparse fossil finds, lack of in situ stone tools, and possibility that the fossils were modified by crocodiles rather than hominins (Pante et al. 2016) all suggest that additional documentation is needed to confirm hominin modification. By 2.0 Ma, sites are found in a broader array of environmental settings. Kanjera South provides the earliest evidence of hominin activities in an open habitat within a grassland-dominated ecosystem, and the coeval site of FwJj20 provides evidence of hominin site formation in riparian gallery forest. The fact that these two sites were formed near opposite ends of the habitat spectrum suggests that by this time activities requiring stone tool use occurred in a diverse array of settings, and were not simply centered on riparian woodlands or riparian woodland/grassland ecotones.

Lithic materials were transported over short distances at the earliest sites

Sites 2.6 to 2.3 Ma were formed near sources of stone raw material, and where transport distances can be estimated they were routinely short, generally a few meters to perhaps a few hundred meters (Plummer 2004, Harmand 2009a, b, Rogers and Semaw 2009, Goldman-Neuman and Hovers 2012). The costs of search, acquisition, and transport of lithic materials therefore seems to have been relatively low. Sites 2.0 Ma and younger tend to combine generally complex raw material selectivity patterns with on average longer transport distances (Plummer 2004, Braun et al. 2008a, Harmand 2009a, b, Goldman-Neuman and Hovers 2012, Potts 2012). For example, approximately 30 percent of the artifact sample from Kanjera South was transported at least 13 km, and these non-local raw materials were more carefully and thoroughly flaked than lower-quality locally available stones (Plummer 2004; Braun et al. 2008a, 2009a,b).

Raw material selectivity at the earliest sites was more variable

The earliest sites exhibit variable degrees of raw material selectivity, from relatively low to more complex (Stout et al. 2005, Goldman-Neuman and Hovers 2012). It is unclear what this variability in selectivity, ranging from simply choosing cobbles with relatively homogenous groundmass to choosing specific raw materials and specific cobble shapes, represents. Some researchers entertain the possibility that it reflects differing levels of cognitive sophistication or technical skills between different groups (or species) of hominins (e.g., Delagnes and Roche 2005, Harmand 2009a).

There may be more variation in technological competency in the early Oldowan site sample

It is clear that many sites in the 2.6–2.3 Ma time interval exhibit the same understanding of stone fracture mechanics and competency in flake production as the assemblages used to define the Oldowan from ca. 1.8 Ma sites at Bed I Olduvai Gorge, Tanzania (Leakey 1971, Ludwig and Harris 1998, Semaw et al. 1997, Semaw 2000, Semaw et al. 2003, Hovers 2009). This has led some to argue that Oldowan flake production follows a fairly straightforward set of rules for approximately one million years (2.6–1.6 Ma), whereby hominins with a good understanding of conchoidal fracture removed flakes from cobbles of varying sizes, lithologies, and shapes, creating assemblages dominated by cores and unmodified flakes (Rogers and Semaw 2009, Stout et al. 2010). Cores were exhausted when knapping surfaces lost their necessary convexity (de la Torre 2004, Delagnes and Roche 2005), and modification of flakes (e.g., by retouching) was carried out at low frequencies or not at all (Braun and Plummer 2013). Sites where hominins had access to larger sized cobbles (e.g., Gona, West Turkana, Hadar) have longer reduction sequences than sites where clasts used in artifact production were small (e.g., Omo, Kanjera South). The primary exception to this view of uniform knapping competency is Lokalalei 1, which has a small lithic assemblage with a high frequency of step fractures, small flakes, and impact damage from repeated percussion that failed to remove flakes (Kibunjia 1994, Delagnes and Roche 2005, Harmand 2009a). The knapping sequences from this assemblage have been characterized as "opportunistic" (Delagnes and Roche 2005), and in part the flaking mistakes can be attributed to hominin selection of rounded cobbles without the natural angles used elsewhere as platforms to initiate fracture (Harmand 2009a). This lack of foresight restricted their available reduction strategies. Comparison of the Lokalalei 1 reduction strategies with the "organized" (*sensu* Delagnes and Roche 2005) reduction of the Lokalalei 2C cores, which were selected for their natural angles and flaked sequentially, provides one of the strongest contrasts in reduction systems in the 2.6 to 2.3 Ma bracket of sites.

Stone tool use may have focused on plant processing at the earliest sites

Initially, researchers argued that the few cutmarked large mammal bones surface collected from sites at Gona indicated that butchery was a prominent, perhaps predominant, function of stone tool use (Dominguez-Rodrigo et al. 2005). Similarly, Rogers and Semaw (2009) argued that the primary function of the earliest stone tools was to produce flakes for butchery. Recently the strong linkage between artifact production and animal butchery has been questioned. There are no in situ assemblages of bones from the earliest sites that preserve unequivocal evidence of butchery (Dominguez-Rodrigo and Martinez-Navarro 2012). Except for the Bouri sample, which we feel needs reevaluation, none of the fossils from the early Oldowan exhibit unambiguous evidence for marrow processing. In fact, it has been argued that many Oldowan sites, including sites postdating 2.0 Ma, are palimpsests of hominin and carnivore activity, where hominins were primarily using stone tools for non-butchery related activities, such as plant processing (e.g., Dominguez-Rodrigo 2009). Similarly, Hovers (2012) argues that early Oldowan tool use may have focused on plant processing. Increased exploitation of butchered carcasses may reflect a behavioral shift after 2.0 Ma that was linked to increasingly sophisticated raw material selectivity and longer lithic transport distances. In this view, more routine processing of animal tissue may have been an "add-on" postdating 2.0 Ma to the existing Oldowan technological and behavioral repertoire focused on plant processing. Certainly sites at 2.0 Ma do provide evidence of hominin processing of fauna, as well evidence of plant processing. Hominins butchered substantially complete small antelope carcasses at Kanjera South, and consumed tissue from larger animals as well (Ferraro et al. 2013). Use-wear on quartz and quartzite artifacts from Kanjera South also documents the processing of animal tissue as well as the processing of a variety of plant tissues, including underground storage organs, wood, and herbaceous plants. At FwJj20, hominins butchered terrestrial mammals as well as aquatic fauna such as fish, turtles, and crocodiles (Braun et al. 2010, Archer et al. 2014).

Conclusions drawn from the Oldowan record

We feel that the existing data indicate both similarities and differences in the behaviors leading to the formation of "early" and "later" Oldowan sites. But the strength of the arguments for these perceived differences is not strong, as the small number of Oldowan sites across both space and time makes generalization about hominin behavior problematic. Therefore, arguments about hominin behavior derived from the data at hand need to be tempered with an appreciation of the potentially strong sampling biases involved.

Several robust conclusions about early Oldowan behavior do seem warranted. The efficient production of stone flakes using good quality raw materials is evident at all sites save Lokalalei 1. Hominins making stone tools did so frequently enough to master conchoidal fracture, and they expended energy selecting specific cobble forms and/or raw material types. Energy was also being expended in lithic transport, both from conglomerate sources to where site accumulations were formed, and in the transport of flaked pieces and flakes to these sites and away from them (Semaw et al. 2003, Delagnes and Roche 2005, Goldman-Neuman and Hovers 2012). This energetic investment suggests that the tasks artifacts were used for had some fitness benefit, even if transport costs were lower in the early Oldowan sites. At some sites (e.g., Lokalalei 2C, Hadar 894) the artifact assemblages are large, falling well within the size range of later Oldowan sites, and show that activities requiring the production of a large number of flakes were at times spatially concentrated on the landscape. If Oldowan lithic technology did have some fitness benefit at an early date, as it appears it did, its slow geographic spread from Gona to the Omo Shungura Formation, Hadar, and Lokalalei all at ca. 2.3 Ma, and the gap between 2.3 and 2.0 Ma almost certainly reflect sampling error. This seems even more likely given that the early Oldowan occurrences fall into narrow stratigraphic intervals (see above) and probably represent brief (decade[s] or smaller) intervals of time (Hovers 2012). The specific tasks artifacts were used for in the early Oldowan remains somewhat ambiguous, though it is likely that they were used for plant processing, and at least occasionally for butchery given the cutmarked fossils from Gona.

Toward a More Integrated Paleoanthropology

One of the goals of paleoanthropological research should be to integrate the record of hominin behavior derived from archeological investigation with paleontological evidence of hominin paleobiology. As reviewed above, Plio-Pleistocene hominin taxonomy is not straightforward, and even within the most commonly used framework of early *Homo* taxa (*Homo habilis*, *H. rudolfensis*, *Homo ergaster*/early *H. erectus*; Wood 1992, Anton et al. 2014), it is not clear which taxon other than *Homo ergaster*/early *H. erectus* would likely have made and used stone tools (Plummer 2004). However, recent research has pointed out some major adaptive trends in early *Homo* postcranial anatomy and energy budgets which ultimately should be relatable to the archeological record. The strongest case for postcranial change in hominins between 3–1.5 Ma is an increase in body mass in *Homo* starting around 2 Ma (Pontzer 2012, Anton et al. 2014) which in turn can be related to a larger home range size (Anton et al. 2002) and an increase in energy expenditure (Aiello and Key 2002, Leonard and Robertson 1997, Steudel-Numbers 2006). An expansion of the daily energy budget in early *Homo* likely reflects a greater reliability in calorie acquisition than the australopiths, and may be linked to increased reproductive investment in early *Homo* as well (Pontzer 2012). These inferences are consistent with the finding that the human lineage has experienced an acceleration in metabolic rate relative to the other living hominoids, providing energy for larger brains and faster reproduction relative to what is predicted for australopiths (Pontzer et al. 2016). Food sharing and enhanced body fat deposition in humans may have evolved with this metabolic strategy as way to mitigate risk-associated dependence on patchily distributed nutrient dense foods.

At ca. 2 Ma, Kanjera South and FwJj20 are the oldest Oldowan sites with large, well-preserved faunal samples, and provide data consistent with the view that *Homo* was foraging for nutrient-dense foods across a broad spectrum of habitats, perhaps reflecting increased average home range size. Lithic technology at Kanjera South was part of a stone tool dependent foraging strategy to acquire high quality foods (Braun and Plummer 2013, Ferraro et al. 2013, Lemorini et al. 2014, Plummer and Bishop 2016). Hominins selectively collected and transported hard lithologies and deployed technological strategies to more intensively utilize them. They consumed foods including animal carcasses and USOs that would have required stone tool use to acquire and/or process, were of high nutritional value, and in the case of whole gazelle carcasses and dense USO patches, came in large packets. The utilization of aquatic resources at FwJj20 suggests expansion of the hominin diet to include aquatic resources. Evidence for the shift towards the acquisition of nutritious, hard-to-acquire foods in packets large enough to be shared at both sites fits well with the evidence for increasing hominin body mass and the argument for metabolic rate acceleration outlined above. The importance of Oldowan technology in hominin foraging is also borne out by the slightly younger sites of Ain Hanech, Algeria; FLK Zinj, Tanzania; and Swartkrans, South Africa. These sites document early access to fleshy carcasses of medium-sized mammals across the full north-south extent of the African continent (Dominguez-Rodrigo et al. 2007; Pickering et al. 2008; Sahnouni et al. 2013, Bunn and Gurtov 2014).

The record as it exists now seems to demonstrate an intensification of resource transport and increase in faunal utilization in the later Oldowan, as suggested by Hovers (2012) and

Potts (2012). However, both the human paleontological record, as well as the archeological record, are characterized by small sample sizes in the interval of interest from 3–2 Ma. Until the fossil and archeological records are better sampled, there will be uncertainty in the actual timing of the increase in body size in *Homo*, as well as the shift towards artifact-dependent foraging of nutrient dense foods. An investment needs to be made to find more hominin fossils, and also investigate Oldowan archeology through space and time, with an eye towards better understanding the evolution of human foraging systems, and changes in the diet of *Homo*. It would be particularly valuable to redouble efforts to understand the behaviors that led to the formation of the earliest Oldowan sites. Research could profitably address the following questions about the early Oldowan.

Is the first appearance datum (FAD) of ca. 2.6 Ma for the Oldowan accurate? Did Oldowan technology emerge in the Gona region alone at its outset, or were early Oldowan sites more widely distributed? Both of these questions will require research in appropriately aged sediments in more regions of Africa and even Eurasia, as reports of finds as old as 2.6 Ma may ultimately demonstrate that artifact-wielding hominins were outside of Africa at an early date (Dennell and Roebroeks 2005, Dennell 2011, Malassé et al. 2016).

Is the apparent expansion of hominin habitat usage ca. 2 Ma real? Are artifact scatters generally smaller, less dense, and stratigraphically constrained at older sites? The uniform occurrence of early Oldowan sites in riparian settings very close to raw material sources, and the "more ephemeral" nature of some of these occurrences, may indicate that the behaviors carried out at these sites were more spatially and or temporally constrained than those carried out at later Oldowan sites.

Did Oldowan artifact function vary over space and time? Was there a greater emphasis on plant processing prior to 2 Ma? Or was butchery and meat consumption routinely carried out but masked by a preservation bias against bone? Artifact function throughout the Oldowan needs more attention, but particularly for early Oldowan sites that generally lack well-preserved fossils. Discovery of Oldowan sites in new locales with better fossil preservation, use-wear analysis of artifacts, and attempts to collect phytoliths and/or starch grains from artifact edges may provide helpful data about artifact function (Dominguez-Rodrigo et al. 2001, Mercader et al. 2008, Lemorini et al. 2014). A careful zooarchaeological analysis of the Lokalalei 1 assemblage would be useful, as at 2.34 Ma it is quite old, it is small bovid dominated like the Kanjera South assemblages, and some fossils are reported to have stone tool damage (Kibunjia 1994).

The relative profitability of foraging with or without the use of stone tools may have varied by ecosystem and season, as has been considered when investigating chimpanzee tool use (Sanz and Morgan 2013). Determining the season of death of animals found butchered at Oldowan sites would be useful (Pike-Tay and Cosgrove 2002, Lee-Thorp and Sponheimer 2006) to demonstrate wherever possible whether carcass processing occurred in both wet and dry seasons (and thus probably year round), or was concentrated during the dry season, when ungulate carcasses may have been more readily available, and hominin populations under greater dietary stress (Blumenschine 1987, Foley 1987, Laden and Wrangham 2005). Year-round butchery of fauna would provide an indirect measure of the importance of stone tools to hominin foraging ecology, and an assessment of the stage of carcass acquisition (largely fleshed from hunting/aggressive scavenging, or largely defleshed if passively scavenged), would provide a sense of its nutritional value, and of hominin ability to compete with large carnivores. An enlarged zooarchaeological sample throughout the temporal range of the Oldowan is needed to assess whether hominins were utilizing larger packets of nutrient-rich carcasses starting around 2 Ma, or utilizing them more frequently, versus earlier in time. This would be expected if resource transport and intensification of tool-dependent foraging for high-quality resources was on the rise with body size increase in at least one species of *Homo*.

Conclusion

The Oldowan is the first widespread and persistent technology in the hominin evolutionary record. However, the features most frequently ascribed evolutionary significance (transport of artifacts and fauna, large mammal butchery) best characterize the later (2 Ma and younger) Oldowan record. A great deal has been published about the technology of the sites from 2.6–2.0 Ma, but the actual activities hominins utilized artifacts for, and the significance of stone tools in the day-to-day existence of artifact-producing hominins, needs further investigation. An important goal of paleoanthropologists working in the Plio-Pleistocene should be to integrate, as much as is possible, the hard tissue record of hominin growth, development and behavior, with the behavioral evidence provided from investigation of early hominin archeological sites. Continued field work and analysis of hominin fossils and Oldowan archeological materials is necessary for this goal to be achieved.

Acknowledgments

TP would like to thank Dr. Gerd Müller, Dr. Eva Lackner, and Dr. Jeffrey Schwartz for the invitation to participate in this engaging and thought-provoking conference.

References

Aiello, L. C., and C. Key. 2002. "Energetic Consequences of Being a *Homo Erectus* Female." *American Journal of Human Biology* 14(5):551–565.

Antón, S. C. 2003. "Natural History of *Homo Erectus*." *American Journal of Physical Anthropology* 122(S37):126–170.

Antón, S. C., R. Potts, and L. C. Aiello. 2014. "Evolution of Early *Homo*: An Integrated Biological Perspective." *Science* 345(6192):1236828.

Archer, W., D. R. Braun, J. W. K. Harris, J. T. McCoy, and B. G. Richmond. 2014. "Early Pleistocene Aquatic Resource Use in the Turkana Basin." *Journal of Human Evolution* 77:74–87.

Asfaw, B., T. White, O. Lovejoy, B. Latimer, S. Simpson, and G. Suwa. 1999. "*Australopithecus Garhi*: A New Species of Early Hominid from Ethiopia." *Science* 284(5414):629–635.

Bamford, M. K. 2011. "Late Pliocene Woody Vegetation of Area 41, Koobi Fora, East Turkana Basin, Kenya." *Review of Palaeobotany and Palynology* 164(3):191–210.

Behrensmeyer, A. K., R. Potts, T. Plummer, L. Tauxe, N. Opdyke, and T. Jorstad. 1995. "The Pleistocene Locality of Kanjera, Western Kenya: Stratigraphy, Chronology and Paleoenvironments." *Journal of Human Evolution* 29(3):247–274.

Beyene, Y., S. Katoh, G. WoldeGabriel, W. K. Hart, K. Uto, M. Sudo, M. Kondo, M. Hyodo, P. R. Renne, G. Suwa, B. Asfaw. 2013. "The Characteristics and Chronology of the Earliest Acheulean at Konso, Ethiopia." *Proceedings of the National Academy of Sciences* 110(5):1584–1591.

Blumenschine, R. J., 1987. "Characteristics of an Early Hominid Scavenging Niche." *Current Anthropology* 28(4):383–407.

Bobe, R., and A. K. Behrensmeyer. 2004. "The Expansion of Grassland Ecosystems in Africa in Relation to Mammalian Evolution and the Origin of the Genus *Homo*." *Palaeogeography, Palaeoclimatology, Palaeoecology* 207(3):399–420.

Braun, D. R., J. W. K. Harris, N. E. Levin, J. T. McCoy, A. I. R. Herries, M. K. Bamford, L. C. Bishop, B. G. Richmond, and M. Kibunjia. 2010. "Early Hominin Diet Included Diverse Terrestrial and Aquatic Animals 1.95 Ma in East Turkana, Kenya." *Proceedings of the National Academy of Sciences* 107(22):10002–10007.

Braun, D. R., and T. W. Plummer. 2013. "Oldowan Technology at Kanjera South: Technological Diversity on the Homa Peninsula." In *Africa: Cradle of Humanity: Recent Discoveries*, edited by M. Sahnouni, 131–145. CNRPAH, Algeria.

Braun, D. R., T. Plummer, P. Ditchfield, J. V. Ferraro, D. Maina, L. C. Bishop, and R. Potts. 2008a. "Oldowan Behavior and Raw Material Transport: Perspectives from the Kanjera Formation." *Journal of Archaeological Science* 35(8):2329–2345.

Braun, D. R., T. Plummer, J. V. Ferraro, P. Ditchfield, and L. C. Bishop. 2009a. "Raw Material Quality and Oldowan Hominin Toolstone Preferences: Evidence from Kanjera South, Kenya." *Journal of Archaeological Science* 36(7):1605–1614.

Braun, D. R., T. W. Plummer, P. W. Ditchfield, L. C. Bishop, and J. V. Ferraro. 2009b. "Oldowan Technology and Raw Material Variability at Kanjera South." In *Interdisciplinary Approaches to the Oldowan*, edited by E. Hovers and D. R. Braun, 99–110. Netherlands: Springer.

Braun, D. R., J. C. Tactikos, J. V. Ferraro, S. L. Arnow, and J. W. K. Harris. 2008b. "Oldowan Reduction Sequences: Methodological Considerations." *Journal of Archaeological Science* 35(8):2153–2163.

Brugal, J., H. Roche, and M. Kibunjia. 2003. "Faunes et paléoenvironnements des principaux sites archéologiques plio-pléistocènes de la formation de Nachukui (Ouest-Turkana, Kenya)." *Comptes Rendus Palevol* 2(8):675–684.

Bunn, H. T., and A. N. Gurtov. 2014. "Prey Mortality Profiles Indicate That Early Pleistocene *Homo* at Olduvai Was an Ambush Predator." *Quaternary International* 322:44–53.

Cerling, T.E., S. A. Andanje, S. A. Blumenthal, F. H. Brown, K. L. Chritz, J. M. Harris, J. A. Hart, F. M. Kirera, P. Kaleme, L. N. Leakey, and M. G. Leakey. 2015. "Dietary Changes of Large Herbivores in the Turkana Basin, Kenya from 4 to 1 Ma." *Proceedings of the National Academy of Sciences* 112(37):11467–11472.

Chavaillon, J. 1976. "Evidence for the Technical Practices of Early Pleistocene Hominids, Shungura Formation, Lower Omo Valley, Ethiopia." In *Earliest Man and Environments in the Lake Rudolf Basin*, edited by Y. C. Glynn, L. I. Isaac, and R. E. Leakey, 565–573. Chicago, IL: University of Chicago Press.

de Heinzelin, J., J. D. Clark, T. White, W. Hart, P. Renne, G. WoldeGabriel, Y. Beyene, and E. Vrba. 1999. "Environment and Behavior of 2.5-Million-Year-Old Bouri Hominids." *Science* 284(5414):625–629.

Delagnes, A., J-R. Boisserie, Y. Beyene, K. Chuniaud, C. Guillemot, and M. Schuster. 2011. "Archaeological Investigations in the Lower Omo Valley (Shungura Formation, Ethiopia): New Data and Perspectives." *Journal of Human Evolution* 61(2):215–222.

Delagnes, A., and H. Roche. 2005. "Late Pliocene hominid Knapping Skills: The Case of Lokalalei 2C, West Turkana, Kenya." *Journal of Human Evolution* 48(5):435–472.

de la Torre, I. 2004. "Omo Revisited: Evaluating the Technological Skills of Pliocene Hominids." *Current Anthropology* 45(4):439–465.

deMenocal, P. B. 2004. "African Climate Change and Faunal Evolution during the Pliocene–Pleistocene." *Earth and Planetary Science Letters* 220:3–24.

Dennell, R., and W. Roebroeks. 2005. "An Asian Perspective on Early Human Dispersal from Africa." *Nature* 438(7071):1099–1104.

Dennell, R. W., M. Martinón-Torres, and J. M. Bermúdez de Castro. 2011. "Hominin Variability, Climatic Instability and Population Demography in Middle Pleistocene Europe." *Quaternary Science Reviews* 30(11):1511–1524.

Ditchfield, P., J. Hicks, T. Plummer, L. C. Bishop, and R. Potts. 1999. "Current Research on the Late Pliocene and Pleistocene Deposits North of Homa Mountain, Southwestern Kenya." *Journal of Human Evolution* 36(2):123–150.

Domínguez-Rodrigo, M. 2009. "Are All Oldowan Sites Palimpsests? If So, What Can They Tell Us about Hominid Carnivory?" In *Interdisciplinary Approaches to the Oldowan*, edited by E. Hovers and D. R. Braun, 129–147. Netherlands, Springer.

Domínguez-Rodrigo, M., R. Barba, and C. P. Egeland. 2007. *Deconstructing Olduvai: A Taphonomic Study of the Bed I Sites*. Dordrecht, Netherlands: Springer.

Domínguez-Rodrigo, M., and B. Martinez-Navarro. 2012. "Taphonomic Analysis of the Early Pleistocene (2.4 Ma) Faunal Assemblage from AL 894 (Hadar, Ethiopia)." *Journal of Human Evolution* 62(3):315–327.

Domínguez-Rodrigo, M., T. R. Pickering, and H. T. Bunn. 2010. "Configurational Approach to Identifying the Earliest Hominin Butchers." *Proceedings of the National Academy of Sciences* 107(49):20929–20934.

Domínguez-Rodrigo, M., T. R. Pickering, and H. T. Bunn. 2012. "Experimental Study of Cut Marks Made with Rocks Unmodified by Human Flaking and Its Bearing on Claims of ~3.4-Million-Year-Old Butchery Evidence from Dikika, Ethiopia." *Journal of Archaeological Science* 39(2):205–214.

Domínguez-Rodrigo, M., T. R. Pickering, S. Semaw, and M. J. Rogers. 2005. "Cutmarked Bones from Pliocene Archaeological Sites at Gona, Afar, Ethiopia: Implications for the Function of the World's Oldest Stone Tools." *Journal of Human Evolution* 48(2):109–121.

Feibel, C. S., F. H. Brown, and I. McDougall. 1989. "Stratigraphic Context of Fossil Hominids from the Omo Group Deposits: Northern Turkana Basin, Kenya and Ethiopia." *American Journal of Physical Anthropology* 78(4):595–622.

Ferraro, J. V. 2007. *Broken Bones and Shattered Stones: On the Foraging Ecology of Oldowan Hominins*. Ann Arbor, MI: ProQuest.

Ferraro, J. V., T. W. Plummer, B. L. Pobiner, J. S. Oliver, L. C. Bishop, D. R. Braun, P. W. Ditchfield, J. W. Seaman III, K. M. Binetti, J. W. Seaman Jr., and F. Hertel. 2013. "Earliest Archaeological Evidence of Persistent Hominin Carnivory." *PloS one* 8(4):e62174.

Foley, R. 1987. *Another Unique Species: Patterns in Human Evolutionary Ecology*. New York: Longman Sc & Tech.

Gabunia, L., S. C. Antón, D. Lordkipanidze, A. Vekua, A. Justus, and C. C. Swisher. 2001. "Dmanisi and Dispersal." *Evolutionary Anthropology: Issues, News, and Reviews* 10(5):158–170.

Gathogo, P. N., and F. H. Brown. 2006. "Stratigraphy of the Koobi Fora Formation (Pliocene and Pleistocene) in the Ileret Region of Northern Kenya." *Journal of African Earth Sciences* 45(4):369–390.

Goldman-Neuman, T., and E. Hovers. 2012. "Raw Material Selectivity in Late Pliocene Oldowan Sites in the Makaamitalu Basin, Hadar, Ethiopia." *Journal of Human Evolution* 62(3):353–366.

Harmand, S. 2009a. "Variability in Raw Material Selectivity at the Late Pliocene Sites of Lokalalei, West Turkana, Kenya." In *Interdisciplinary Approaches to the Oldowan*, edited by E. Hovers and D. R. Braun, 85–97. Netherlands: Springer.

Harmand, S. 2009b. "Raw Materials and Techno-economic Behaviors at Oldowan and Acheulean Sites in the West Turkana Region, Kenya." *Lithic Materials and Paleolithic Societies*, edited by B. Adams and B. Blades, 1–14. Chichester, UK: John Wiley & Sons.

Harmand, S., J. E. Lewis, C. S. Feibel, C. J. Lepre, S. Prat, A. Lenoble, X. Boës, R. L. Quinn, M. Brenet, A. Arroyo, and N. Taylor. 2015. "3.3-Million-Year-Old Stone Tools from Lomekwi 3, West Turkana, Kenya." *Nature* 521(7552):310–315.

Harris, J. W. K. 1983. "Cultural Beginnings: Plio-Pleistocene Archaeological Occurrences from the Afar, Ethiopia." *African Archaeological Review* 1(1):3–31.

Harris, J. W. K., and S. Capaldo. 1993. "The Earliest Stone Tools: Their Implications for an Understanding of the Activities and Behaviour of Late Pliocene Hominids." In *The Use of Tools by Human and Non-Human Primates,* edited by A. Berthelet and J. Chavaillon, 197–224. Oxford: Clarendon Press.

Harrison, T. 2011. "Hominins from the Upper Laetolil and Upper Ndolanya Beds, Laetoli." In *Paleontology and Geology of Laetoli: Human Evolution in Context,* edited by T. Harrison, 141–188. Dordrecht, Netherlands: Springer.

Hovers, E. 2009. "Learning from Mistakes: Flaking Accidents and Knapping Skills in the Assemblage of AL 894 (Hadar, Ethiopia)." In *The Cutting Edge: New Approaches to the Archaeology of Human Origins. Stone Age Institute Press Publications, Series 3,* edited by K. D. Schick and N. P. Toth, 137–148. Bloomington, IN: Stone Age Institute Press.

Hovers, E. 2012. "Invention, Reinvention and Innovation: The Makings of Oldowan Lithic Technology." In *Origins of Human Innovation and Creativity*, edited by S. Elias, In *Developments in Quaternary Science* edited by J. J. M van der Meer, 16: 51–68. Oxford, UK: Elsevier.

Hovers, E. 2015. "Archaeology: Tools Go Back in Time." *Nature* 521(7552):294–295.

Howell, F. Clark, P. Haesaerts, and J. de Heinzelin. 1987. "Depositional Environments, Archeological Occurrences and Hominids from Members E and F of the Shungura Formation (Omo Basin, Ethiopia)." *Journal of Human Evolution* 16(7–8):665–700.

Isaac, G. L., and Koobi Fora Research Project. 1997. "Koobi Fora Research Project Vol. 5: Plio-Pleistocene Archaeology."

Kibunjia, M. 1994. "Pliocene Archaeological Occurrences in the Lake Turkana Basin." *Journal of Human Evolution* 27(1):159–171.

Kimbel, W. H., R. C. Walter, D. C. Johanson, K. E. Reed, J. L. Aronson, Z. Assefa, C. W. Marean, G. G. Eck, R. Bobe, E. Hovers, and Y. Rak. 1996. "Late Pliocene *Homo* and Oldowan Tools from the Hadar Formation (Kada Hadar Member), Ethiopia." *Journal of Human Evolution* 31(6):549–561.

Kingston, J. D. 2007. "Shifting Adaptive Landscapes: Progress and Challenges in Reconstructing Early Hominid Environments." *American Journal of Physical Anthropology* 134:20–58.

Kuman, K. and A. S. Field. 2009. "The Oldowan Industry from Sterkfontein Caves, South Africa." In *The Cutting Edge: New Approaches to the Archaeology of Human Origins,* edited by K. D. Schick and N. P. Toth, 151–170. Bloomington, IN: Stone Age Institute Press.

Laden, G. and R. Wrangham. 2005. "The Rise of the Hominids as an Adaptive Shift in Fallback Foods: Plant Underground Storage Organs (USOs) and Australopith Origins." *Journal of Human Evolution* 49(4):482–498.

Laland, K. N., J. Odling-Smee, and S. Myles. 2010. "How Culture Shaped the Human Genome: Bringing Genetics and the Human Sciences Together." *Nature Reviews Genetics* 11(2):137–148.

Leakey, L. S. B., P. V. Tobias, and J. R. Napier. 1964. "A New Species of the Genus *Homo* from Olduvai Gorge." *Nature* 202(4927):7–9.

Leakey, M. D. 1971. *Olduvai Gorge. Vol. 3, Excavations in Beds I and II, 1960–63.* Cambridge: Cambridge University Press.

Lee-Thorp, J., and M. Sponheimer. 2006. "Contributions of Biogeochemistry to Understanding Hominin Dietary Ecology." *American journal of physical anthropology* 131(S43):131–148.

Lemorini, C., T. W. Plummer, D. R. Braun, A. N. Crittenden, P. W. Ditchfield, L. C. Bishop, F. Hertel, J.S. Oliver, F. W. Marlowe, M. J. Schoeninger, and R. Potts. 2014. "Old Stones' Song: Use-Wear Experiments and Analysis of the Oldowan Quartz and Quartzite Assemblage from Kanjera South (Kenya)." *Journal of Human Evolution* 72:10–25.

Leonard, W. R., and M. L. Robertson. 1997. "Comparative Primate Energetics and Hominid Evolution." *American Journal of Physical Anthropology* 102(2):265–281.

Lepre, C. J., H. Roche, D. V. Kent, S. Harmand, R. L. Quinn, JP. Brugal, PJ. Texier, A. Lenoble, and C. S. Feibel. 2011. "An Earlier Origin for the Acheulian." *Nature* 477(7362):82–85.

Levin, N. E., J. Quade, S. W. Simpson, S. Semaw, and M. Rogers. 2004. "Isotopic Evidence for Plio–Pleistocene Environmental Change at Gona, Ethiopia." *Earth and Planetary Science Letters* 219(1):93–110.

Ludwig, B. V., and J. W. K. Harris. 1998. "Towards a Technological Reassessment of East African Plio-Pleistocene Lithic Assemblages." In *Early Human Behaviour in Global Context: The Rise and Diversity of the Lower Palaeolithic Record,* edited by M. D. Petraglia and R. Korisettar, 84–107. London: Routledge.

Magill, C. R., G. M. Ashley, and K. H. Freeman. 2013. "Ecosystem Variability and Early Human Habitats in Eastern Africa." *Proceedings of the National Academy of Sciences* 110(4):1167–1174.

Malassé, A. D., A. M. Moigne, M. Singh, T. Calligaro, B. Karir, C. Gaillard, A. Kaur, V. Bhardwaj, S. Pal, S. Abdessadok, and C. C. Sao. 2016. "Intentional Cut Marks on Bovid from the Quranwala Zone, 2.6 Ma, Siwalik Frontal Range, Northwestern India." *Comptes Rendus Palevol* 15(3):317–339.

McCoy, J. T. 2009. *Ecological and Behavioral Implications of New Archaeological Occurrences from Upper Burgi Exposures at Koobi Fora, Kenya.* Ann Arbor, MI: ProQuest.

McPherron, S. P., Z. Alemseged, C. W. Marean, J. G. Wynn, D. Reed, D. Geraads, R. Bobe, and H. A. Béarat. 2010. "Evidence for Stone-Tool-Assisted Consumption of Animal Tissues before 3.39 Million Years Ago at Dikika, Ethiopia." *Nature* 466(7308):857–860.

McPherron, S. P., Z. Alemseged, C. Marean, J. G. Wynn, D. Reed, D. Geraads, R. Bobe, and H. Béarat. 2011. "Tool-Marked Bones from before the Oldowan Change the Paradigm." *Proceedings of the National Academy of Sciences* 108(21):E116–E116.

Mercader, J., Bennett, T., and M. Raja. 2008. "Middle Stone Age Starch Acquisition in the Niassa Rift, Mozambique." *Quaternary Research,* 70(2):283–300.

Merrick, H. V. 1976. "Recent Archaeological Research in the Plio-Pleistocene Deposits of the Lower Omo, Southwestern Ethiopia." In *Human Origins: Louis Leakey and the East African Evidence,* edited by G. L Isaac and E. R. McCown, 461–481. Menlo Park, CA: WA Benjamin Advanced Book Program.

Merrick, H. V., and J. P. S. Merrick. 1976. "Archaeological Occurrences of Earlier Pleistocene Age from the Shungura Formation." In *Earliest Man and Environments in the Lake Turkana Basin*, edited by Y. Coppens, F. C. Howell, G. L. Isaac, and R. E. F. Leakey, 574–584. Chicago: University of Chicago Press.

Panger, M. A., A. S. Brooks, B. G. Richmond, and B. Wood. 2002. "Older than the Oldowan? Rethinking the Emergence of Hominin Tool Use." *Evolutionary Anthropology: Issues, News, and Reviews* 11(6):235–245.

Pante, M. C., J. K. Njau, T. Keevil, M. Muttart, R. J. Blumenschine, and S. R. Merritt. 2016. "A Quantitative Reassessment of Feeding Trace Morphology and Implications for the Earliest Cut Marked Bones." *PaleoAnthropology* A22.

Pickering, T. R., C. P. Egeland, M. Domínguez-Rodrigo, C. K. Brain, and A. G. Schnell. 2008. "Testing the 'Shift in the Balance of Power' Hypothesis at Swartkrans, South Africa: Hominid Cave Use and Subsistence Behavior in the Early Pleistocene." *Journal of Anthropological Archaeology* 27(1):30–45.

Pike-Tay, A., and R. Cosgrove. 2002. "From Reindeer to Wallaby: Recovering Patterns of Seasonality, Mobility, and Prey Selection in the Palaeolithic Old World." *Journal of Archaeological Method and Theory* 9(2):101–146.

Plummer, T. W. 2004. "Flaked Stones and Old Bones: Biological and Cultural Evolution at the Dawn of Technology." *American Journal of Physical Anthropology* 125(S39):118–164.

Plummer, T. W., and L. C. Bishop. 2016. "Oldowan Hominin Behavior and Ecology at Kanjera South, Kenya." *Journal of Anthropological Sciences* 94:1–12.

Plummer, T. W., L. C. Bishop, P. W. Ditchfield, J. V. Ferraro, J. D. Kingston, F. Hertel, and D. R. Braun. 2009a. "The Environmental Context of Oldowan Hominin Activities at Kanjera South, Kenya." In *Interdisciplinary Approaches to the Oldowan*, edited by E. Hovers and D. R. Braun, 149–160. Netherlands: Springer.

Plummer, T. W., L. C. Bishop, P. Ditchfield, and J. Hicks. 1999. "Research on Late Pliocene Oldowan Sites at Kanjera South, Kenya." *Journal of Human Evolution* 36(2):151–170.

Plummer, T. W., P. W. Ditchfield, L. C. Bishop, J. D. Kingston, J. V. Ferraro, D. R. Braun, F. Hertel, and R. Potts. 2009b. "Oldest Evidence of Toolmaking Hominins in a Grassland-Dominated Ecosystem." *PLoS One* 4(9):e7199.

Pontzer, H. 2012. "Ecological Energetics in Early *Homo*." *Current Anthropology* 53(S6):S346–S358.

Pontzer, H., M. H. Brown, D. A. Raichlen, H. Dunsworth, B. Hare, K. Walker, A. Luke, L. R. Dugas, R. Durazo-Arvizu, D. Schoeller, and J. Plange-Rhule. 2016. "Metabolic Acceleration and the Evolution of Human Brain Size and Life History." *Nature* 533(7603):390–392.

Potts, R. 1991. "Why the Oldowan? Plio-Pleistocene Toolmaking and the Transport of Resources." *Journal of Anthropological Research* 47(2):153–176.

Potts, R. 1998. "Environmental Hypotheses of Hominin Evolution." *American Journal of Physical Anthropology* (27):93–136.

Potts, R. 2007. "Environmental Hypotheses of Pliocene Human Evolution." In *Hominin Environments in the East African Pliocene: an Assessment of the Faunal Evidence*, edited by R. Bobe, Z. Alemseged, and A. K. Behrensmeyer, 25–49. Dordrecht, Netherlands: Springer.

Potts, R. 2012. "Environmental and Behavioral Evidence Pertaining to the Evolution of Early *Homo*." *Current Anthropology* 53(S6):S299–S317.

Quade, J., N. Levin, S. Semaw, D. Stout, P. Renne, M. Rogers, and S. Simpson. 2004. "Paleoenvironments of the Earliest Stone Toolmakers, Gona, Ethiopia." *Geological Society of America Bulletin* 116(11–12):1529–1544.

Quinn, R. L., Christopher J. Lepre, C. S. Feibel, J. D. Wright, R. A. Mortlock, S. Harmand, JP. Brugal, and H. Roche. 2013. "Pedogenic Carbonate Stable Isotopic Evidence for Wooded Habitat Preference of Early Pleistocene Tool Makers in the Turkana Basin." *Journal of Human Evolution* 65(1):65–78.

Roche, H. 1996. "Remarques sur les plus anciennes industries en Afrique et en Europe." In *The First Humans and Their Cultural Manifestations*, edited by F. Facchini, 55–68. Forli: A.B.A.C.O. s.r.l.

Roche, H., A. Delagnes, J. P. Brugal, C. Feibel, M. Kibunjia, V. Mourre, and P. J. Texier. 1999. "Early Hominid Stone Tool Production and Technical Skill 2.34 Myr Ago in West Turkana, Kenya." *Nature* 399(6731):57–60.

Roche, H., and J. J. Tiercelin. 1980. "Industries lithiques de la formation plio-pléistocène d'Hadar: Campagne 1976." In *Proceedings, VIIIth Panafrican Congress of Prehistory and Quaternary Studies*, edited by R. E. F. Leakey and B. A. Ogot, 194–199. Nairobi: Louis Leakey Memorial Institute for African Prehistory.

Rogers, M. J., and S. Semaw. 2009. "From Nothing to Something: The Appearance and Context of the Earliest Archaeological Record." In *Sourcebook of Paleolithic Transitions*, edited by M. Camps and P. Chauhan, 55–171. New York, NY: Springer.

Rose, M. D. 1984. "A Hominine Hip Bone, KNM-ER 3228, from East Lake Turkana, Kenya." *American Journal of Physical Anthropology* 63(4):371–378.

Sahnouni, M., J. Rosell, J. van der Made, J. M. Vergès, A. Ollé, N. Kandi, Z. Harichane, A. Derradji, and M. Medig. 2013. "The First Evidence of Cut Marks and Usewear Traces from the Plio-Pleistocene Locality of El-Kherba (Ain Hanech), Algeria: Implications for Early Hominin Subsistence Activities circa 1.8 Ma." *Journal of Human Evolution* 64(2):137–150.

Sanz, C. M., J. Call, and C. Boesch. 2013. *Tool Use in Animals: Cognition and Ecology*. Cambridge: Cambridge University Press.

Sanz, C. M., and D. B. Morgan. 2013. "Ecological and Social Correlates of Chimpanzee Tool Use." *Philosophical Transactions of the Royal Society of London B: Biological Sciences* 368(1630):20120416.

Schick, K., and N. Toth. 2006. "An Overview of the Oldowan Industrial Complex: The Sites and the Nature of Their Evidence." In *The Oldowan: Case Studies into the Earliest Stone Age*, edited by K. D. Schick, N. P. Toth, and S. Semaw, 3–42. Bloomington, IN: Stone Age Institute Press.

Schwartz, J. H., and I. Tattersall. 2015. "Defining the Genus *Homo*." *Science* 349(6251):931–932.

Semaw, S. 2000. "The World's Oldest Stone Artefacts from Gona, Ethiopia: Their Implications for Understanding Stone Technology and Patterns of Human Evolution between 2.6–1.5 Million Years Ago." *Journal of Archaeological Science* 27(12):1197–1214.

Semaw, S. 2006. "The Oldest Stone Artifacts from Gona (2.6–2.5 Ma), Afar, Ethiopia: Implications for Understanding the Earliest Stages of Stone Knapping." In *The Oldowan: Case Studies into the Earliest Stone Age*, edited by K. D. Schick, N. P. Toth, and S. Semaw, 43–75. Bloomington, IN: Stone Age Institute Press.

Semaw, S., P. Renne, J. W. K. Harris, C. S. Feibel, R. L. Bernor, N. Fesseha, and K. Mowbray. 1997. "2.5-Million-Year-Old Stone Tools from Gona, Ethiopia." *Nature* 385(6614):333–336.

Semaw, S., M. J. Rogers, J. Quade, P. R. Renne, R. F. Butler, M. Dominguez-Rodrigo, D. Stout, W. S. Hart, T. Pickering, and S. W. Simpson. 2003. "2.6-Million-Year-Old Stone Tools and Associated Bones from OGS-6 and OGS-7, Gona, Afar, Ethiopia." *Journal of Human Evolution* 45(2):169–177.

Steudel-Numbers, K. L. 2006. "Energetics in *Homo Erectus* and Other Early Hominins: The Consequences of Increased Lower-Limb Length." *Journal of Human Evolution* 51(5):445–453.

Stout, D., and T. Chaminade. 2007. "The Evolutionary Neuroscience of Tool Making." *Neuropsychologia* 45(5):1091–1100.

Stout, D., J. Quade, S. Semaw, M. J. Rogers, and N. E. Levin. 2005. "Raw Material Selectivity of the Earliest Stone Toolmakers at Gona, Afar, Ethiopia." *Journal of Human Evolution* 48(4):365–380.

Stout, D., S. Semaw, M. J. Rogers, and D. Cauche. 2010. "Technological Variation in the Earliest Oldowan from Gona, Afar, Ethiopia." *Journal of Human Evolution* 58(6):474–491.

Stout, D., N. Toth, K. Schick, and T. Chaminade. 2008. "Neural Correlates of Early Stone Age Toolmaking: Technology, Language and Cognition in Human Evolution." *Philosophical Transactions of the Royal Society of London B: Biological Sciences* 363(1499):1939–1949.

Susman, R. L. 1988. "Hand of *Paranthropus Robustus* from Member 1, Swartkrans: Fossil Evidence for Tool Behavior" *Science* 240(4853):781–784.

Susman, R. L. 1991. "Who Made the Oldowan Tools? Fossil Evidence for Tool Behavior in Plio-Pleistocene Hominids." *Journal of Anthropological Research* 47(2):129–151.

Toth, N. 1987. "Behavioral Inferences from Early Stone Artifact Assemblages: An Experimental Model." *Journal of Human Evolution* 16(7):763–787.

Toth, N., K. Schick, and S. Semaw. 2006. "A Comparative Study of the Stone Tool-Making Skills of *Pan, Australopithecus*, and *Homo Sapiens*." In *The Oldowan: Case Studies into the Earliest Stone Age,* edited by K. D. Schick, N. P. Toth, and S. Semaw, 155–222. Bloomington, IN: Stone Age Institute Press.

Villmoare, B., W. H. Kimbel, C. Seyoum, C. J. Campisano, E. N. DiMaggio, J. Rowan, D. R. Braun, J. R. Arrowsmith, and K. E. Reed. 2015. "Early *Homo* at 2.8 Ma from Ledi-Geraru, Afar, Ethiopia." *Science* 347(6228):1352–1355.

Wood, B. 1992. "Origin and Evolution of the Genus *Homo*." *Nature* 355(6363):783–790.

14 Human Brain Evolution: History or Science?

Dietrich Stout

Introduction

In 1998, Robin Dunbar published an influential review of the "Social Brain Hypothesis" (SBH) in the journal *Evolutionary Anthropology*. Although the idea that social complexity played a central role in primate brain evolution had been around at least since the 1960s (Jolly 1966) and gathered considerable support in the 1980s (Byrne and Whiten 1988), it was Dunbar's application of comparative methods from evolutionary biology (Harvey and Pagel 1991) that established the SBH as the consensus view in anthropology and beyond. By operationalizing intelligence ("information-processing capacity") as a ratio of neocortex to rest-of-brain volume and social complexity as average group size, Dunbar was able to show a strong correlation between the two over a wide range of primate species. This was taken as evidence that brain size constrains group size, so that evolutionary increases in group size (for whatever reason) would generate concomitant selective pressure for brain size increase to handle the increase in social complexity. At the same time, Dunbar showed that neocortex ratio was *not* correlated with various measures of ecological complexity (e.g., percent fruit in diet, range size). This straightforward and decisive result convinced many skeptics that social complexity was not merely important, but was almost exclusively responsible for generating the selective pressures leading to primate brain expansion.

Dunbar then went a step further to consider implications for human evolution specifically. This led to what is now popularly known as "Dunbar's Number" of 150: the group size predicted by a modern human neocortex ratio. Although humans obviously have social groupings at various scales of organization, from nuclear families to nations, Dunbar argued that 150 is the approximate number of stable interpersonal relationships a typical human can maintain at any one time. This number is thought to recur in everything from the size of Hutterite farming communities to the length of British Christmas card mailing lists and has even attracted the attention of social media software designers, some of whom are explicitly building "Dunbar's Number" into their systems (Bennett 2013). The success of a hominoid (ape) regression equation in predicting human group size suggests continuity between human and nonhuman primates and supports the view that human brain evolution has been a straightforward extension of a primate trend.

Following this logic, the hominoid regression has also been used to interpolate group sizes for extinct hominin species (using neocortex ratios predicted from whole brain volumes inferred from fossil crania). Aiello and Dunbar (1993) used these interpolated group sizes in combination with comparative primate data on the relationship between group size and time spent grooming to date the origin of language to approximately 250,000 years ago, arguing that increasing hominin group sizes would have necessitated language as a time-saving, "cheap" form of social grooming. Dunbar (2003), using a revised fossil data set, increased this estimate to 500,000 years ago and suggested this earlier date may actually have marked the emergence of "musical chorusing" prior to the development of grammatical speech. Fossil cranial capacities were similarly used to date the emergence of religion (to archaic *Homo sapiens*) based on a correlation between frontal lobe volume and achievable level of intentionality across living catarrhines and the auxiliary argument that supernatural proscriptions require at least level 4 intentionality. Thus, for example: "I have to *believe* that you *suppose* that there are supernatural beings who can be made to *understand* that you and I *desire* that things should happen in a particular way," where italics mark levels of intentionality (Dunbar 2003, p. 177).

These conclusions are stunning. If accepted, they provide specific dates for two of the most evolutionarily significant, hotly debated, and archaeologically invisible events in human prehistory. More importantly, this is accomplished within a single, elegantly simple theoretical framework that explains not just the when, but also the why of these evolutionary events, providing answers to what others have described as evolutionary "questions we will never answer" (Lewontin 1998). The pragmatic power of this comparative, evolutionary biological approach to answer questions that have stymied archaeologists and paleontologists is undeniably appealing. For many, its theoretical emphasis on continuity between humans and other animals is equally attractive and might be seen to place the study of human evolution within a broader, more truly scientific framework. Along these lines, Gowlett et al. (2012) argue paleoanthropology's focus on providing detailed reconstructions of the past has produced a dearth of theoretical content and ceded "big picture" evolutionary interpretation to other disciplines. To remedy this, they propose that a comparative approach based on the SBH can compensate for inherently limited archaeological data:

If we can understand the broad primate-wide rules that govern the behavior, ecology, and demography of primates (including humans), then we may be better able to identify the sequence of changes that have taken place since the last common ancestor that led, step by (nonteleological) step, to ourselves. (p. 695)

This comparative method is seen as an antidote to a "what you see is what there was" (p. 693) archaeological paradigm unduly limited by the availability of actual material evidence of past behavior.

Obviously, the validity and efficacy of this approach to human evolution depends on the validity and detail of the comparative models that are employed. As one such model, the SBH has led to a great deal of productive research, and this has naturally included important criticisms, extensions, and revisions that must be taken into account and will be discussed below. But there is a more fundamental issue to be considered: Should we in fact expect human evolution to have conformed to "broad primate-wide rules" in the first place? To put it more broadly, to what extent is the evolution of any one particular species actually predictable based on general principles? This is a fundamental and debated question in evolutionary theory that should be addressed by any attempt to apply comparative methods to elucidate the course of human evolution.

Science Versus History?

"Science" means many things to many people, but prediction has a central place in many definitions. Indeed there can be little dispute that the scientific method is fundamentally based upon testing the predictions (i.e., implications) of hypotheses. But it does not stop there. In the natural sciences especially, there is also the idea that one core objective of science is to discover universal laws that allow accurate prediction of physical phenomena. The paradigm example is Newton's Law of Universal Gravitation: $F = G(m_1 m_2)/r2$. This is a remarkably brief and powerful summary of the way the world works that, among many other things, allowed accurate prediction of the existence and position of Neptune prior to telescopic confirmation. Matt Cartmill (1990, 2002) has argued that scientific (including evolutionary) explanation *requires* an appeal to such lawful "if . . . then" relationships. More specifically, he argues that scientific explanations must combine laws and narratives to form a *modus ponens* argument: "If A, then B (the law); A (the narrative); therefore B (the explained event)" (2002, p. 190). He concludes (p. 194) that if students of human evolution "never go beyond narrative to seek recurrent patterns, then we are not doing science. We're just telling stories." Note that Cartmill is not questioning the validity of "historical science" in the conventional sense of using the scientific method to test hypotheses about the occurrence of particular events in the past (Cleland 2002). Rather, he is concerned with the next step of using such historical data to induce generalizable laws of evolution. Although it is not cited by Cartmill, the SBH purports to have identified one such recurrent pattern in primate evolution that can be used to construct properly scientific explanations of human brain evolution and thus to remedy the particularistic storytelling of paleoanthropologists.

Stephen J. Gould was also very concerned to avoid "storytelling" in evolutionary science, but from a very different perspective. Gould and Lewontin (1979) critiqued what they referred to as the "adaptationist programme" in evolutionary biology by asserting the importance of historical and contextual factors such as phyletic heritage, developmental and structural constraint, and (what would later [Gould and Vrba 1982] be termed) exaptation. In essence Gould and Lewontin were questioning the validity of "adaptive optimization" as a universal evolutionary law, and the storytelling they wished to avoid was the uncritical use of this "law" to explain particular traits without due consideration of alternative causes. On the face of it, this might appear to simply be a critique of one particular (putative) law rather than a comment on the nature of evolutionary explanation. However, the underlying philosophical stance goes deeper than that.

Throughout his work, Gould championed the idea of evolutionary contingency. As he (Gould 1995, p. 36) rather bluntly put it:

> Apply all the conventional "laws of nature" type explanations you wish . . . and we will still be missing a fundamental piece of "what is life?" The events of our complex natural world may be divided into two broad realms—repeatable and predictable incidents of sufficient generality to be explained as consequences of natural law, and uniquely contingent events that occur, in a world full of both chaos and genuine ontological randomness as well because complex historical narratives happened to unfurl along the pathway actually followed, rather than along any of the myriad equally plausible alternatives. . . . Contingent events, though unpredictable at the onset of a sequence are as explainable as any other phenomenon after they occur. The explanations, as contingent rather than law-based, do require a knowledge of the particular historical sequence that generated the result . . . many natural sciences, including my own of paleontology, are historical in this sense, and can provide information if the preserved archive be sufficiently rich.

This would seem to be a fundamentally different view on the proper role of generalization, prediction, and law in evolutionary explanation to that espoused by Cartmill. Is Gould's version of contingent evolution really scientific? Or does it simply provide historical narratives of "one thing after another" without any actual explanation?

It seems clear, as Cartmill contends, that scientific explanation must be based on known causal relationships. It is not enough to say that y follows x, we must also say why. Of course this is true of history and narrative as well. Historians don't simply say "Germany invaded Poland and then Britain declared war"; they explain why this particular act produced that particular response rather than another. The point at issue seems to be what exactly qualifies as a "known causal relationship" or law. According to Cartmill (2002), such laws are (1) discovered through the identification of recurrent patterns, (2) universally applicable, and (3) sufficient to allow prediction. To the extent that historical narratives attempt to explain unique events—"Britain declared war on Germany in 1939"—that are not part of a recurrent pattern and could not be predicted from first principles, they should not be considered scientific. Importantly, Cartmill does not deny that such unique events also occur in evolutionary history (and may in fact be commonplace); he merely asserts that these events are scientifically inexplicable. For example, he states (2002, p. 196–197) that "I doubt that we will ever know why our ancestors became bipedal" because human bipedalism is unlike that of any other bipedal animal and must have resulted from "some unique coincidence of factors—some contingency—that does not conform to any recurring regularity of evolution." Such pessimism about one of the core questions in human evolution will not sit well with many paleoanthropologists. Does Gould's approach to contingency offer a viable, scientific alternative?

The positions of Gould and Cartmill are actually closer than either would likely admit. Both recognize that many evolutionary events are unpredictably contingent on particular circumstances, and that explanation requires an appeal to general causal principles. Where they differ is on what counts as a "general causal principle." Gould's best known contributions to evolutionary theory, including work on punctuated equilibria (Eldredge and Gould 1972), evolutionary developmental biology (Gould 1977), and evolutionary constraint (Gould and Lewontin 1979), were clearly attempts to identify recurrent, universally applicable patterns in biological evolution—whether or not he would have been comfortable calling them "laws." The key difference is that these are regularities of process or mechanism whereas Cartmill prefers to focus on regularities of outcome (i.e., convergent evolution). This focus reflects an implicit commitment to the primacy of natural selection as *the* causal principle for evolutionary explanation and restricts the scope of such explanation to recurring adaptations. Dunbar (1998, p. 179) was more explicit about this when he argued that "large brains will evolve only when the selection factor in their favor is sufficient" and that mechanisms other than selection should not be regarded as causal. The implication for Cartmill is that a trait like human bipedalism is unique and inexplicable to the extent that its adaptive function appears different from that of other bipeds. For Gould, on the other hand, an event plus a known mechanism (selective or otherwise) constitutes an explanation, even if the particular outcome is unique and contingent. From this perspective, human bipedalism may be unique in many ways, perhaps including its adaptive functions, but can still be explained in terms of general processes operating in a specific context.

As discussed by Fitch (2012), such an approach to the question of human bipedalism starts with the recognition that " 'Bipedalism' as a monolithic entity is too broad to allow a single, simple causal explanation" (p. 624). For example, bipedal walking and running may have evolved at different times and for different adaptive reasons. Furthermore, each would have evolved from a historically unique set of initial conditions including "the number of leg bones and physiological constraints of bone strength, muscle properties, the rhythm of breathing, and neural control of balance" (p. 624) as well as capacities for phenotypic accommodation that would constrain optimization in some respects while facilitating variation and adaptation in others (Laland et al. 2015). Parallels for aspects of this complex evolutionary process can be found and studied across other species (e.g., biomechanical trade-offs, anatomical plasticity) allowing for a generalizing, comparative, and "law"-based approach. Nevertheless, the particular history of human bipedalism will remain unique. In principle, this is no different than explaining the particular orbital path of Neptune using Newton's Law of Universal Gravitation—it is just that the complex interaction of the causes and conditions involved means the task bears more similarity to explaining the path of a falling leaf in a windstorm.

Clearly, simple explanations are greatly to be preferred where possible, and, regrettably, we remain a long way from explaining the complex evolution of human bipedalism. If human brain evolution turns out to have a simpler explanation than bipedalism, then we would certainly want to take advantage of this. However, we must also remain cautious of the dangers of oversimplification. It is an empirical, rather than epistemological, question as to whether the simplifying and generalizing theoretical framework offered by the SBH is superior to more particularistic and historical archaeological and paleontological approaches.

Problems for Social Brain Theory

Although it is still conventionally referred to as a hypothesis, the use of proposed relations between brain size and social complexity to explain everything from human social structure to the origin of language implies that the SBH actually has the status of an established scientific theory. That is: "a comprehensive explanation of some aspect of nature that is supported by a vast body of evidence" such that it "can be used to make predictions about natural events or phenomena that have not yet been observed" though of course remaining "subject to continuing refinement" (National Academy of Sciences 2008, p. 11). Indeed, decades of research have left little doubt that there is an important evolutionary relationship between brain size and social complexity (variously measured). Nevertheless, there are important reasons to doubt both the causal primacy and universal applicability of this relationship, even if we restrict ourselves to the Primate order where the hypothesis was first formulated. These issues suggest that the SBH may need revision more substantial than "continuing refinement."

Interconnectedness of brain, behavior, and life history variation

The SBH is distinguished from other accounts by its strong claim that the evolution of brain size is driven primarily by the selective pressures of social complexity. In Primates, this claim has been supported by various statistical analyses showing that social group size (as a proxy for complexity) is the best predictor of (relative) neocortex size across an array of ecological and life history variables (e.g., Dunbar 1998, Dunbar and Shultz 2007). The fundamental challenge for this approach is the fact that brain size, body size, life span, developmental length, range size, activity pattern, diet, and sociality are *all* highly correlated with one another as part of species' integrated life history and adaptive strategies (Charvet and Finlay 2012). Using correlation and path analyses to identify one particular variable in this web of complex covariation as causally primary is problematic for two closely related reasons.

First, because so many of the variables under consideration co-vary so closely, comparisons of their relative predictive power will be highly sensitive to differences in error variation within each variable (Isler and Van Schaik 2014). That is, the variable measured with the highest accuracy will tend to be the most successful. Compounding this problem is the fact that many of variables under consideration, including both group size and neocortex size, are actually intended as proxies for something else (social complexity, "computational power"). Thus, we could reasonably expect substantial differences in error variance for both theoretical (e.g., if group size captures social complexity better than percentage of fruit in diet captures foraging complexity) and practical (e.g., if a given variable is easier to measure with precision and accuracy and/or its value is inferred from a larger, more representative sample due to differential research effort) reasons.

One useful way to address such concerns has been to consider a greater range of proxy variables. In addition to group size, social complexity has also been operationalized as number of females in the group, grooming clique size, frequency of coalitions, male mating strategies, prevalence of social play, and frequency of tactical deception, all of which show a relation to relative brain size (reviewed by Dunbar and Shultz 2007). However, a similar effort has not been made to diversify measures of "ecological complexity," or to directly test all of these alternative metrics against one another. When alternative measures of ecological complexity (notably "technical innovation" frequency [Navarrete et al. 2016]) have been included in models, results do not always support the primacy of social complexity as a driving force in Primate brain evolution.

Second, there is the problem of scaling. It is well known that brain size scales with body size across species. What is not well understood are the implications this fact has for the use of various different measures of "brain size" in comparative studies. As Deaner et al. (2000) showed some time ago, different scaling methods (e.g., residuals vs. ratios, scaling to body mass vs. brain subdivisions) produce different results and there is no clear theoretical motivation for preferring one method to another. The SBH is most strongly supported by Dunbar's (1998) preferred scaling measure: the neocortex/rest-of-brain ratio. Other measures generally confirm an association between "brain size" and group size (Dunbar and Shultz 2007, Deaner, Nunn, and van Schaik 2000, Pérez-Barbería, Shultz, and Dunbar 2007), but may not support the strong claims of the SBH that group size is the *best* predictor of brain size (Charvet and Finlay 2012, Deaner, Nunn, and van Schaik 2000) and therefore that "the key selection pressure promoting the evolution of large brains is explicitly social" (Dunbar and Shultz 2007, p. 1345).

An important issue for the use and interpretation of neocortex ratios is that this measure does not control the effect of body size (Dunbar 1992). As Aiello and Dunbar (1993) show, there is a tight correlation between neocortex ratio and total brain size in primates. This is simply a reflection of more widespread brain scaling relationships documented by Finlay and colleagues (Finlay and Darlington 1995, Finlay and Uchiyama 2015, Yopak et al. 2010). Since total brain size scales with body size, it is unsurprising that body size is a good predictor of neocortex ratio (figure 14.1a). This is not necessarily a problem as there are good reasons to suppose that cognitive complexity might scale with absolute brain and

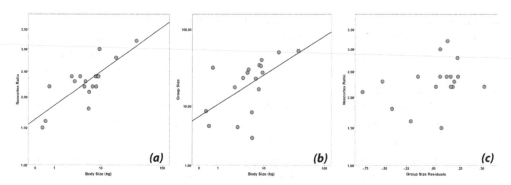

Figure 14.1
Relationships between body size, neocortex ratio (a) and group size (b). There is no relationship between group
size and neocortex ratio if effects of body size are controlled using residuals (c). Data from Schillaci (2015).

body size (Deacon 1997). If this is so, rigorously "controlling" for body size would actually remove the effect of interest. However, this positive scaling relationship does need to
be explicitly considered in comparative analyses.

For the SBH, it is important to note that body size is also positively associated with
group size (figure 14.1b) in all but the largest primates (Janson and Goldsmith 1995).
Should this confounding effect of body size be controlled (Schillaci 2015)? Perhaps not,
if the association between body and group size is mediated by neocortical constraints on
group size (cf. Dunbar and Shultz 2007). However, the actual reasons for body size/group
size correlation are unknown and may be ecological in nature (e.g., related to resource
distribution, feeding competition, and travel costs [Chapman and Valenta 2015]). An
analogous issue confronts the use of range size as a proxy for ecological complexity since
range size is also correlated with body size. In Dunbar's (1998) "head-to-head" comparison of range size vs. group size as predictors for neocortex ratio, range size was controlled
for body size (using residuals) but group size was not. Not surprisingly, group size fared
better. If both variables are treated in the same way (body size controlled/not controlled)
the apparent advantage of group size is reduced or eliminated, and range size may appear
as a better predictor (e.g., Deaner, Nunn, and van Schaik 2000). In at least one recent data
set (Schillaci 2015) a strong association between neocortex ratio and group size (df = 17,
$\beta = 0.662$, $F = 12.504$, $p = 0.003$, $r^2 = 0.439$) is completely eliminated if group size residuals
are used (figure 14.1c: df = 17, $\beta = 0.292$, $F = 1.495$, $p = 0.239$, $r^2 = 0.085$).

Theoretically, it is not clear whether one or both proxy variables should be controlled
for body size. Do the cognitive challenges of range size scale to "ecological grain effects"
that need to be removed (Dunbar 1992)? Or is it instead the case that "bigger is different
as far as cognition goes" (Deacon 1997, p. 163). Similarly, is the relation between group
size and body size driven by cognitive or ecological constraints? We don't know the answer
to these questions, but they inform decisions about appropriate scaling methods that have

a huge impact on comparative results. The bottom line, once again, is that all of the behavioral, ecological, and anatomical variables under consideration are intimately connected as elements of species' cohesive life history and adaptive strategies. It is theoretically unlikely that one particular element of this complex web will always have been the primary cause of evolutionary changes in brain size, and attempts to identify such a "key element" are methodologically compromised by sensitivity to differences in data quality and analytical methods.

This brings us back to the issue of "causation" in evolutionary explanation. As alluded to earlier, Dunbar (1998) rejects constraint-based (energetic, developmental) hypotheses of brain evolution by arguing that constraints simply represent obstacles to be overcome by the causal force of selection. Thus, constraint-based explanations "do not do not tell us why brains actually evolved as they did" (p. 179). However, it is clear that selection acts on the actual (net) fitness effects of a trait, and that these are determined by both costs and benefits. The causal reason why "brains evolved as they did" can just as easily be a change in costs as a change in benefits. Thus, the observed pattern of variation across species might in fact be "explained" by constant brain size benefits interacting with variable life history constraints (van Schaik, Isler, and Burkart 2012).

Interestingly enough, Cartmill (2002, p. 197) also makes this suggestion, arguing that "Evidently, bigger brains are advantageous in all sorts of ways of life. . . . So why aren't other mammals as smart as we are? Perhaps they can't afford it." Indeed, the generalized utility of enlarged brains has been supported by research on primate "general intelligence" (Reader, Hager, and Laland 2011, Reader and Laland 2002) and mammalian invasion of new habitats (Sol et al. 2008). This suggests an interesting rapprochement between the universalizing approach of Cartmill and the historicism of Gould: Brain evolution is indeed explained by a general "law" (all else equal, bigger brains are better), but with particular effects that are contingent on a diverse set of context-dependent benefits and constraints. These variables may themselves interact in lawful ways (e.g., Isler and Van Schaik 2014), but with sufficient complexity and sensitivity to conditions that individual outcomes can only really be explained on a case-by-case basis as contingent historical sequences (cf. Gould 1995).

Explaining the evolution of individual species

Another major problem for the SBH is its inability explain apparent "grade shifts" in the relation between social complexity and brain size (van Schaik, Isler, and Burkart 2012, Isler and Van Schaik 2014). The existence of such shifts has been apparent from the outset (Dunbar 1992), and Dunbar (1998, p. 185) suggested that they reflect qualitative differences in social organization such that "apes require more computing power to manage the same number of relationships that monkeys do, and monkeys in turn require more than prosimians do." This argument may in fact be accurate, and it has been extended to explain brain size variation associated with different mating systems across a wide array of mammals and birds (Dunbar and Shultz 2007). Nevertheless, it is an ad hoc accommodation of observed discontinuities in the data and does not provide any causal explanation or predictive criteria for the occurrence of grade shifts.

This is problematic for the SBH in two ways. First, a substantial amount of the brain size variation we are interested in explaining is attributed to grade shifts, and there is no evidence that the forces driving any or all of these shifts were "explicitly social." Byrne (1997), for example, suggested that the discontinuity between apes and monkeys is more likely due to selection on "technical intelligence." In the original data set of Dunbar (1992), this unexplained hominoid shift (+0.83 to intercept) is not much less than the total range of variation in neocortex ratios (1.12) within hominoids.

Second, it is not clear where to draw the line between grade shifts and unexplained residual variation. At the level of individual species, such as unexpectedly large-brained aye-ayes (van Schaik, Isler, and Burkart 2012), these explanations are actually equivalent: The expected relationship has broken down and some additional explanation is needed. In larger, species-rich taxonomic groups, such as the paraphyletic monkey grade (e.g., Dunbar 1998), it is possible to show that a reliable group size-brain size relationship is indeed present (but see previous section regarding interpretation) even if the intercept has shifted relative to other groups. However, problems arise with species-poor taxonomic groups such as the apes. With only three hominoid data points (Dunbar 1998), it is difficult to tell if there really is a reliable correlation ($r^2=0.89$, $p=0.216$). Are we dealing with a grade shift followed by SBH business-as-usual, or has the predictive group size-brain size relationship simply broken down for this taxonomic group? This is a critical point for applications of the SBH to human evolution, as it is the 3-species hominoid regression that is used to predict hominin group sizes (Gowlett, Gamble, and Dunbar 2012).

Unfortunately, there is no way around the paucity of extant hominoid species, but it is at least possible to include two more: orangutans and bonobos. Orangutan neocortex ratio values have been available for some time (e.g., Dunbar 1993) but the species is typically excluded from SBH analyses, perhaps due to uncertainty about the relevant group size. Indeed, recent research repudiates the classic characterization of orangutans as "solitary" and instead recognizes the existence of social "clusters" of related females together with associated males and immatures (Mitra Setia et al. 2009). Here we use a group size value of 31, calculated as the average size (6) of the two female clusters described by Singleton and van Schaik (2002) plus the number of associated males and immatures expected from population sex ratios (ibid.). Group sizes for other species are from Lehmann et al. (2007) and neocortex ratios are from the MRI data of Rilling and Insel (1999).

Regression results (figure 14.2, solid line) fail to show any clear relationship between neocortex ratio and group size (slope: $B=0.084$, $t(5)=0.073$, $p=0.947$; regression: $r^2=0.262$, $F(1, 4)=1.064$, $p=0.378$), and yield a predicted human group size of 62. This suggests that the SBH does not apply to hominoids and cannot be used to predict hominin group sizes, including Dunbar's Number of 150 for modern humans. Of course it is possible that these particular results are unreliable due to inaccuracies in the data sets used. But this is exactly the problem. As discussed previously, the presence of an association between neocortex ratio and group size is robust across data sets and analysis methods but the strength and shape of this association is more sensitive. For example, it has recently been shown that accounting for intraspecific variation in group size can substantially alter regression slopes (Sandel et al. 2016). This calls into question the reliability of model-based predictions for individual species, a problem that can only be compounded by using small numbers of species to predict the extreme values associated with humans. Results obtained here illustrate the fragility of the empirical argument that human brain evolution has been a straightforward extension of a hominoid trend.

In the absence of such empirical support, the SBH provides no theoretical justification for assuming that humans occupy the same "grade" as other hominoids. This is again illustrated by figure 14.2. Even at a glance, it is readily apparent that the regression is heavily leveraged by two species: gibbons (small group size) and gorillas (small neocortex). A reasonable argument could be made for excluding either one (e.g., derived pair-bonding in gibbons, derived cerebellar expansion in gorillas), with major implications for the inferred hominoid regressions (figure 14.2, dashed lines) and human group size predictions (87 vs. 9,913). In essence, these ad hoc accommodations are hypotheses about additional, unpredicted grade shifts (represented by single species) and must be evaluated with reference to arguments and evidence external to the SBH model. Similarly, there is no SBH-internal reason to assume that one or more such grade shifts did not also occur over human evolution, and no way to test the proposition absent detailed archaeological and paleontological evidence of what actually happened in the past.

Archaeology and the Contingent Brain

Any simple regression of real-world biological data can be expected to leave some unexplained residual variation. In the case of the SBH, this residual variation is both substantial and theoretically interesting. While it is possible to address this variation through ad hoc accommodation, as discussed above, the preferable alternative is to develop a more comprehensive model that can support systematic explanations. Such a model might not be simple enough to allow reliable prediction from one or two key variables, but would allow principled explanation of sufficiently detailed individual cases.

Expensive, cultural brains

The most comprehensive account of brain-size evolution currently available is that of van Schaik and colleagues, which brings together two core elements: the "expensive brain" (Isler and Van Schaik 2014) and the "cultural brain" (van Schaik, Isler, and Burkart 2012).

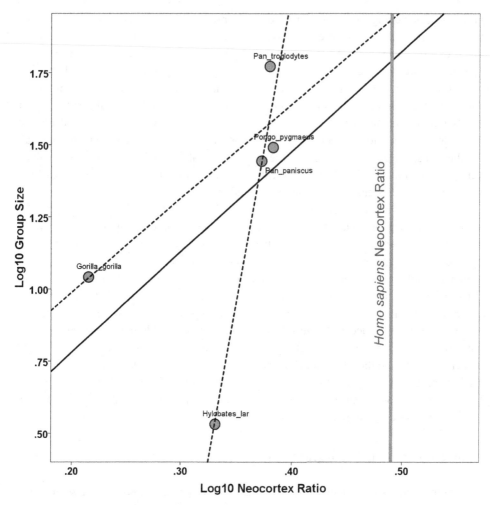

Figure 14.2
Relationship between neocortex ratio and group size in hominoids. Solid line (*n.s.*) is for whole sample, dashed lines exclude either gorillas or gibbons. Grey line indicates human neocortex ratio. Orangutan group size calculated from Singleton and van Schaik (2002); other species as reported by Lehmann et al. (2007). Neocortex ratios are from the MRI data of Rilling and Insel (1999).

The underlying assumption is that, all else equal, bigger brains are generally advantageous (i.e., potentially favored by a large number of different selective pressures). This is supported by comparative evidence of a correlation between brain size and behavioral flexibility or "general intelligence" (Reader and Laland 2002, Reader, Hager, and Laland 2011) and is consistent with the domain-general nature (Duncan 2010) of the large-scale functional networks that occupy much of human (Power et al. 2011) and monkey (Neubert et al. 2014) neocortex. Conserved developmental mechanisms appear to favor the disproportionate expansion of these flexible association networks (Buckner and Krienen 2013, Finlay and Uchiyama 2015) in response to any selection on brain size, which may in turn help explain the convergent evolution of general intelligence (e.g., spanning foraging, sociality, and tool-use) in taxa ranging from birds to cetaceans (Van Horik, Clayton, and Emery 2012). At the same time, the functional plasticity of developing cortex (e.g., Bedny et al. 2011), raises the possibility that even species differences in "modular" abilities (e.g., Amici et al. 2012) may arise through divergent developmental selection (Heyes 2003, Dehaene and Cohen 2007, Heyes and Frith 2014) on more general shared substrates.

Given that any individual could likely benefit in some way from more neural tissue, the "expensive brain" framework seeks to explain interspecific variation in brain size with respect to *net* fitness effects that take energetic and life history constraints into account. Thus, larger brains can only evolve if mortality is low enough to reward investment in slowly developing "embodied capital" (Kaplan et al. 2000) and a sufficient energy budget can be found through increased intake and/or reallocation. Importantly, many of these relationships are inherently multidirectional. For example, brain enlargement initially funded by a shift to a higher quality diet might produce general cognitive benefits with further impacts on both foraging productivity (Genovesio, Wise, and Passingham 2014) and predation avoidance (Byrne and Bates 2007), offsetting metabolic costs, reducing mortality and allowing further brain expansion. Holloway (1967) already emphasized the importance of addressing such feedback or "deviation amplification" relationships in human brain evolution, as well as the difficulty of identifying a primary cause or "initial kick" in such a complex interacting system. An additional implication is that many different causes can have the same effect, so that the long-term trend toward brain expansion along the human line need not imply a similar constancy in selective context.

To this already complex framework, the "cultural brain" (van Schaik and Burkart 2011, van Schaik, Isler, and Burkart 2012) adds the possibility of gene-culture coevolution. Modeling indicates that, if baseline conditions of frequency, learning ability, and skill complexity are met, social learning can increase the mean fitness of a population and lead to cumulative cultural evolution (Henrich and McElreath 2003). This generates yet another potential feedback relationship, in which increasingly complex, socially learned skills both fund and require greater investment in neural tissue, as well as requiring/promoting social tolerance (van Schaik and Burkart 2011), slower life histories, and extensive resource transfers (Kaplan et al. 2000). Humans are seen as an extreme extension of this trend, characterized by reliance on a technological niche (Boyd, Richerson, and Henrich 2011, Stout and Khreisheh 2015, Sterelny 2007) subsiding redistributive "biocultural reproduction" (Bogin, Bragg, and Kuzawa 2014). The cultural brain thus places great importance on sociality but differs from the SBH by explaining fitness benefits within the model (benefits of increasing group size are attributed to exogenous predation pressure by the SBH [Dunbar and Shultz 2007]). More broadly, the cultural brain helps to erode a problematic dichotomy between the "ecological" and the "social" that is difficult to recognize either in the real lives of primates (Rapaport and Brown 2008) or in the brain (Stout 2010).

Implications for the integrated study of human brain evolution

Accepting that there are many possible evolutionary pathways to brain enlargement, it follows that a causal explanation of human brain evolution cannot simply be inferred from its end product. Insofar as outcomes are contingent upon specific conditions, a properly historical account is required. A comparative approach can reveal the law-like regularity of these contingencies, but it is up to the historical sciences of archaeology, paleontology, and geology to provide the narrative element of Cartmill's (2012) *modus ponens* argument structure. This is not to suggest that the aim is to reconstruct a complete, narrative prehistory (cf. Gowlett, Gamble, and Dunbar 2012). Rather, the focus should be on providing rigorous evidence of change through time on the critical contingencies identified by the comparative model. These include brain and body size, diet quality, foraging efficiency, life history, cooperation, sharing, and cultural skill accumulation (Schuppli et al. 2016, Isler and Van Schaik 2014). It is a striking theoretical validation that this list of key issues is nothing new to archaeologists (e.g., Isaac 1971, Washburn 1960). Much remains to be done, but decades of research on these issues have already led to substantial empirical and methodological progress (reviewed by Antón, Potts, and Aiello 2014). Further progress will require ever more thorough integration across the evolutionary and behavioral sciences.

Indeed, evolutionary theory itself has undergone a renovation over the past 25 years, leading to calls for a new, "extended" evolutionary synthesis (Laland et al. 2015). Students of human evolution may now appreciate a broader range of evolutionary processes, including reciprocal causation (organisms as active agents in evolution), inclusive inheritance (more than just genes), and developmental bias (including phenotypic accommodation). This has led to integrated accounts of human evolution that emphasize the role of developmental plasticity, evolvability, niche construction, and cultural evolution (Antón, Potts, and Aiello 2014, Fuentes 2015). Such accounts confront researchers with an increasingly complex web of interacting causes to consider, but also suggest new opportunities for inference from available material remains to the biological and behavioral variables of interest. Examples include links between behavior and plastic anatomical responses (Skinner et al. 2015, Hecht et al. 2015); anatomy, technology, and foraging efficiency (Marzke et al. 2015, Zink and Lieberman 2016); artifacts and socially facilitated learning (Stout and Khreisheh 2015, Morgan et al. 2015, Fragaszy et al. 2013); and development, neuroanatomy, and cognition (Byrge, Sporns, and Smith 2014, Hublin, Neubauer, and Gunz 2015). Particularly promising is the adoption of experimental methods from the behavioral and neural sciences to better understand the implications of reconstructed Paleolithic behaviors (review in Stout and Hecht 2015), and the study of modern individual variation across genetics, anatomy, behavior, and cognition as a window on evolutionary relationships (Bruner et al. 2017, Hopkins et al. 2015, Thornton and Lukas 2012). Formal evolutionary modeling (e.g., Morgan 2016) will be an important conceptual tool as the complexity of interacting processes under consideration increasingly exceeds the scope of informal linguistic arguments.

Despite such causes for optimism, however, we remain a long way from a providing a detailed causal explanation of human brain evolution. Isler and van Schaik (2014) do suggest a plausible scenario in which cooperative hunting assumed major importance in hominin subsistence as early as 2.5 million years ago, leading to improved diet quality and stability and thus to sharing, cooperative breeding, and reduced mortality. This would have initiated a coevolutionary feedback loop between brain, behavior, technology, and life history ultimately leading to the modern human condition. Like Holloway (1967), however, Isler and van Schaik recognize the inherent difficulty of identifying the "initial kick" for such an evolutionary feedback loop. Archaeological evidence of early cooperative hunting is quite scant, being limited to a few cut marks on large animal bones (Semaw et al. 2003, De Heinzelin et al. 1999) and, depending on how big an initial kick we think is needed, we might also consider more subtle causes, such as dietary shifts (Ungar 2012) and/or technology assisted food processing (Zink and Lieberman 2016) as logical antecedents. Given this potential for regress, we may never be able to point to a singular "first cause" of hominin brain expansion. What we can do, and what is ultimately more interesting, is learn quite a lot about the timing, order, and causal relations of the broad span of human evolution. This has been and should continue to be the project of hominin paleontology and Paleolithic archaeology, with a renewed focus on tailoring fieldwork and methodological development to address questions and insights derived from comparative and evolutionary biology. Some high priorities for attention include evidence of cooperative hunting and sharing (Stiner, Barkai, and Gopher 2009), diet quality (Braun et al. 2010), life history variation (Hublin, Neubauer, and Gunz 2015), technological accumulation (Perreault et al. 2013), and the acquisition and social transmission of skills (Stout and Khreisheh 2015).

Conclusion

Sadly, comparative evidence reviewed here does not support a simple, mono-causal explanation of human brain evolution. This does not, however, mean that accounts of human brain evolution are stuck "just telling stories" without scientific content. Rather, it is possible to recognize evolutionary processes as both contingent *and* regular, both narrative *and* lawful. Indeed, it is in the very nature of scientific explanation to combine specific conditions with general principles. The amount of historical detail required depends on the complexity and sensitivity of the causal relations involved, as well as the explanatory detail desired. Since we desire a through explanation of the evolutionary history of a complex trait (brain size) for one particular species (*Homo sapiens*), the level of required detail is high. Fortunately, the historical sciences of archaeology and paleontology have a long tradition of pursuing such detail. If properly informed by progress in comparative and evolutionary biology, continued research in these disciplines will yield increasingly detailed scientific explanations of human brain evolution.

Acknowledgments

Thanks to Adrian Jaeggi for advice on primate comparative data sources and orangutan social structure, to Jeffrey Schwartz, Gerd Müller, Eva Lackner, and all the KLI staff for organizing the workshop, and to the participants for lively discussion.

References

Aiello, Leslie C., and R. I. M. Dunbar. 1993. "Neocortex Size, Group Size, and the Evolution of Language." *Current Anthropology* 34(2):184–193. doi: 10.2307/2743982.

Amici, Federica, Bradley Barney, Valen E Johnson, Josep Call, and Filippo Aureli. 2012. "A Modular Mind? A Test Using Individual Data from Seven Primate Species." *PLoS One* 7(12):e51918.

Antón, Susan C., Richard Potts, and Leslie C. Aiello. 2014. "Evolution of Early *Homo*: An Integrated Biological Perspective." *Science* 345(6192):1236828.

Bedny, Marina, Alvaro Pascual-Leone, David Dodell-Feder, Evelina Fedorenko, and Rebecca Saxe. 2011. "Language Processing in the Occipital Cortex of Congenitally Blind Adults." *Proceedings of the National Academy of Sciences* 108(11):4429–4434.

Bennett, Drake. 2013. "The Dunbar Number, from the Guru of Social Networks." *Bloomberg Businessweek*. http://www.businessweek.com/articles/2013-01-10/the-dunbar-number-from-the-guru-of-social-networks. Accessed: March 15, 2016.

Bogin, Barry, Jared Bragg, and Christopher Kuzawa. 2014. "Humans Are Not Cooperative Breeders but Practice Biocultural Reproduction." *Annals of Human Biology* 41(4):368–380.

Boyd, Robert, Peter J. Richerson, and Joseph Henrich. 2011. "The Cultural Niche: Why Social Learning Is Essential for Human Adaptation." *Proceedings of the National Academy of Sciences* 108(Supplement 2):10918–10925. doi: 10.1073/pnas.1100290108.

Braun, David R., John W. K. Harris, Naomi E. Levin, Jack T. McCoy, Andy I. R. Herries, Marion K. Bamford, Laura C. Bishop, Brian G. Richmond, and Mzalendo Kibunjia. 2010. "Early Hominin Diet Included Diverse Terrestrial and Aquatic Animals 1.95 Ma in East Turkana, Kenya." *Proceedings of the National Academy of Sciences* 107(22):10002–10007. doi: 10.1073/pnas.1002181107.

Bruner, Emiliano, Todd M Preuss, Xu Chen, and James K Rilling. 2016. "Evidence for Expansion of the Precuneus in Human Evolution." *Brain Structure and Function*:1–8.

Buckner, Randy L., and Fenna M. Krienen. 2013. "The Evolution of Distributed Association Networks in the Human Brain." *Trends in Cognitive Sciences* 17(12):648–665.

Byrge, Lisa, Olaf Sporns, and Linda B. Smith. 2014. "Developmental Process Emerges from Extended Brain–Body–Behavior Networks." *Trends in Cognitive Sciences* 18(8):395–403. doi: http://dx.doi.org/10.1016/j.tics.2014.04.010.

Byrne, Richard. 1997. "The Technical Intelligence Hypothesis: An Additional Evolutionary Stimulus to Intelligence?" In *Machiavellian Intelligence II: Extensions and Evaluations*, edited by Andrew Whiten and Richard Byrne, 289–311. Cambridge: Cambridge University Press.

Byrne, Richard, and Lucy Bates. 2007. "Sociality, Evolution and Cognition." *Current Biology* 17:R714–R723.

Byrne, Richard, and Andrew Whiten. 1988. *Machiavellian Intelligence: Social Expertise and the Evolution of Intellect in Monkeys, Apes, and Humans*. Oxford: Oxford University Press.

Cartmill, Matt. 1990. "Human Uniqueness and Theoretical Content in Paleoanthropology." *International Journal of Primatology* 11(3):173–192.

Cartmill, Matt. 2002. "Paleoanthropology: Science or Mythological Charter?" *Journal of Anthropological Research* 58(2): 183–201.

Chapman, Colin A, and Kim Valenta. 2015. "Costs and Benefits of Group Living Are Neither Simple nor Linear." *Proceedings of the National Academy of Sciences* 112(48):14751–14752.

Charvet, Christine J., and Barbara L. Finlay. 2012. "Embracing Covariation in Brain Evolution: Large Brains, Extended Development, and Flexible Primate Social Systems." *Progress in Brain Research* 195:71–87. doi: 10.1016/B978-0-444-53860-4.00004-0.

Cleland, Carol. 2002. "Methodological and Epistemic Differences between Historical Science and Experimental Science." *Philosophy of Science* 69(3):447–451. doi: 10.1086/342455.

De Heinzelin, Jean, J. Desmond Clark, Tim White, William Hart, Paul Renne, Giday WoldeGabriel, Yonas Beyene, and Elisabeth Vrba. 1999. "Environment and Behavior of 2.5-million-year-old Bouri Hominids." *Science* 284(5414):625–629.

Deacon, Terrence W. 1997. *The Symbolic Species: The Co-evolution of Language and the Brain.* New York: W.W. Norton.

Deaner, Robert O., Charles L. Nunn, and Carel P. van Schaik. 2000. "Comparative Tests of Primate Cognition: Different Scaling Methods Produce Different Results." *Brain, Behavior & Evolution* 55(1):44. doi: 10.1159/000006641.

Dehaene, Stanislas, and Laurent Cohen. 2007. "Cultural Recycling of Cortical Maps." *Neuron* 56(2):384–398. doi: http://dx.doi.org/10.1016/j.neuron.2007.10.004.

Dunbar, R. I. M., and Susanne Shultz. 2007. "Evolution in the Social Brain." *Science* 317(5843):1344–1347.

Dunbar, Robin I. M. 1992. "Neocortex Size as a Constraint on Group Size in Primates." *Journal of Human Evolution* 22(6):469–493.

Dunbar, Robin I. M. 1993. "Coevolution of Neocortical Size, Group Size and Language in Humans." *Behavioral and Brain Sciences* 16(4):681–694.

Dunbar, Robin I. M. 1998. "The Social Brain Hypothesis." *Evolutionary Anthropology: Issues, News, and Reviews* 6(5):178–190.

Dunbar Robin I. M. 2003. "The Social Brain: Mind, Language, and Society in Evolutionary Perspective." *Annual Review of Anthropology*, 32(1), 163–181.

Duncan, John. 2010. "The Multiple-Demand (MD) System of the Primate Brain: Mental Programs for Intelligent Behaviour." *Trends in Cognitive Sciences* 14(4):172–179.

Eldredge, N., and Gould, S. J. 1972 "Punctuated Equilibria: An Alternative to Phyletic Gradualism." In *Models in Paleobiology*, edited by T. J. M. Schopf, 82–115. San Francisco: Freeman, Cooper & Co.

Finlay, Barbara, and Richard Darlington. 1995. "Linked Regularities in the Development and Evolution of Mammalian Brains." *Science* 268:1578–1584.

Finlay, Barbara L., and Ryutaro Uchiyama. 2015. "Developmental Mechanisms Channeling Cortical Evolution." *Trends in Neurosciences* 38(2):69–76.

Fitch, W. Tecumseh. 2012. "Evolutionary Developmental Biology and Human Language Evolution: Constraints on Adaptation." *Evolutionary Biology* 39(4):613–637.

Fragaszy, D. M., D. Biro, Y. Eshchar, T. Humle, P. Izar, B. Resende, and E. Visalberghi. 2013. "The Fourth Dimension of Tool Use: Temporally Enduring Artefacts Aid Primates Learning to Use Tools." *Philosophical Transactions of the Royal Society B: Biological Sciences* 368(1630). doi: 10.1098/rstb.2012.0410.

Fuentes, Agustín. 2015. "Integrative Anthropology and the Human Niche: Toward a Contemporary Approach to Human Evolution." *American Anthropologist* 117(2):302–315.

Genovesio, Aldo, Steven P. Wise, and Richard E. Passingham. 2014. "Prefrontal–Parietal Function: From Foraging to Foresight." *Trends in Cognitive Sciences* 18(2):72–81.

Gould, S. J., and R. C. Lewontin. 1979. "The Spandrels of San Marco and the Panglossian Paradigm: A Critique of the Adaptationist Programme." *Proceedings of the Royal Society of London. Series B, Biological Sciences* 205(1161):581–598.

Gould, Stephen Jay. 1977. *Ontogeny and Phylogeny*. Cambridge, MA: Harvard University Press.

Gould, Stephen Jay. 1995. "'What Is Life?' as a Problem in History." In *What Is Life? The Next Fifty Years*, edited by M. P. Murphey and L. A. J. O'Neill, 25–39. Cambridge: Cambridge University Press.

Gould, Stephen Jay, and Elisabeth S. Vrba. 1982. "Exaptation-A Missing Term in the Science of Form." *Paleobiology* 8(1):4–15.

Gowlett, J., C. Gamble, and R. Dunbar. 2012. "Human Evolution and the Archaeology of the Social Brain." *Current Anthropology* 53(6):693–722.

Harvey, Paul H., and Mark D. Pagel. 1991. *The Comparative Method in Evolutionary Biology*. Oxford: Oxford University Press.

Hecht, E. E., D. A. Gutman, N. Khreisheh, S. V. Taylor, J. Kilner, A. A. Faisal, B. A. Bradley, T. Chaminade, and D. Stout. 2014. "Acquisition of Paleolithic Toolmaking Abilities Involves Structural Remodeling to Inferior Frontoparietal Regions." *Brain Structure and Function* 220(4): 2315–2331.

Henrich, Joseph, and Richard McElreath. 2003. "The Evolution of Cultural Evolution." *Evolutionary Anthropology: Issues, News, and Reviews* 12(3):123–135. doi: 10.1002/evan.10110.

Heyes, Cecilia. 2003. "Four Routes of Cognitive Evolution." *Psychological Review October* 110(4):713–727.

Heyes, Cecilia, and Chris D. Frith. 2014. "The Cultural Evolution of Mind Reading." *Science* 344(6190):1243091.

Holloway, Ralph L. 1967. "The Evolution of the Human Brain: Some Notes Toward a Synthesis between Neural Structure and the Evolution of Complex Behavior." *General Systems* 12:3–19.

Hopkins, William D., Lisa Reamer, Mary Catherine Mareno, and Steven J. Schapiro. 2015. "Genetic Basis in Motor Skill and Hand Preference for Tool Use in Chimpanzees (*Pan Troglodytes*)." *Proceedings of the Royal Society B: Biological Sciences* 282(1800):20141223. doi: 10.1098/rspb.2014.1223.

Hublin, Jean-Jacques, Simon Neubauer, and Philipp Gunz. 2015. "Brain Ontogeny and Life History in Pleistocene Hominins." *Philosophical Transactions of the Royal Society of London B: Biological Sciences* 370(1663):20140062.

Isaac, Glynn. 1971. "The Diet of Early Man: Aspects of Archaeological Evidence from Lower and Middle Pleistocene Sites in Africa." *World Archaeology* 2(3):278–299.

Isler, Karin, and Carel P. Van Schaik. 2014. "How Humans Evolved Large Brains: Comparative Evidence." *Evolutionary Anthropology: Issues, News, and Reviews* 23(2):65–75. doi: 10.1002/evan.21403.

Janson, Charles H., and Michele L. Goldsmith. 1995. "Predicting Group Size in Primates: Foraging Costs and Predation Risks." *Behavioral Ecology* 6(3):326–336. doi: 10.1093/beheco/6.3.326.

Jolly, Alison. 1966. "Lemur Social Behavior and Primate Intelligence." *Science* 153(3735):501–506. doi: 10.1126/science.153.3735.501.

Kaplan, Hillard, Kim Hill, Jane Lancaster, and A. Magdalena Hurtado. 2000. "A Theory of Human Life History Evolution: Diet, Intelligence, and Longevity." *Evolutionary Anthropology: Issues, News, and Reviews* 9(4):156–185.

Laland, Kevin N., Tobias Uller, Marcus W. Feldman, Kim Sterelny, Gerd B. Müller, Armin Moczek, Eva Jablonka, and John Odling-Smee. 2015. "The Extended Evolutionary Synthesis: Its Structure, Assumptions and Predictions." *Proceedings of the Royal Society of London B: Biological Sciences* 282(1813). doi: 10.1098/rspb.2015.1019.

Lehmann, Julia, A. H. Korstjens, and R. I. M. Dunbar. 2007. "Group Size, Grooming and Social Cohesion in Primates." *Animal Behaviour* 74(6):1617–1629.

Lewontin, R. C. 1998. "The Evolution of Cognition: Questions We Will Never Answer." In *An Invitation to Cognitive Science, Volume 4: Methods, Models, and Conceptual Issues*, edited by D. Scarborough and S. Sternberg, 107–132. Cambridge, MA: MIT Press.

Marzke, Mary W., Linda F. Marchant, William C. McGrew, and Sandra P. Reece. 2015. "Grips and Hand Movements of Chimpanzees during Feeding in Mahale Mountains National Park, Tanzania." *American Journal of Physical Anthropology* 156(3):317–326.

Mitra Setia, T., R. A. Delgado, S. S. Utami Atmoko, I. Singleton, and C. P. Van Schaik. 2009. "Social Organization and Male-Female Relationships." In *Orangutans, Geographic Variations in Behavioral Ecology and Conservation*, edited by Serge A. Wich, S. Suci Utami Atmoko, Tatang Mita Setia amd Carel P. van Schaik, 245–253. New York: Oxford University Press.

Morgan, Thomas J. H. 2016. "Testing the Cognitive and Cultural Niche Theories of Human Evolution." *Current Anthropology* 57(3):370–377.

Morgan, T. J. H., N. T. Uomini, L. E. Rendell, L. Chouinard-Thuly, S. E. Street, H. M. Lewis, C. P. Cross, C. Evans, R. Kearney, and I. de la Torre. 2015. "Experimental Evidence for the Co-evolution of Hominin Tool-Making Teaching and Language." *Nature Communications* 6:1–6. doi:10.1038/ncomms7029

National Academy of Sciences. 2008. *Science, Evolution, and Creationism*. Washington, DC: National Academies Press.

Navarrete, Ana F., Simon M. Reader, Sally E. Street, Andrew Whalen, and Kevin N. Laland. 2016. "The Coevolution of Innovation and Technical Intelligence in Primates." *Philosophical Transactions of the Royal Society of London B* 371(1690):20150186.

Neubert, Franz-Xaver, Rogier B. Mars, Adam G. Thomas, Jerome Sallet, and Matthew F. S. Rushworth. 2014. "Comparison of Human Ventral Frontal Cortex Areas for Cognitive Control and Language with Areas in Monkey Frontal Cortex." *Neuron* 81(3):700–713.doi.org/10.1016/j.neuron.2013.11.012.

Pérez-Barbería, F. Javier, Susanne Shultz, and Robin I. M. Dunbar. 2007. "Evidence for Coevolution of Sociality and Relative Brain Size in Three Orders of Mammals." *Evolution* 61(12):2811–2821.

Perreault, Charles, P. Jeffrey Brantingham, Steven L. Kuhn, Sarah Wurz, and Xing Gao. 2013. "Measuring the Complexity of Lithic Technology." *Current Anthropology* 54(S8):S397–S406.

Power, Jonathan D., Alexander L. Cohen, Steven M. Nelson, Gagan S. Wig, Kelly Anne Barnes, Jessica A. Church, Alecia C. Vogel, Timothy O. Laumann, Fran M. Miezin, and Bradley L. Schlaggar. 2011. "Functional Network Organization of the Human Brain." *Neuron* 72(4):665–678.

Rapaport, Lisa G., and Gillian R. Brown. 2008. "Social Influences on Foraging Behavior in Young Nonhuman Primates: Learning What, Where, and How to Eat." *Evolutionary Anthropology: Issues, News, and Reviews* 17(4):189–201.

Reader, Simon M., Yfke Hager, and Kevin N. Laland. 2011. "The Evolution of Primate General and Cultural Intelligence." *Philosophical Transactions of the Royal Society of London B: Biological Sciences* 366(1567):1017–1027.

Reader, Simon, and Kevin Laland. 2002. "Social Intelligence, Innovation and Enhanced Brain Size in Primates." *Proceedings of the National Academy of Sciences* 99:4436–4441.

Rilling, James K., and Thomas R. Insel. 1999. "The Primate Neocortex in Comparative Perspective Using Magnetic Resonance Imaging." *Journal of Human Evolution* 37:191–223.

Sandel, Aaron A., Jordan A. Miller, John C. Mitani, Charles L. Nunn, Samantha K. Patterson, and László Zsolt Garamszegi. 2016. "Assessing Sources of Error in Comparative Analyses of Primate Behavior: Intraspecific Variation in Group Size and the Social Brain Hypothesis." *Journal of Human Evolution* 94:126–133. doi: doi .org/10.1016/j.jhevol.2016.03.007.

Schillaci, Michael. 2015. "Letter to the Editor: Body Mass as a Confounding Variable When Predicting Group Size from Orbit Diameter and Neocortex Ratio." *American Journal of Physical Anthropology* 158(1):170–171.

Schuppli, Caroline, Sereina M. Graber, Karin Isler, and Carel P. van Schaik. 2016. "Life History, Cognition and the Evolution of Complex Foraging Niches." *Journal of Human Evolution* 92:91–100. doi: doi.org/10.1016/j .jhevol.2015.11.007.

Semaw, Sileshi, Michael J. Rogers, Jay Quade, Paul R. Renne, Robert F. Butler, Manuel Dominguez-Rodrigo, Dietrich Stout, William S. Hart, Travis Pickering, and Scott W. Simpson. 2003. "2.6-Million-Year-Old Stone Tools and Associated Bones from OGS-6 and OGS-7, Gona, Afar, Ethiopia." *Journal of Human Evolution* 45:169–177.

Singleton, Ian, and Carel P. van Schaik. 2002. "The Social Organisation of a Population of Sumatran Orang-utans." *Folia Primatologica* 73(1):1–20.

Skinner, Matthew M., Nicholas B. Stephens, Zewdi J. Tsegai, Alexandra C. Foote, N. Huynh Nguyen, Thomas Gross, Dieter H. Pahr, Jean-Jacques Hublin, and Tracy L. Kivell. 2015. "Human-like Hand Use in *Australopithecus Africanus*." *Science* 347(6220):395–399.

Sol, Daniel, Sven Bacher, Simon M. Reader, and Louis Lefebvre. 2008. "Brain Size Predicts the Success of Mammal Species Introduced into Novel Environments." *The American Naturalist* 172(S1):S63–S71.

Sterelny, Kim. 2007. "Social Intelligence, Human Intelligence and Niche Construction." *Philosophical Transactions of the Royal Society of London B: Biological Sciences* 362(1480):719–730. doi: 10.1098/rstb.2006.2006.

Stiner, Mary C., Ran Barkai, and Avi Gopher. 2009. "Cooperative Hunting and Meat Sharing 400–200 kya at Qesem Cave, Israel." *Proceedings of the National Academy of Sciences* 106(32):13207–13212.

Stout, D., and E. E. Hecht. 2015. "Neuroarchaeology." In *Human Paleoneurology*, edited by Emiliano Bruner, 145–175. New York: Springer.

Stout, Dietrich. 2010. "The Evolution of Cognitive Control." *Topics in Cognitive Science* 2(4):614–630. doi: 10.1111/j.1756-8765.2009.01078.x.

Stout, Dietrich, and Nada Khreisheh. 2015. "Skill Learning and Human Brain Evolution: An Experimental Approach." *Cambridge Archaeological Journal* 25(04):867–875. doi: doi:10.1017/S0959774315000359.

Thornton, Alex, and Dieter Lukas. 2012. "Individual Variation in Cognitive Performance: Developmental and Evolutionary Perspectives." *Philosophical Transactions of the Royal Society of London B: Biological Sciences* 367(1603):2773–2783. doi: 10.1098/rstb.2012.0214.

Ungar, Peter S. 2012. "Dental Evidence for the Reconstruction of Diet in African Early *Homo*." *Current Anthropology* 53(S6):S318–S329. doi: 10.1086/666700.

Van Horik, J., N. Clayton, and N. Emery. 2012. "Convergent Evolution of Cognition in Corvids, Apes and Other Animals." In *The Oxford Handbook of Comparative Evolutionary Psychology*, edited by Jennifer Vonk, Todd Kennedy Shackelford, and Peter E. Nathan, 80–101. Oxford: Oxford University Press.

van Schaik, Carel P., and Judith M. Burkart. 2011. "Social Learning and Evolution: The Cultural Intelligence Hypothesis." *Philosophical Transactions of the Royal Society of London B: Biological Sciences* 366(1567):1008–1016.

van Schaik, Carel P., Karin Isler, and Judith M. Burkart. 2012. "Explaining Brain Size Variation: From Social to Cultural Brain." *Trends in Cognitive Sciences* 16(5):277–284.

Washburn, S. L. 1960. "Tools and Human Evolution." *Scientific American* 203(3):3–15.

Yopak, Kara E., Thomas J. Lisney, Richard B. Darlington, Shaun P. Collin, John C. Montgomery, and Barbara L. Finlay. 2010. "A Conserved Pattern of Brain Scaling from Sharks to Primates." *Proceedings of the National Academy of Sciences* 107(29):12946–12951. doi: 10.1073/pnas.1002195107.

Zink, Katherine D., and Daniel E. Lieberman. 2016. "Impact of Meat and Lower Palaeolithic Food Processing Techniques on Chewing in Humans." *Nature* 531(7595):500–503. doi: 10.1038/nature16990.

15 Brain Size and the Emergence of Modern Human Cognition

Ian Tattersall

Introduction

For reasons previously explored both by this author (Tattersall 1997, 2015) and by the editor of this volume (Schwartz 2006, 2016), paleoanthropology has been mired since the mid-twentieth century in the beguiling notion that evolution in the hominid family (hominin subfamily/tribe, if you prefer; the difference is notional) has consisted essentially of the gradual burnishing by natural selection of a central lineage that culminated in *Homo sapiens*. Yet accretions to the hominid fossil record over the same period have, in contrast, consistently shown that hominid phylogeny instead involved vigorous evolutionary experimentation. Over the seven-million-odd years of our family's existence, new species and lineages were regularly thrown out onto the ecological stage, to be triaged in competition with organisms both closely and distantly related. Extinction rates were high to match. Further, it is by now well established that all this took place in the context of constantly oscillating climates and habitats (deMenocal 2011), to which steady, perfecting adaptation would not have been possible, even in principle.

The diversity we see in the material record is, in fact, entirely in keeping with what we might expect from what we know of evolutionary process; and although paleoanthropological theory itself has significantly lagged, the practical need to accommodate the steady beat of new fossil discoveries over recent decades has meant that the number of extinct hominid species most paleoanthropologists are willing to recognize has significantly increased since the late twentieth century (there are 26 in the tentative phylogeny in figure 15.1, updated to 2016, as compared to a mere 11 in the earliest version of this diagram, drawn two decades earlier: Tattersall 1993). Nevertheless, the gradualist interpretive framework has tenaciously lingered, leading to the widespread application in practice of a strictly minimalist systematic approach that has often been justified by spectacularly contorted reasoning (see Spoor et al. [2007], and Lordkipanidze et al. [2014] for classic examples).

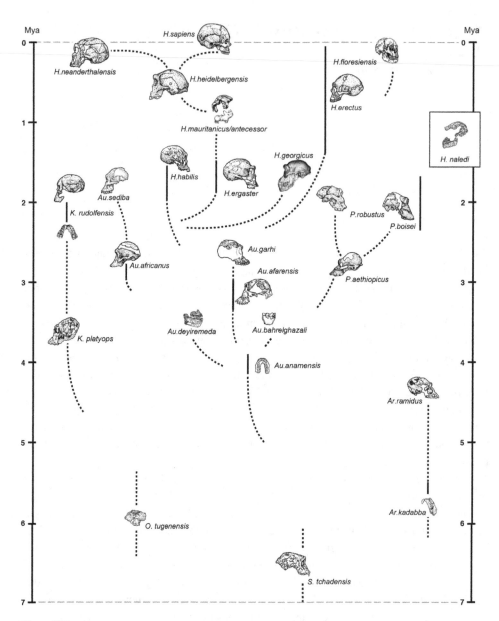

Figure 15.1
Provisional genealogical tree of the hominid family, showing both high diversity and that typically several hominid species have coexisted at any one time. Drawn by Patricia Wynne.

Hominid Brain Size Increase

The most seductive of the factors that superficially appear to sustain the gradualist model of human phylogeny is the seemingly inexorable increase in hominid brain volumes following the origin of the genus *Homo* (at least when this taxon is envisaged as a morphologically coherent entity; see Collard and Wood [2014] for a recent review). Very crudely, two million years (myr) ago, hominid brain sizes were only marginally larger than those of living apes; a million years ago they had on average doubled; and in the last million years they have doubled again. Within the gradualist framework, this dramatic increase certainly seems by itself to support the progressive intra-lineage interpretation of hominid phylogeny. But look closer, and the illusion disappears. For, even using Collard and Wood's (2014) limited concept of *Homo* (excluding such forms as *habilis*, *rudolfensis*, and *georgicus*, and in my view also *floresiensis* and *naledi*), the genus that contains and is defined by *Homo sapiens* is beginning to appear as a complex clade, rather than as a time-transgressive succession of intergrading species (Schwartz and Tattersall 2015, Tattersall 2016). Given this reality, it is clear that we can no longer continue to think in gradualist terms of the evolution of the various body systems—the "foot," the "brain," and so forth—in isolation from one another. It is the whole individual into which those systems are bundled, and the species to which such individuals belong, that are the relevant units of analysis.

As yet, we do not have anything like an accurate estimate of species diversity within the genus *Homo*, even as restricted by Collard and Wood; we are even farther from knowing what the stratigraphic and brain-volume ranges of the various species comprising it might have been. But it is already evident that brain volume increase proceeded independently in at least three different lineages within the genus. In Africa, in the period following about 1.7 myr ago, brain sizes in fossils attributed to *Homo* varied from 691 to 825 ml (nearly all values quoted here are from Holloway, Broadfield, and Yuan 2004). At around 1.0 myr ago, hominid brains were in the 995–1,067 ml range. In the period following about 600 thousand years (kyr) ago, they already stood at 1,225–1,325 ml; and very early *Homo sapiens* at Herto in Ethiopia, dated to~160 kyr ago, boasted a cranial capacity of 1,450 ml. In eastern Asia, early *Homo erectus* from Sangiran in Java comes in with brain volumes of 950–1,059 ml. And while the rather later Trinil holotype skullcap is estimated only at 940 ml, the indeterminately dated but probably significantly later specimens from Sambungmacan range from 917 to 1,006 ml. Finally, the ~40 kyr crania from Ngandong average 1,150 ml for six individuals. This pattern of brain size increase toward the recent is also seen in the Neanderthal clade in Europe, where the >430 kyr "pre-Neanderthal" sample from Atapuerca Sima de los Huesos comes in at a mean brain volume of 1,220 ml, while a sample of fully fledged *Homo neanderthalensis* (<200 kyr) averages 1,487 ml.

In Africa, Asia, and Europe, then, we see independent trends toward increasing brain volume with the passage of time. The trend is strongest among the systematically diverse hominids of Africa, and weakest within the single species *Homo erectus* in eastern Asia.

But overall, the signal is clear enough to indicate that the tendency to brain enlargement within *Homo* was a propensity of the genus as a whole, rather than specifically of the lineage leading to *Homo sapiens*. Significantly, this means that we are not necessarily justified in generalizing any specific cognitive property of our own (observable) living species to extinct members of its genus, at least from a strictly biological (as opposed to behavioral) perspective.

What, then, might have been driving the metabolically expensive average increase in brain volumes over the history of the genus *Homo*? Stout (this volume) elegantly demonstrates the shortcomings of the popular "Social Brain Hypothesis" (SBH: Dunbar 2003) as an explanation of such expansion. Through viewing neocortex-to-rest-of-brain ratios as proxies for general intelligence, and social group sizes as a measure of social complexity, the SBH proposes that enlarging brain size is directly associated with increasing social complexity, as an extension of a trend seen among the primates as a whole. However, seductive as this notion may be in storytelling terms, Stout convincingly shows how it fails to close the circle. Direct environmental causes as drivers of increasing intelligence are equally dubious: Dunbar (1998) found no significant correlation between neocortex ratio and any ecological measure he examined, while Hart and Sussman (2009) have very persuasively argued, by analogy with modern open-country primates such as baboons and macaques, that group sizes among the small-brained early hominids would—contrary to Dunbar's prediction—plausibly have been very large. What is more, while it might seem intuitively obvious (to members of a big-brained species) that the "high intelligence" presumptively associated with a large brain would have been advantageous for small-bodied and physically defenseless hominids living in predator-infested woodland and bushland environments, it is hard even to speculate about precisely what this would have meant in terms of behavioral repertoire.

At the level of intra-group dynamics, the invention of projectile weapons—which potentially, at least, made hominids each other's most dangerous enemies—might well have placed individual wiliness at a premium. But in a lineage whose members are also so clearly distinguished today by their extreme prosocial and cooperative tendencies, it is hard to see inter-individual or even inter-group conflict as a long-term selective dynamic inexorably driving brain sizes (and by extension intelligence) higher. At a higher hierarchical level, an equivalent explanation might (more convincingly) involve interspecies competition (or potentially, in the hominid context, interspecies conflict). And indeed, in in a world that was typically populated at any one time by several hominid species (at around 2 myr ago, for example, we have long had evidence for at least four just in the vicinity of Kenya's Lake Turkana, while Haile-Selassie et al. [2016] have recently pointed out something similar nearby for the period between about 3.8 and 3.3 myr ago), it is plausible in principle that larger-brained hominid species would regularly have outcompeted those with smaller brains—as, on the face of it, the empirical evidence also seems to indicate they did. Still, even this does not self-evidently give us the full picture because, from the very earliest phases of hominid history, several different kinds of hominid were routinely able to coexist (see figure 15.1), in a pattern that continued almost to the present day. Indeed, it was not until a species emerged that possessed a qualitatively new form of cognition that all the hominid competition abruptly disappeared (Tattersall 2012).

A very recent hypothesis (Piantosi and Kidd, 2016) posits that "extreme intelligence" in the human lineage might have arisen "through a positive evolutionary feedback loop" (much as the SBH also proposes). According to the authors, "Humans must be born unusually early to accommodate larger brains, but this gives rise to unusually helpless neonates. Caring for these children . . . requires more intelligence—thus even larger brains" (Piantadosi and Kidd 2016, p. 1). Still, although the extended dependency period of human infants is perhaps the most striking of all the many unusual features of human life history, the notion that ever-more-helpless newborns require ever-more intelligent parents to raise them shares with all feedback models the difficulty of establishing how the cycle started—and in this case, why it occurred independently in three lineages of *Homo*, and nowhere else. It also signally fails to explain the recent diminution in human brain size, or the qualitative nature of the switch to symbolic cognition (discussed later). The exact nature of the behavioral factor that evidently made metabolically expensive larger brains advantageous, driving brain size increase across the entire spectrum of the genus *Homo*, thus remains obscure for the moment.

Symbolic Cognition

Members of the living species *Homo sapiens* are distinguished from their closest extant relatives in many ways, but most strikingly in the manner in which they process the information provided by their senses. What modern human beings alone do (as far as we can tell) is to deconstruct their experience of the worlds both around and within them into a vocabulary of mental symbols. Those symbols can be recombined to produce new statements about those inner and outer worlds not only as they are, but as they *might* be. The ability to create and manipulate discrete mental representations in this way is the basis of our imagination and creativity, allowing us to conceive that the world might be other than our senses tell us it is. And the resulting capacity resonates in every domain of human experience. In combination with our clever hands—evidently a much earlier acquisition in our evolution—our ability to reason symbolically has allowed our species *Homo sapiens* to impact its environment in an entirely unprecedented way. Many other organisms can do one or another of the unusual things—make tools or utter words, for example—that our symbolic capacity, in combination with various fortuitously present peripheral structures (nimble hands, modern vocal tract), permits. But no other organism can do everything that this capacity of ours allows, or can function with such originality and inventiveness. Of course, many other creatures also significantly influence their environments (indulge in "niche construction"). For example, in tearing up vegetation for social as well as dietary reasons, gorillas keep plants at the early stage of succession that provides the young, tender shoots that they prefer. But no other organism (perhaps with the single exception of the first oxygen-producing cyanobacteria) has ever radically altered the world around it as *Homo sapiens* has (often equally unwittingly) done.

None of this, of course, means that we *Homo sapiens* are the only clever beings around, or even the only ones that can recognize and use symbols. For example, our close relatives the bonobos and chimpanzees can add symbols to make to make and understand simple statements such as "put . . . bottle . . . in . . . refrigerator." But the additive algorithm is ultimately limiting; and for all that apes are hugely intelligent in the larger scheme of things, we remain the only organisms that can shuffle symbols, according to rules, to mentally transcend the world that Nature presents to us. And we are equally the only extant organisms able to express the ideas our brains produce in structured articulate language—a quality that itself maps very closely on to our processes of thought (Hinzen 2012).

Clearly, in cognitive terms we *Homo sapiens* are qualitatively distinct from every other organism we can observe behaving in the natural world. We are not simply doing what chimpanzees are doing, only better or faster. We are manipulating and communicating information in a new and entirely unique fashion; and it is evident that when we compare ourselves to other primates we cannot treat raw "intelligence" as a single, continuous variable. It is *how* our minds process information that is the key element in our human cognitive uniqueness, and not the quantity of information processed, or how quickly our brains crunch it.

Still, there can be no rational question that *Homo sapiens* is intimately nested into the great Tree of Life that unites all living beings. Which, in turn, means that today's symbolic and linguistic people are ultimately descended from an ancestor that was neither of those things. So, at some point along the unique evolutionary trajectory of *Homo sapiens*, a cognitive and linguistic barrier was breached. A new *kind* of cognition was achieved, one that was not simply an improvement on the ancestral form. As I shall argue in the next section, this breaching was both recent and abrupt, a fact that renders only partially valid, at best, any behavioral generalizations about ourselves that might invoke deep evolutionary roots, or our similarities with other primates. Of course, our remarkable capacities are unquestionably underwritten by a history of evolutionary innovation that stretches back hundreds of millions of years to the first vertebrate brains; and because of this we share a huge amount of neural structure and process with ever-widening circles of our vertebrate relatives (and beyond). We could not be the kind of creature we are in the absence of any element of this complex neural heritage that took so long to accumulate. Nonetheless, all of our conscious mental processes are significantly modulated by our recently acquired ability to make symbolically transformed mental associations; and this associative overlayering of the more ancient intuitive and emotional brain has changed the cognitive rules by which we human beings operate.

It is important not to forget this, because we are closely bound by the ways in which we ourselves perceive the world. We find it difficult or more likely impossible to conceive of any mindset that is not our own, and the closer to us other forms are phylogenetically, the harder the exercise becomes. It is even very clear that not all modern people perceive the world in exactly the same way, as our fraught contemporary politics proves. And symbolic cognition makes culture of such supreme importance that, while all members of *Homo sapiens* are initially capable of absorbing the norms of any society, by the time they achieve adolescence members of different cultures may have strikingly divergent and often irreconcilable worldviews. Nonetheless, under most conditions individual *Homo sapiens* can operate from day to day on the assumption that the developmentally normal people around them are living perceptually in the same world that they are. We can reasonably suppose that others feel and calculate pretty much as we do, and we can anticipate their reactions accordingly. But on the other side of the cognitive divide, things are clearly different. We cannot know, or even make reliable assumptions about, what apes are feeling, still less about what they are "thinking." For all the similarities between us, the apes and we are clearly processing information about ourselves and our surroundings in very different ways. And this plain fact should give us pause when we are tempted to view even our closest extinct relatives as (simpler) versions of us. Crafty and resourceful as our ancestors may have been, and no matter how similar physically they may have been to us, whether or not they had breached the symbolic divide is almost certainly a better guide than phylogenetic propinquity to the way in which they perceived and interacted with the world. This makes it worthwhile to look briefly at the archaeological record with the aim of determining at what point that record begins to show evidence of behaviors that we might reasonably associate with the symbolic manipulation of information.

The Fossil and Archaeological Records

Over the course of human evolution, episodes of biological and technological innovation have consistently been out of phase (see discussion by Tattersall 2012). In other words, as far as can be told, behavioral innovations have always been made *within* preexisting species, rather than marking the (often fuzzy) boundaries between species. As judged from the lithic record (obviously a very oblique indicator of wider cognitive status, but the one that reliably preserves), innovations in this domain were characteristically rare, and separated by long periods of stasis or only minor refinement. Indeed, the pattern of constant technological change that conditions our experience (and expectations) today appears to be highly atypical in the larger human evolutionary context.

The time gap between the initial introduction of Mode 1 flaked stone tools and the appearance of shaped Mode 2 handaxes was probably considerably in excess of a million years (Lepre et al. 2011, Mc Pherron et al. 2010, Harmand et al. 2015); longer still was to elapse before the appearance of prepared-core techniques, the next major conceptual advance in lithic technology, which occurred at~300 kyr ago (Klein 2009). Over this extended period numerous hominid species came and went (figure 15.1); but although the lithic record strongly suggests that by its end hominids had become considerably more complex cognitively than they had been at the beginning, all the identifiable refinements in terms of their actual activities were strictly functional, rather than symbolic in nature. Resourceful and intelligent as, for example, *Homo heidelbergensis* clearly was (as, for example, the putative maker of the earliest known shelters, compound tools, throwing spears, and so forth: Thieme 1997, de Lumley 2016), there is little substantive evidence to suggest that any Middle Pleistocene hominid was manipulating information in a symbolic manner. Indeed, the only object known in the period preceding the introduction of prepared-core tools that might potentially encode symbolic significance is a~500 kyr-old mollusk shell from Trinil, in Java, that is incised with a pattern of zig-zag and parallel lines (Joordens et al. 2015). Yet while this curious object—presumably produced by a *Homo erectus*—indicates considerable manual dexterity on the part of its engraver, it remains a floating point, bereft of evidence that it was produced within a symbolic social tradition. The same may be also be said of the next putatively symbolic object, produced within both the tenure and the territory of *Homo heidelbergensis*: the~250 kyr "Venus" of Berekhat Ram, on the Golan Heights: a vaguely anthropomorphic lump of welded volcanic ash that some believe was deliberately grooved to appear more human-like (Goren-Inbar 1986, Nowell and d'Errico, 2000). Even *Homo neanderthalensis*, a well-documented hominid with a brain as large as our own, failed to produce a material record that is uncontestably symbolic. Fairly early on in the history of this species (which appeared < 200 kyr ago), it left evidence of eagle talons that had been grooved for stringing (Radovcic et al. 2015); and at the very end (>38.5 kyr) it is associated with a curious grid-like deep engraving (Rodriguez-Vidal et al. 2014). But once again these are isolated expressions, and there is precious little evidence that either was embedded within a more extensive symbolic context.

Perhaps even more remarkably, exactly the same is true for the earliest fossil *Homo sapiens*, known from deposits in Ethiopia dating between about 200 and 160 kyr ago (Clark et al. 2003, McDougall et al. 2005). At the site of Herto, lithics from sediments that yielded early anatomical humans are unremarkable for their period, and in some respects they are distinctly archaic, including the latest known occurrence in Africa of Mode 2 hand axes (Clark et al. 2003). It was not until around 100 kyr ago, when anatomically recognizable *Homo sapiens* had been in existence for a hundred thousand years, and the African Middle Stone Age lithic tradition had been established for yet longer, that we begin to pick up convincing evidence that symbolic activities had finally become incorporated into the behavioral repertoire of *Homo sapiens*.

A certain amount of controversy attends the issue of exactly which categories of artifact may be taken with confidence as evidence of symbolic mental processes on the part of their makers; but most observers are willing to accept deliberately pieced and ochre-stained marine gastropod shells (found in both Mediterranean and southern African sites; e.g., Bouzouggar et al. 2007, Vanhaeren et al. 2006, Henshilwood et al. 2004) as strong evidence for symbolic bodily decoration in the >100 kyr to 75 kyr period. More definitively, several smoothed ochre plaques >70 kyr old from South Africa's Blombos Cave (Henshilwood et al. 2002) bear almost certainly symbolic geometric engravings. Additionally, there is evidence in this time frame (and possibly considerably earlier), both from Blombos and from the Pinnacle Point cave complex to its east, for a multistage fire-hardening technology that most observers believe was complex enough to have necessitated symbolic reasoning on the part of its inventors (Brown et al. 2009).

Of course, individual objects or artifact categories are frequently arguable as evidence of the cognitive status of the individuals who produced them. But what seems most noteworthy of all in the <100 kyr period is the inflection we begin to see in the pace of change and innovation among those early African populations of *Homo sapiens*. Previously, significant technological change as reflected in the lithic record had been highly sporadic, and extremely rare. Hominids appear typically to have responded to the sometimes dramatic and sudden swings in climate and habitat that typified the Pleistocene by adapting old tools to new uses, rather than by inventing new types of utensil to respond to new exigencies. But, halfway through the African Middle Stone Age, we see in contrast an entirely new picture emerging. From this point on, the material record begins to show an unprecedented pattern of innovation and change. What is more, at some time before 100 kyr ago, an earlier (and presumably nonsymbolic) iteration of *Homo sapiens* had forayed out of Africa and into the Levant without establishing itself there, or contriving to dislodge the resident Neanderthals (Bar-Yosef 1998). But once members of our species had begun to behave in a manner suggesting they had acquired modern symbolic cognition, they rapidly moved out of Africa and spread throughout the entire Old World, displacing all endemic hominid species in the process. Possibly as early as 80 kyr ago (Wu et al. 2015), and certainly not long thereafter, *Homo sapiens* was already in China. By around 40 kyr ago our species had not only supplanted the resident *Homo erectus* in Asia, and *Homo neanderthalensis* in Europe, but it had begun in both regions to create representational art on rock walls (Pike et al. 2012, Aubert et al. 2014)—providing us with the most eloquent evidence possible that they had achieved a cognitive level entirely on a par with our own. By 29 kyr ago, elaborate technologies had already allowed these originally tropical hominids to establish themselves north of the Arctic Circle (Pitulko et al. 2004); and the newly neophiliac *Homo sapiens* was well and truly embarked on the frenzied course of continual change and innovation that so dramatically marks our own lives today.

The abrupt nature of the event that gave rise to our symbolic cognitive style immediately rules out the possibility that our unprecedented form of consciousness was driven into existence by natural selection. This is because—to the extent that it ever functions as an agent for change, rather than as a promoter of homeostasis (see Tattersall 2015)—natural selection would have worked far too slowly to have produced an evolutionary pattern of the kind we see in the record just discussed (Tattersall 2012). In turn, this implies very strongly that we human beings are not biologically fine-tuned to be creatures of a specific kind, or to display particular behavior patterns; it also rather neatly explains why the nature of the "human condition" that philosophers and others have so long sought to define is so elusive.

The Role of Language

When *Homo sapiens* came into existence as an anatomically recognizable entity, its activities seem to have been broadly comparable to those of its large-brained contemporaries such as the Neanderthals. Yet, at some point during the African Middle Stone Age, something happened to *Homo sapiens* to provoke this established species into behaving in an entirely new and unprecedented way. The change involved was both a profound and an entirely unpredictable one, for clearly the symbolic cognitive state is not simply an extrapolation of earlier trends towards greater "intelligence" of the kind that might in theory be promoted by natural selection. As already emphasized, the cognitive shift involved was a *qualitative* one; and indeed, the only reason we have to believe that it *could* ever have happened, is that it so self-evidently *did* happen.

So how did that transformation come about? The principal contender for the element that spurred "high intelligence" in the modern human lineage has undoubtedly been the refinement of "theory of mind" (the ability to understand what is going on in the increasingly complex minds of others), the factor that underwrites the Social Brain Hypothesis that Dietrich Stout so efficiently dismantles elsewhere in this volume. Significantly, however, in addition to the deficiencies elaborated by Stout, the SBH is selection-driven and gradualist by nature, and thus cannot account for the sudden revolution in cognitive style that evidently occurred when humans, evidently within the tenure of *Homo sapiens*, became the unprecedented cognitive entity they so clearly are. So we need to seek another mechanism, one that supports the discontinuity the archaeological record compels us to infer.

A virtually instantaneous cognitive transformation of the kind that evidently occurred in our lineage could only have occurred under two conditions. First, the underwriting biology must already have been present, since without that enabling structure no behavioral transformation would have been possible. Second, the new potential inherent in that structure must have been recruited via the action of a stimulus that had a more or less instantaneous effect. That stimulus must have been a behavioral one, since the biology was necessarily already there. The most parsimonious hypothesis that fulfills both of these conditions—the one requiring the fewest steps—is that the neural structure, the physical ability to make the complex associations in the brain that make symbolic reasoning possible, was acquired in the extensive developmental reorganization that gave rise to the distinctive and highly autapomorphic skeletal structure of *Homo sapiens*. Although it most likely involved a structurally minor alteration in gene regulation, that genomic/transcriptomic change clearly had cascading developmental effects throughout the skeletal system. And while the skeleton is all that preserves in the material record, there is no reason in principle to believe that the effects of this modification would not have extended to other systems, including the neural one, giving rise to what has been called the "language-ready brain" (Boeckx and Benitez-Burraco 2014).

As to the behavioral innovation that recruited the new capacity for symbolic thought, by far the most plausible candidate is—as the "language-ready" brain implies—the spontaneous invention of externalized language at some point following about 100 kyr ago—in evolutionary terms, not very long after the origin of our species. It is not difficult, at least in principle, to imagine that, in an isolate of *Homo sapiens* living in Africa in this time frame, a small subset of individuals in some remote dusty encampment (perhaps most likely juveniles at play) began to attach meanings to specific spoken sounds, starting a self-reinforcing feedback in their brains between those spoken symbols and their mental representations, as they combined and recombined them to produce the beginnings of what we now experience as thought (Tattersall, 2012).

Structured externalized language is also a particularly attractive driver of symbolic cognition for a whole variety of other reasons. First, as Hinzen (2012, p. 647) has elegantly pointed out, externalized language and internalized symbolic thought are intimately intertwined and are not "two independent domains of inquiry." Second, although many have considered articulate language far too complex to have emerged other than slowly under natural selection (e.g., Pinker and Bloom 1993, Lieberman 2016), Noam Chomsky and his collaborators have recently argued very convincingly (Bolhuis et al. 2014, 2015, Berwick and Chomsky 2016) that the algorithmic basis for language is in fact an extremely simple one, rendering it much easier to understand the sudden emergence suggested by the material record. Direct empirical evidence for the process involved is inaccessible since all modern human societies already possess spoken language; however, an analogously rapid and spontaneous linguistic emergence has indeed been observed by modern linguists in the case of similarly structured sign language (Senghas et al. 2005). Third, the geometry of events adumbrated here has the signal advantage of explaining how the highly derived vocal tract necessary for the expression of modern articulate language happened to be available precisely when it became needed for that purpose. For it was already in place, having been acquired initially as a simple byproduct of the retraction of the face below the front of the braincase that is the most obvious cranial apomorphy of *Homo sapiens*. And finally, externalized spoken language would have been poised to spread rapidly among members of a species whose members already possessed a language-ready brain, explaining how events were able to move so rapidly following the initial appearance of symbolic cognition.

Though highly speculative, this scenario of the emergence of language and symbolic cognition is entirely in line with the repurposing of existing cognitive capacities that we have seen was typical throughout human evolution. Similarly, it is completely consistent with what we know of evolutionary mechanisms in general. In other words, as non-predictable and extraordinary as the cognitive product undoubtedly was, and is, the evolutionary processes that produced it were altogether routine.

The Shrinking Human Brain

There is an intriguing coda to the story of human brain size and cognition. Thanks to the fact that *Homo neanderthalensis* boasted a large brain but apparently did not reason symbolically, we know that our modern intellectual prowess is not a simple consequence of large brain mass. As reported by Holloway et al. (2004), the mean volume of 26 Neanderthal cranial vaults is 1,487 ml, a figure substantially higher than the 1,330-ml mean volume the same authors report for a large and geographically varied sample of modern *Homo sapiens* braincases. But it is noteworthy that, using figures from the same source, the mean endocranial volume of 29 Upper Pleistocene *Homo sapiens* individuals is 1,499 ml: a figure virtually identical to that of the Neanderthal sample. Over the last few tens of thousands of years, it appears, members of our species have lost on average some 12.7 percent of their brain mass.

What might explain this? Hawks (2011) found that this diminution could not be associated with a reduction in body mass, and it is doubtful that the effect can be traced to the effects of climate (Beals, Smith, and Dodd 1984). Equally dubious are the notions that we have "dumbed down" due to the individual protection provided by complex social structures (Bailey and Geary 2009), or that we have reduced our brains through "self-domestication" (Wrangham 2011). So what has been happening? A more plausible scenario might refer back to the "expensive tissue" hypothesis advanced by Aiello and Wheeler (1995). Our large brains consume up to 25 percent of all the metabolic energy we use, and such expenditure demands justification. Other things being equal, then, brain tissue in excess of requirements is hugely uneconomic as well as being expendable, and with the passage of time elimination of excess neural mass might be expected.

Over the history of the genus *Homo* the archaeological record clearly shows that, on average, complexities of behavior seem to have increased along with brain size, if not necessarily exactly in step. Given the strong and independent trends toward brain enlargement seen in all *Homo* lineages, it seems reasonable to suppose that, typically, this increasing behavioral complexity was more or less directly underwritten by expanding brain mass. In other words, the intuitive ancestral cognitive style likely depended on a "brute force" information-processing algorithm, whereby cognitive output corresponded more or less to raw brain volume. In contrast, it seems entirely possible that the symbolic algorithm is energetically a more efficient one, using the modern brain's associative abilities to produce more complex and flexible outputs using lesser amounts of brain tissue and thereby permitting today's human beings to get by on a more economical smaller brain.

This scenario makes the "incredible shrinking human brain" appear less anomalous in terms of cognitive output, while simultaneously accommodating the relationship between expanding brain size and the presumed increasing intelligence that we infer from the material record throughout the bulk of the history of the genus *Homo*. It takes the origin of modern human cognition out of the constricting frame of reference imposed by the reduction of virtually all evolutionary phenomena to special cases of gradual change via natural selection—a formulation that most paleoanthropologists still accept, or at least are strongly influenced by. And it emphasizes the emergent and adventitious nature of our astonishing and unprecedented cognitive system.

Acknowledgments

I should like to thank Jeffrey Schwartz for inviting me to contribute to this volume even though circumstances prevented me from attending what was clearly a fascinating and stimulating workshop. Whether or not he might have approved of them, I would like to dedicate the speculations in this essay to our dear friend and irreplaceable colleague Bob Sussman, who prematurely died while it was being written.

References

Aiello, L., and P. Wheeler. 1995. "The Expensive-Tissue Hypothesis: The Brain and the Digestive System in Human and Primate Evolution." *Current Anthropology* 36:199–221.

Aubert, M., A. Brumm, M. Ramli, T. Sutikna, E. W. Saptomo, B. Hakim, M. J. Morwood, G. D. van den Bergh, L. Kinsley and A. Dosseto. 2014. "Pleistocene Cave Art from Sulawesi, Indonesia." *Nature* 514:223–227.

Bailey, D. H., and D. C. Geary. 2009. "Hominid Brain Evolution: Climatic, Ecological, and Social Competition Models." *Human Nature* 20:67–79.

Bar-Yosef, Y. 1998. "The Chronology of the Middle Paleolithic of the Levant." In *Neandertals and Modern Humans in Western Asia*, edited by T. Akazawa, K. Aoki, and O. Bar-Yosef, 39–56. New York: Plenum Press.

Beals, K. L., C. L. Smith, and S. M. Dodd. 1984. "Brain Size, Cranial Morphology, Climate and Time Machines." *Current Anthropology* 25:301–330.

Berwick, R. C. and N. Chomsky. 2016. *Why Only Us. Language and Evolution.* Cambridge, MA: MIT Press.

Boeckx, C., and A. Benitez-Burraco. 2014. "The Shape of the Human Language-Ready Brain." *Frontiers in Psychology* 5:282. doi: 10.3389/fpsyg.2014.00282.

Bolhuis, J. J., I. Tattersall, N. Chomsky, and R. Berwick. 2014. "How Could Language Have Evolved?" *PLoS Biology* 12(8):e101934.

Bolhuis, J. J., I. Tattersall, N. Chomsky, and R. Berwick. 2015. "UG or Not to Be, That Is the Question." *PLoS Biology* 13(2):e1002063.

Bouzzougar, A., N. Barton, M. Vanhaeren, F. d'Errico, S. Colcutt, T. Higham, E. Hodge, S. Parfitt, E. Rhodes, J.-L. Schwenninger, C. Stringer, E. Turner, S. Ward, A. Moutmir and A. Stambouli. 2007. "82,000-Year-Old Shell Beads from North Africa and Implications for the Origins of Modern Human Behavior." *Proceedings of the National Academy of Sciences, USA* 104:9964–9969.

Brown, K. S., C. W. Marean, A. I. R. Herries, Z. Jacobs, C. Tribolo, D. Braun, D. L. Roberts M. C. Meyer, and J. Bernatchez. 2009. "Fire as an Engineering Tool of Early Modern Humans." *Science* 325:859–862.

Clark, J. D., Y. Beyene, G. WoldeGabriel, W. K. Hart, P. R. Renne, H. Gilbert, A. Defleur, G. Suwa, S. Katoh, K. R. Ludwig, J.-R. Boisserie, B. Asfaw, and T. D. White. 2003. "Stratigraphic, Chronological and Behavioural Contexts of Pleistocene *Homo sapiens* from Middle Awash, Ethiopia." *Nature* 423:747–752.

Collard, M., and B. Wood. 2014. "Defining the genus *Homo*." In *Handbook of Paleoanthropology*, 2nd ed., Vol. 3, edited by W. Henke and I. Tattersall, 2107–2144. Heidelberg: Springer.

d'Errico, F., and A. Nowell. 2000. "A New Look at the Berekhat Ram Figurine: Implications for the Origins of Symbolism." *Cambridge Archaeological Journal* 10:123–167.

de Lumley, H. (editor). 2016. *Terra Amata: Nice, Alpes-Maritimes, France, Tome V, Comportement et Mode de Vie des Chasseurs acheuléens de Terra Amata*. Paris: CNRS Editions.

deMenocal, P. 2011. "Climate and Human Evolution." *Science* 331:540–542.

Dunbar, R. I. M. 1998. "The Social Brain Hypothesis." *Evolutionary Anthropology* 6:178–190.

Dunbar, R. I. M. 2003. "The Social Brain: Mind, Language and Society in Evolutionary Perspective." *Annual Reviews of Anthropology* 32:163–181.

Goren-Inbar, N. 1986. "A Figurine from the Acheulean Site of Berekhat Ram." *Mitekufat Haeven* 19:7–12.

Haile-Selassie, Y., S. M. Melillo, and D. F. Su. 2016. "The Pliocene Hominin Diversity Conundrum: Do More Fossils Mean Less Clarity?" *Proceedings of the National Academy of Sciences USA* 113:6364–6371.

Harmand, S., J. E. Lewis, C. S. Feibel, C. J. Lepre, S. Prat, A. Lenoble, X. Boes, R. L. Quinn, M. Brenet, A. Arroyo, N. Taylor, S. Clement, G. Daver, J.-P. Brugal, L. Leakey, R. A. Mortlock, J. D. Wright, S. Lokorodi, C. Kirwa, D. V. Kent, and H. Roche. 2015. "3.3-Million-Year-Old Stone Tools from Lomekwi 3, West Turkana, Kenya." *Nature* 521:310–316.

Hart, D., and R. W. Sussman. 2009. *Man the Hunted: Primates, Predators, and Human Evolution*. Expanded edition. New York: Westview/Perseus.

Hawks, J. 2011. "Selection for Smaller Brains in Holocene Human Evolution." *arXiv* 1102.5604v1.

Henshilwood, C. S., F. d'Errico, R. Yates, Z. Jacobs, C. Tribolo, G. A. T. Duller, N. Mercier, J. C. Sealy, H. Valladas, I. Watts, A. G. Wintle. 2002. "Emergence of Modern Human Behavior: Middle Stone Age Engravings from South Africa." *Science* 295:1278–1280.

Henshilwood, C., F. d'Errico, M. Vanhaeren, K. van Niekerk, and Z. Jacobs. 2004. "Middle Stone Age Shell Beads from South Africa." *Science* 304:404.

Hinzen, W. 2012. "The Philosophical Significance of Universal Grammar." *Language Sciences* 34:635–649.

Holloway, R. L., D. C. Broadfield, and M. S. Yuan. 2004. *The Human Fossil Record, Vol. 3: Hominid Endocasts, the Paleoneurological Evidence*. New York: Wiley-Liss.

Joordens J. C. A., F. d'Errico, F. P. Wesselingh, S. Munro, J. de Vos, J. Wallinga, C. Ankjaergaard, T. Reimann, J. R. Wijbrans, K. F. Kuiper, H. J. Mucher, H. Coqueugniot, V. Prie, I. Joosten, B. van Os, A. S. Schulp, M. Panuel, V. van der Haas, W. Lustenhouwer, J. J. G. Reijmer and W. Roebroeks. 2015. "*Homo erectus* at Trinil on Java Used Shells for Tool Production and Engraving." *Nature* 518:228–231.

Klein, R. G. 2009. *The Human Career*, 3rd ed. Chicago: University of Chicago Press.

Lepre, C. J., H. Roche, D. V. Kent, S. Harmand, R. L. Quinn, J.-P. Brugal, P.-J. Texier, A. Lenoble, and C. S. Feibel. 2011. "An Earlier Origin for the Acheulian." *Nature* 477:82–85.

Lieberman, P. 2016. "The Evolution of Language and Thought." *Journal of Anthropological Sciences* 94:127–146.

Lordkipanidze, D., M. Ponce de León, A. Margvelashvili, Y. Rak, G. P. Rightmire, A. Vekua, and C. P. E. Zollikofer. 2014. "A Complete Skull from Dmanisi, Georgia, and the Evolutionary Biology of Early *Homo.*" *Science* 342:326–331.

McPherron, S., Z. Alemseged, C. W. Marean, J. G. Wynne, D. Reed, D. Geraads, R. Bobe, and H. A. Béarat. 2010. "Evidence for Stone-Tool-Assisted Consumption of Animal Tissues before 3.39 Million Years Ago at Dikika, Ethiopia." *Nature* 466:857–860.

Piantadosi, S. T., and C. Kidd. 2016. "Extraordinary Intelligence and the Care of Infants." *Proceedings of the National Academy of Sciences USA*, doi:10.1073/pnas.1506752113.

Pike, A. W. G., D. L. Hoffmann, M. Garcia-Diez, P. B. Pettit, J. Alcolea, R. De Balbin, C. Gonzalez-Sainz, C. de las Heras, J. A. Lasheras, R. Montes, and J. Zilhao. 2012. "U-series Dating of Paleolithic Art in 11 Caves in Spain." *Science* 336: 1409–1413.

Pinker, S., and P. Bloom. 1990. "Natural Language and Natural Selection." *Behavior and Brain Sciences* 13(4):707–784.

Pitulko, V. V., P. A. Nikolsky, E. Y. Girya, A. E. Basilyan, V. E. Tumskoy, S. A. Koulakov, S. N. Astakhov, E. Y. Pavlova, and M. A. Anisimov. 2004. "The Yana RHS Site: Humans in the Arctic before the Last Glacial Maximum." *Science* 303:52–56.

Radovčić, D., A. O. Sršen, J. Radovčić, J., and D. W. Frayer. 2015. "Evidence for Neandertal Jewelry: Modified White-Tailed Eagle Claws at Krapina." *PLoS ONE* 10(3):e0119802.

Rodriguez-Vidal, J., F. d'Errico, F. G. Pacheco, R., Blasco, J. Rosell, R. P. Jennings, A. Quaffelec, G. Finlayson, D. A. Fa, J. M. G. Lopez, J. S. Carrion, J. J. Negro, S. Finlayson, L. M. Caceres, M. A. Bernal, S. F. Jimenez and C. Finlayson. 2014. "A Rock Engraving Made by Neanderthals in Gibraltar." *Proceedings of the National Academy of Sciences USA.* doi: 10.1073/pnas.1411529111.

Roebroeks, W., and P. Villa. 2011. "On the Earliest Evidence for Habitual Fire Use in Europe." *Proceedings of the National Academy of Sciences USA* 108:5209–5214.

Schwartz, J. H. 2006. "Race and the Odd History of Human Paleontology." *The Anatomical Record Part B: The New Anatomist* 289B(6):225–240.

Schwartz, J. H. 2016. "What Constitutes *Homo sapiens*? Morphology versus Received Wisdom." *Journal of Anthropological Sciences* 94:65–80. doi 10.4436/JASS.94028.

Schwartz, J. H., and I. Tattersall. 2015. "Defining the Genus *Homo.*" *Science* 349:931–932.

Senghas, A., S. Kita, and A. Özyürek. 2005. "Children Creating Core Properties of Language: Evidence from an Emerging Sign Language in Nicaragua." *Science* 305:1779–1782.

Spoor, F., M. G. Leakey, P. N. Gathogo, F. H. Brown, S. C. Anton, I. McDougall, C. Kiarie, F. K. Manthi, and L. N. Leakey. 2007. "Implications of New Early *Homo* Fossils from Ileret, East of Lake Turkana, Kenya." *Nature* 448:688–691.

Stout, D. 2016. "Human Brain Evolution: History or Science?" This volume.

Tattersall, I. 1993. *The Human Odyssey: Four Million Years of Human Evolution.* Englewood Cliffs, NJ: Macmillan.

Tattersall, I. 1997. *The Fossil Trail: How We Know What We Think We Know About Human Evolution.* New York: Oxford University Press.

Tattersall, I. 2012. *Masters of the Planet: The Search for Our Human Origins*. New York: Palgrave Macmillan.

Tattersall, I. 2015. *The Strange Case of the Rickety Cossack and Other Cautionary Tales from Human Evolution*. New York: Palgrave Macmillan.

Tattersall, I. 2016. "The Genus *Homo*." *Inference: International Review of Science* 2(1). http://inference-review.com/article/the-genus-homo

Thieme, H. 1997. "Lower Palaeolithic Hunting Spears from Germany." *Nature* 385:807–810.

Vanhaeren, M., F. d'Errico, C. Stringer, S. L. James, J. A. Todd, and H. K. Mienis. 2006. "Middle Paleolithic Shell Beads in Israel and Algeria." *Science* 312:1785–1788.

White, T. D., B. Asfaw, D. DeGusta, H. Gilbert, G. D. Richards, G. Suwa, and F. C. Howell. 2003. "Pleistocene *Homo sapiens* from Middle Awash, Ethiopia." *Nature* 423:742–747.

Wrangham, R. 2011. "Chimpanzees, Bonobos and the Self-Domestication Hypothesis." *American Journal of Primatology* 73:33–45.

Wu, L., M. Martinon-Torres, Y.-J. Cai, S. Xing, H.-W. Tong, S.-W. Pei, M. J. Sier, X.-H. Wu, R. L. Edwards, H. Cheng, Y.-Y. Li, X.-X. Yang, J. M. Bermudez de Castro and X.-J. Wu 2015. "The Earliest Unequivocally Modern Humans in Southern China." *Nature* 526:596–600.

16 Sex, Reproduction, and Scenarios of Human Evolution

Claudine Cohen

Introduction

Reproduction, in combination with selection and variation, is fundamental to the biological process of evolution. Reproductive physiology and behavior play an obvious role in the modes and tempo of evolution of all heterosexual animals, including humans.

Here I will consider historically and critically different scenarios of human evolution that have been, and still are, based on human reproductive characteristics and behavior. I will first focus on Darwin's concept of sexual selection as exposed in his ([1871] 1877) opus, *The Descent of Man, and Selection in Relation to Sex*, and consider the reception of his ideas as a major source of evolutionary scenarios in cultural anthropology, sociobiology, and evolutionary psychology. Second, I will consider several anatomical and physiological features that characterize human reproduction, in particular *concealed ovulation*—a major feature of women's physiology—and examine its importance in determining human reproductive behavior, as well as its consequences for establishing social forms and norms. I will finally analyze two other human characteristics: the immaturity of the human child at birth, and the long duration of women's lives after their period of reproduction. Through the examination of these different human reproductive features, I will underline the essential role that cultural and social frameworks associated with reproductive behavior and sexual practices, played in the evolution of Mankind.

Darwin's Sexual Selection and Its Anthropological Meaning

When, in 1868, almost ten years after the publication of his *Origin of Species* (Darwin 1859), Darwin undertook writing about human evolution, he chose to consider it from a perspective that was markedly different than those of his contemporaries, that is, Huxley's comparative anatomy (1863), Haeckel's embryology (1868) and Lyell' (1863) and Mortillet's (1867) archeology. Instead, in the resultant monograph, *The Descent of Man*, Darwin ([1871] 1877) reflected on the role of sex and reproduction in evolution, and proposed the new concept of sexual selection, which he saw as a major element in the evolution of all sexed animals, including humans (Cohen, 2009).

Darwin first observed that, in animals, some anatomical features characterize one sex and not the other. From this he concluded that these features, which do not appear to have any particular adaptive role, are actually related to reproductive choice (*ibid*. chapters VIII–XVIII). Sexual selection explains, for example, the colorful feathers and elaborate singing of male birds, as well as the development, in the males of a number of mammalian species, of antlers or horns. These ornaments have a double role because they allow males to compete with other males and conquer females. Indeed, in most animal species, the less morphologically endowed females (e.g., in birds, less elaborate and colorful plumage) pick and choose their reproductive mates from among the more adorned males (*ibid*. chapter XV).

From this, Darwin (*ibid*. XIX–XXI) argued that, through the selection of a number of anatomical and also physiological, psychological, and social features, mate choice and sexual selection were central to human evolution: specific features belonging only to one sex, likely emerged long ago via sexual selection. These elements include features that characterize body appearance (size and position of teeth, hair and skin color), as well as behavior (sound of the voice and ability to sing, strength, aggressiveness, and ability to fight in males; beauty, kindness, and generosity in women). Darwin distinguished two kinds of sexual selection: Either a fight takes place between two individuals of the same sex (generally males), while females remain passive (this being what is currently referred to as *intrasexual* selection [Andersson 1994]), or it takes place when females choose the most attractive partner (which is *intersexual* selection [*ibid*.]). According to Darwin ([1871] 1877, VI, p. 160), the latter form of selection (by females) was dominant during the earliest stages of human prehistory, while the former finally became more prominent, implying male dominance until the present times in our societies.

In conceiving the role of sexual selection during the early times of human evolution, Darwin posited that females were probably able to choose the males with whom they copulated, which, through repeated acts of selection, had the effect of modifying males in body and behavior. As he wrote: "Man in all probability owes his beard, and perhaps some other characters, to inheritance from an ancient progenitor who thus gained his ornaments. But this form of selection may have occasionally acted during later times; for in utterly barbarous tribes women have more power in choosing, rejecting, and tempting their lovers, or of afterwards changing their husbands, than might have been expected"(*ibid*. chapter XX, p. 597).

However Darwin limits the preeminence of human female choice to the most ancient, and thus most primitive, period of our genus' evolutionary history. Only later on would other expressions of sexual selection emerge. As a result of the "Paw of battle" selecting the strongest and the bravest males, it is likely that males started choosing their mates in accordance with their predilections. As Darwin expressed this:

The most vigorous and strongest males—those who could best defend a family and hunt for its food, those who had the best arms and possessed more goods, such as dogs and other animals, could raise more children than individuals who were poorer and weaker in the same tribe. No doubt these men could generally choose the more attractive women. (*ibid.* XX: 595)

As a consequence,

women are everywhere conscious of the value of their own beauty; and when they have the means, they take more delight in decorating themselves with all sorts of ornaments than men do. They borrow the plumes of male birds, with which nature has decked this sex, in order to charm the females. (*ibid.* XX: 597)

Through repeated choice, sexual selection can normalize that which had originally been exceptional (*ibid.*). Indeed, over time, it could transform and embellish an entire species. But it would do so in different ways in different human groups, because each culture has its own canons of "beauty" reflecting its particular tastes. Because cultural norms could increase morphological differences between one human group and another, Darwin speculated that sexual selection may well have been the mechanism behind the diversification of human groups into distinct "races" (*ibid.* VII; I, XX: 572–585): "I conclude that of all the causes which have led to the difference in external appearance between the races of Man . . . sexual selection has been the most efficient" (*ibid.* chapter XX, p. 606). Thus in Darwin's views, the emergence of racial differences in humans could well be the result of a cultural process.

Darwin also observed that sexual selection could be transformed, molded, and limited by a number of sociocultural practices. For example, in what he calls "savage tribes," mate-sex choice was and is limited not only by promiscuity, but also by such factors as female infanticide, arranged marriages, and the circumstances in which women can be subjected to different forms of slavery. In "civilized peoples" features such as fortune and social position often determine sexual choices, reproductive capacity, and even beauty (*ibid.* XX, p. 586). In this context, sexual selection plays a role in modifying people's appearance, but does so differently in different social classes: "Our aristocracy, including under this term all wealthy families in which primogeniture has long prevailed, from having chosen during many generations from all classes the more beautiful women as their wives, have become handsomer, according to the European standard, than the middle classes" (*ibid.* XX, p. 587). Here, the notion of selection seems to endorse ideas espoused by Darwin's cousin Francis Galton (1909), according to whom it would be possible to improve the human species by favoring the reproduction of its social elite. Clearly, these analyses strongly imbue the process of sexual selection within cultural and social frameworks.

In *The Descent of Man*, Darwin assumed that "primitive man originally lived in small communities, each man having as many women as he could maintain or he could procure, and he had to defend jealously against any other man." He also hypothesized that young males, who would be expelled from their group and left to survive on their own, would eventually succeed in finding a mate and thus avoid close inbreeding with members of their family. From Sigmund Freud (1918) to René Girard (1972, 1978), this vision of a primeval horde, configured by a dominant father who forced younger males into exile, was extremely influential. Did, as Darwin believed, ancient human societies comprise groups of females and children that were led by a single male? Or were these societies organized around nuclear or extended families? How did the human family become what it is today? By the turn of the twentieth century, these questions became a focal subject of extensive inquiry in cultural anthropology.

Evolution of the Family

Many examples featured in *The Descent of Man* are drawn from contemporary anthropological literature, especially from the works of John Lubbock (1865, 1870), one of the most active proponents at the time of "cultural evolutionism." Darwin, for instance, turned to ethnographic reports of so-called "savage behavior" (*ibid*. XX, p. 599–600) to demonstrate the importance of women's choice during the primitive stages of human evolution. Thus, his evolutionary perspective in *The Descent of Man* converged on positions held by the cultural anthropology of his time (see Stocking 1968), in which gender relationships were considered fundamental to the evolution of societies.

Works by John Lubbock, Edward Tylor (1873–1877), Herbert Spencer (1877–1896), Edward Westermarck (1891), John Ferguson McLennan (1876), Lewis Henry Morgan (1877), Friedrich Engels (1884), and even Sigmund Freud (1918) illustrate this important scientific and ideological trend. Indeed, the proponents of "cultural evolutionism" invoked Darwin's evolutionary concepts in support of their contention that the development of human societies and cultures was both a gradual and a progressive process. These scholars and their followers also relied on the idea that cultural features which existed early on in the evolution of humanity survived in present-day societies which they dubbed "primitive." Cultural evolutionism was rested on racial classifications and hierarchies, and on a concept of evolution viewed as a linear and progressive process (both culturally and biologically), conceiving prehistoric cultures as similar to contemporary "primitive" societies. In particular, these anthropologists viewed social evolution through the lens of change in the structure and organization of the family, from primitive sexual promiscuity to exogamy, and from matriarchy to the appearance of modern patriarchal societies.

One central concept to these constructs was that matriarchy was an early form of human society. This theme was first theorized by Basel jurist Johann Jakob Bachofen (1861) in his 1,300-page monograph, *Das Mutterrecht*, in which he cited ancient Greek histories and myths to argue that women had played a dominant role in the early history of humanity. Subsequently, *gynecocracy* (*gynarchy*), the reign of the Mother, was supplanted by the patriarchal structures that persisted into the present. Bachofen's work was, however, severely criticized by his contemporaries, especially with regard to his method, which took myths as real, historical documents. This notwithstanding, his embrace of matriarchy as the first stage in the evolution of human societies echoed Darwin's emphasis on the dominance of women's sexual choices in primeval human groups, a conception that was taken up by "Darwinian" cultural anthropologists, and which later on achieved great success in feminist prehistoric archaeology.

From the 1920s, these evolutionary speculations were challenged by novel trends in cultural anthropology. Refuting their racist foundations, anthropologists such as Franz Boas (1940) and Claude Lévi-Strauss (1949) strongly criticized this dominant paradigm in anthropology, as he argued that the diversity of human social structures cannot be reduced to a biological evolutionary model and to a simplistic scheme of linear progress. In the field of Prehistory, this critique led to questioning the heuristic value of ethnographical comparison and the use of ethnographical models as windows on prehistoric cultures (see Raphaël 1945, Laming-Emperaire 1962, Leroi-Gourhan 1964).

The Return of Sexual Selection and Hominization Scenarios

However, explanations of human evolution based on biologically determined sex roles and choices were reconfigured, reemerged and regained prominence. Since the 1970s, sex relationships and reproductive behavior have become a major player in the production of evolutionary scenarios. Sociobiology (Wilson 1975, Dawkins 1976) and evolutionary psychology (Miller 2000, Barrett et al. 2002) call for Darwin's sexual selectionism as explanatory of human behaviors and sociocultural practices. For sociobiologists, individuals in social groups (and humans in particular) are essentially bearers of genes whose sole goal is to replicate themselves. Thus, the emergence and diversity of human behavior is reduced to different reproductive strategies (Dawkins 1976). Desire, pleasure, seduction, violence, and culture itself, are nothing but whims of nature that converge on the same result: the maximization of reproductive fitness through opportunities for individuals to reproduce their genes.

If the goal of any sexed organism is to replicate its genes, then men's and women's behaviors "naturally" diverge. Since males produce very small gametes in very large numbers that can be disseminated virtually at will without any consequence to their producers, the choice of whether or not to care for their offspring is separate from the act of giving life. In contrast, the time-constrained production by women of a limited number of ova, their mobilization in anticipation of fertilization, and the consequent pregnancy, involves a much more serious investment of time, energy, and a resources. Further, women are "naturally" responsible for caring for, rearing, and educating children, a responsibility that demands a significant investment of time (including transmitting the group's knowledge and values), energy (including carrying the infant/child), and procurement and preparation of food (Barrett et al. 2002).

These different aspects of human reproduction would generate a natural asymmetry between the two sexes. In terms of the urge to disseminate their gametes and replicate their genes, one consequence for males would be marked activity and mobility as they seek multiple sexual partners. In contrast, female behavior would be influenced by the physiological necessities of ovulation (i.e., the production of one valuable gamete), gestation (nine months carrying an increasing heavy and bulky fetus), childbirth, and lactation, which would *naturally* lead to being sedentary because of attachment to infants and young children. There would thus be a natural division of labor between men, who can travel outside the boundaries of the group and perhaps engage in dangerous tasks, and women, whose activities must be performed at home and in the presence of children, and require only a small investment of time and minimal skill (*ibid.*). From the very beginning to the present, these "natural" conditions would underlie the relational link between men and women.

Although currently popular, this reductionist conceit is highly disputable, and Darwin himself did not support it (Zihlman, 1995), as he appreciated that many factors limit or distort the role and efficacy of sexual selection in human societies. As has been well-argued in many critical analyses (e.g., Sahlins 1977), human reproductive behavior cannot be viewed solely as a biological fact, nor understood as the result of a universal human instinct. In short, human reproductive behavior gains meaning and function from the general cultural, social, symbolic, and economic context in which it takes place.

Such naturalistic and androcentric speculations have also become the target of feminist critique. Beginning in the 1960s, new scenarios refuted the concept of women's "natural" passivity and championed their evolutionary role, insisting on their ability to choose their sexual mates, compete and create social bonds, raise, educate, and socialize their offspring, and beyond, stressing the importance of their social, cultural and technical skills. Numerous anthropological studies have explored the influence of women on human evolution and the emergence of societies (Hrdy 1999, Tanner 1980, Dahlberg 1981, Zihlman 1981, 1995, 1997; also see Haraway 1995 and Cohen 2003, 2016), focusing especially on the significance and role in the shaping of human sexual and social behavior of female physiological features, such as menstruation (Knight 1999, Testart 1987, 2014) and concealed ovulation.

Concealed Ovulation and the Emergence of Kinship Rules

Long ago, Sigmund Freud stressed the importance of the loss of manifestations of estrus in human sexuality, and Simone de Beauvoir (1949 vol. I, p. 70) in turn emphasized "the absurd dependency of the female, in humans, to sex and reproduction." But it is primarily because of the more recent formalization of comparative primate ethology (see Dixson 1998) that the matter of "concealed ovulation," that is the loss of estrus perceptible manifestations in women, has garnered renewed interest.

The sexual behavior of apes is now well known (Dixson 1998). In chimpanzees, with lifespans of ca. 40 years, there is a hierarchy between generations and between sexes.

Females become sexually mature at about ten years of age. Menstrual periods are short and not easily observed. Female menstrual cycles last 36 days, ovulation one day, and tumescence of the vulva in conjunction with the production of male-attracting pheromones ten days. Chimpanzee sexual behavior includes "consorting" (choosing a temporary partner), wherein the female stays with the chosen male. However, after three or fewer weeks, the female breaks this union. Although many males promiscuously copulate with rutting females, only the dominant male does so at the peak of estrus, which coincides with ovulation. Sexual intercourse is intertwined with various kinds of exchange, food in particular. As in many primate species, chimpanzee promiscuity does not lead to stable family groups. In fact, all female-male pairings are possible, except between mother and son. Incest is not prohibited, but it is avoided. In the rare event of incestual intercourse, it is brief, and usually associated with a specific function, such as reassurance. Upon reaching puberty, females leave the mother's group. Gestation is usually 230 days, and the mother gives birth to one newborn (rarely two) typically weighing 4 kg. The neonate is completely dependent on its mother, who first carries it on her belly, then on her back, and stays with it for two years. While females raise the young, males live together in their own territory.

In humans, a woman's sexual attractiveness and receptivity is not tied to periods of rutting associated with ovulation. There are no external manifestations (visual, olfactory) of her hormonal state of sexual receptivity—"concealed ovulation"—and sexual relations are possible at any time, regardless of the timing of ovulation (Alexander et al. 1978). This results in a set of social norms that regulate and organize gender relationships. These features of human female reproductive physiology make clear that chimpanzee behavior cannot be used as a model for early human societies. Indeed, it is the differences that help us understand hominization.

Permanent bipedal gait may have played a role in the concealment of ovulation and the subsequent transformations of human sexual practices and behavior. With upright posture, the vulva is situated under the body, hiding signs of estrus (visual, olfactory) from the male. But while this makes the external manifestations of estrus (e.g., swelling of the vulva) less visible, it highlights other body parts, thereby shifting sexual stimuli from the purely physiological to the social, symbolic, and cultural, which can contribute to attraction and seduction regardless of the timing of ovulation. Since the surface of the human body bears multiple erogenous zones, sexual arousal can be triggered by the perception of females' non primary sexual characters (e.g., prominence of breasts, buttocks, reduction of hair). In the absence of visible estrus, human sexual behavior and reproduction become disconnected (Godelier, 2004). Consequently, the uniquely human manifestations of eroticism and sexual pleasure (see Bataille 1957, 1961), coexist with, and may even replace, the physiological function of procreation (Zwang 2002).

However, as with many other features, human sexual behavior cannot be thought of as an absolute exception in the animal world. In Bonobos (the pygmy chimpanzee), sexual intercourse is possible well beyond the estrous period (see de Waal and Lanting 1997). Males and females engage in non-reproductive sexual activities (masturbation, homosexuality), and mate face-to-face. Sex appears to play a role in sociability, such as reassurance and the reduction of threat. Although the degree to which Bonobo and human sexual practices are similar is still debated, their resemblances illustrate that perhaps in this, as in many other features, humans are not as "unique" as some of us would like to think.

According to anthropologist Owen Lovejoy (1981), the unconstrained availability of sex in humans was the primary cause for the existence of permanent couples, from which versions of family life, education of children, division of labor, and sharing behavior developed. Human sexual practices established the conditions for permanent exchange: women's sexual favors against food brought by men. Concealed ovulation has also been viewed as a "trick"—deception of ovulation—through which women could obtain fidelity, commitment, and permanent presence of men, and secure their survival. Should we assume "a feminine ruse to force the male to protect the female, and make sure that he is the only one to fertilize her, and that the offspring is indeed his own" (Zwang 2002)? This would suppose many calculations on the part of Paleolithic women, whose knowledge about the mechanism of procreation has often been denied (see Malinowki 1927). Would not the necessity to protect pregnant women and very young offspring, as well as the cohesive characteristics of couple relationship, be sufficient to explain the mutual attachment between human mating partners (*ibid.*)?

The acquisition of concealed ovulation has been viewed as a key event leading to the transformation of gender relationships and roles in human social groups. If it was related to the acquisition of the Hominin bipedal gait, its roots lie well before the origin of the genus *Homo*. Understanding concealed ovulation—its origin, causes, and effects—is likely fundamental to understanding human evolution and the emergence of social structure, as is reflected in its being the starting point of several scenarios of " hominization."

One of these scenarios assume that, with sex always being possible, at any moment or place, between any males and females, it became a threat to the cohesion of the social group. That is, the dangers of permanent sexual availability led in all societies to rules and prohibitions that limit sexual exchanges and practices (Godelier 2004). According to this scenario, this could be the origin of kinship rules that determine incest prohibition and the nature of incestuous ties—which, anthropologists claim, are fundamental to all human societies. As Lévi-Strauss (1956, p. 281) wrote:

The universality of the incest prohibition specifies a rule that persons regarded as parent and child, or brother and sister, even in name only, can not have sex, let alone marry. . . . [A]ll the marriage prohibitions have as their only purpose to establish mutual dependency between biological families, or to put it in stronger terms . . . marriage rules express the refusal on the part of society, to admit the exclusive existence of the biological family.

While incest prohibition exists in all human societies, the content of these rules vary considerably, thereby making them a basis for human extensive cultural diversity.

The "Unfinished" Child and the Evolutionary Meaning of Women's Longevity

"A fundamental difference between Humans and their ancestors on the one hand, and the closest primates, Chimpanzees and Bonobos, on the other, is that the breeding of human children is not the responsibility of women only: reproductive units and child rearing in general include the co-men and adult women. . . . What is fundamental (and new) here is that adults of both sexes come to cooperate for the establishment and maintenance of a social group in which children are born and raised, in short a family," writes French anthropologist Maurice Godelier (2004, p. 477).

This familial cooperation is initially determined by the child's needs. With its still-patent cranial sutures, and a head that is too heavy for its body, the human infant long remains unable to move, eat by itself, or express its needs. This early-born, " incomplete " child, as the product of developmental and physiological changes (likely associated with the acquisition of a bipedal gait and the reconfiguring of the pelvis and pelvic canal), probably emerged early on in human evolution with consequences for group structure. In addition, human infants must be breastfed for a long time. Paleoanthropologists have been able to study breast milk consumption, breastfeeding, and weaning practices at different times in the past (Herrscher 2013), through nitrogen, oxygen, and carbon isotope analysis of children's fossilized teeth: they demonstrated that, since the lower Paleolithic, lactation could last for several years. While this long stint of breastfeeding may be a burden for the mother, it also brings with it a number of significant advantages: A prolonged period of intimacy between mother and child reinforces opportunities of exchange, communication, education, and learning the social group's rules and norms. Language, which plays an essential role in mother-child interaction, was probably also an element of the earliest stages of human evolution (Holloway 1986). The strong mutual attraction between mother and child lasts for many years, and is never fully severed. Since a neonate's brain is not fully formed, its growth and plasticity during this long period of mother-offspring association permits increased transmission and acquisition of knowledge, including learning techniques, arts, and symbols, and the potential to innovate. From a social standpoint, the impact of the "incomplete" child likely went beyond the ties binding mother and child, to include the father, siblings, as well as ascendant, descendant, and collateral relatives, such as uncles, aunts, cousins, nephews, and grandparents, with consequences on the entire set of social norms and regulations.

These considerations stress yet another unique aspect of human reproductive physiology: The survival of women beyond their childbearing years. In contrast to great apes, which are capable of reproducing through their ca. 50-year life span, human female fertility lasts less than half of a woman's lifetime. With menopause, fertility ceases, but other physiological functions remain intact. Since humans are unique among primates in this particular biological rhythm and long-term survival after menopause, what might be the function, evolutionary advantage, or benefit for the individual or for the group? Can we offer a plausible explanation for why women live long after their period of fertility?

American anthropologist Kristen Hawkes (1998) argues that women's longevity and the extension of their lives beyond childbearing age are intrinsically (and probably genetically) connected. A consequence of this is that, while they are still physically vigorous, aging women can become *grandmothers*. Taking care of her children's children, the grandmother improves the condition of her own offspring, and at the same time ensures reproductive success to her daughters. She contributes to the care of the nursing mother and of her children, including taking responsibility for feeding infants, which could have the effect of both accelerating the process of weaning and shortening the interval between births. The presence of older women as grandmothers could also have redounded upon the timing of when young mothers could engage in activities that impact survival as well as those that are culturally and socially relevant. It turn, this could yield an indirect reproductive advantage not only for her offspring, but also for the survival and expansion of the group. If this is the case, neither an increase in number of births per individual and, thus, for the group at large, nor sustainable and normal growth and development of children depends entirely on the efficiency of (male) hunting (*contra* e.g., Lee and De Vore 1968, Washburn and Lancaster 1968). Rather, they would be in part the consequence of a transformation of women's physiology, and its impact upon family structures.

These considerations embody implications for the organization of hominin communities. Whereas young female apes leave their mothers to join their male partners' territories (*patrilocality*), the positive role of grandmothers is favored by *matrilocality*, that is, the cohabitation of daughters with mothers (in contrast to cohabitation of the son's mother and his wife). In turn, this leads to a close and harmonious distribution of roles for caring and educating children. In humans, matrilocality permits a better environment for raising young children. Older women, freed from reproductive constraints, can achieve the status of a wise and dominant figure, and contribute to the welfare of their group, by virtue of the knowledge and experience they acquired over a long period of time, one that extends well beyond the cessation of fertility.

This perspective could suggest implications for reconsidering interpretations of feminine representations in Upper Paleolithic art. For more than half a century, prehistoric portable and parietal art was invariably read as reflecting expressions of fertility cults (Breuil 1955). The opulent shapes of the Aurignacian, Gravettian, and Solutrean Upper Paleolithic female figurines or "Venuses," which were carved in ivory, bone, or stone, or modeled in clay and found at sites from the Atlantic shores to the depths of Russia, were invariably interpreted as representing fertility, pregnancy, and childbirth (*ibid.*). And some Venuses do appear to illustrate these themes. Several Russian figurines from the site of Kostienki I in the Don Valley depict a pregnant woman and others the position of giving birth (Abramova 1995). Other figurines seem to recall symbols of fertility, such as the Laussel Venus whose "horn,"

Figure 16.1
The "Venus of Willendorf": This 110-mm-tall feminine figurine, carved in oolitic limestone and painted with red ochre, was unearthed in 1908 at the site of Willendorf, on the Danube River valley, near the town of Krems, Austria, during excavations conducted by archaeologists Josef Szombathy, Hugo Obermayer, and Josef Bayer. It is currently dated between 28,000 and 25,000 BCE. Its discoverers ironically dubbed it "Venus," and the French archaeologist Abbé Henri Breuil expressed his disgust at its "horrible realism." The abundant shape characteristic of such "Venuses" which was depicted in a number of Western European Gravettian statuettes has often been interpreted as embodying a fertility cult. This shape could just as well represent an homage to mature, postmenopausal women, and a recognition of the role grandmothers may have played in these Gravettian societies. (Source: Matthias Kabel, Wikimedia Commons)

with its 13 divisions, might represent a lunar or obstetric calendar (see Cohen 2016). However, while the requirement of fertility is an acceptable hypothesis to account for animal representations in Upper Paleolithic art (the reproduction of game animals being necessary to hunters' survival), it is more questionable in regard to human figures. Rather than representing a magic of fertility, these figurines could well be amulettes protecting women during pregnancy and childbirth (White and Bisson, 1998).

Moreover, in their schematic form, many Aurignacian and Gravettian figurines seem to represent the bodies of mature women who had undergone many pregnancies, and not young women of childbearing age. This interpretation gains credence, for example, in the famous Willendorf Venus (found in the Danube Valley, Austria, by Joseph Szombathy in 1908), whose heavy breasts and generous mid-body can plausibly be interpreted as depicting a post-menopausal condition. If this realm of interpretation is viable, as it appears to be, it might well be that Upper Paleolithic societies venerated grandmothers (and older women in general) for their embodiment of wisdom, for their roles in protecting and unifying familial generations, and contributing to the success of the social group and beyond, of their species.

Reproduction, Fertility, and Upper Palaeolithic Demography

The hypothesis of an important and rapid demographic growth of *Homo sapiens* populations has been cited as a major feature of a "human revolution" at the dawn of the Upper Paleolithic that accounts for the success of the extant species at the expense of the Neanderthals inhabiting Western Europe (Mellars and. Stringer. 1989, Mellars et al. 2007; see also Stringer 2013). Specifically, the Upper Palaeolithic growth of human populations may have been triggered in part by certain conditions, such as a shift to sedentary or semi-sedentary modes of life (as seen in Gravettian "mammoth civilizations" of the Russian plain), or settlement (as exemplified at the site of Ohalo II on the southwest shore of the Sea of Galilee, Israel). Nevertheless, in light of the modes of production and reproduction of hunter-gatherer populations, the idea of general, steady growth of human populations remains in dispute (French 2015).

Ethnographic evidence indicates that living hunter-gatherers put striving to limit and space births above seeking a woman's fertility (Lee 1979). Indeed, even decades earlier, Henri Luquet (1926, p. 124–125) wrote: "In economic terms, increase in the number of hunters increases competition for food to a greater extent, it seems, than it facilitates hunting through mutual aid. Besides, children will only become useful as hunters in the distant future and will remain for years useless mouths." Further, nomadic groups cannot ensure care for children who are unable to sustain themselves. Women, who generally carry heavy burdens, would not be able to move efficiently while also carrying several infants. A balance between the necessities of the group's survival and the physical capacities of women is achieved by spacing births every four years, when children can walk by themselves, express their needs, and help their mothers as well as the group. This scenario, with significant intervals between births, can result from a lengthy period of breastfeeding, which induces lactational amenorrhea (temporary cessation of menstruating; Lee 1979).

In hunter-gatherer societies, a constant rate of reproduction is socially limited by sexual prohibitions requiring abstinence or control of sexual relationships. This can be achieved by regulating access to fertile women, systematically limiting contact between men and women, or reducing fertility through diet, use of contraceptives, abortion, and even infanticide (Vila et al. 2010). The social necessity of reproductive control generally involves establishing prohibitions, taboos, and a hierarchy that, in general, imposes norms and limits to a woman's activities. Indeed, "[i]n societies of modern hunter-gatherers, this control is also achieved through ideology, expressed in the form of myths and ritual practices, which aims precisely to maintain this social order" (Mansur et al. 2007, p. 144) Although these practices and collective ceremonies leave little archeological trace, studies by archaeo-anthropologists on recently extinct, as well as extant, hunter-gatherer societies provide substantial insight into past social structures, cultural habits, and reproductive practices (Godelier 1987, Vila et al. 2001).

Conclusion

Reproduction is vital to the perpetuation of any group by delineating the realms of possibility for survival. However, what has been perpetuated since prehistoric times by human reproduction, are not only genes. It is also social groups and their frameworks, traditions, cultures, prohibitions, and rules. Every human society manages in its own way its reproductive behavior and prescribes its norms, modes, and rhythms of sex relationships that underlie its institutional and ideological structure. Further, kinship rules have organized reproductive behavior, probably since hominids came into existence. Family structures, gender roles, the place and fate of children, patterns of education and transmission, and the group's need for increasing or reducing numbers of births weave a tight network in which individual and collective behaviors are inexorably intertwined.

Consequently, human sexual and reproductive practices and behaviors, cannot be thought of merely in terms of biological determinism, and reduced to the level of genes. Rather, as with other human practices and behaviors, they result from interactions between the biological and the socio-cultural. Scenarios of hominization must take into consideration a number of parameters—biological, environmental, material, but also behavioral, cultural and social—that, together, might account for changes which ultimately distinguished humans from other primates.

Acknowledgments

I express my warmest thanks to Professor Jeffrey Schwartz for inviting me to participate in the symposium that led to this volume, and for his kind encouragement to write this text. I am grateful to all our colleagues, who gave stimulating presentations and insightful comments on multiple aspects of human evolution, and to the team of the Konrad Lorenz Institute, who provided an ideal setting for our exchanges. I also owe many thanks to my friends and colleagues Jean Gayon, Michel Veuille, Alison Wylie, Asumpcio Vila, Jordi Estevez, and Maurice Godelier for their inspiration and willingness to share ideas on different aspects of these issues.

References

Abramova, Zoïa. 1995. *L'art paléolithique d'Europe centrale et de Sibérie*. Grenoble: Jérome Millon.

Alexander, Richard D., and Katherine M. Noonan. 1979. *Concealment of Ovulation, Parental Care, and Human Social Evolution*. In *Evolutionary Biology and Human Social Behavior: An Anthropological Perspective*, edited by Napoleon A. Chagnon and William Irons, 436–453. North Scituate, MA: Duxbury Press.

Andersson, M. 1994. *Sexual Selection*. Princeton: Princeton University Press.

Bachofen, Johann.-Jakob. 1861. *Das Mutterrecht. Eine untersuchung uber die gynaikokratie der alten welt nach ihrer religiosen und rechtlichen natur*. Stuttgart: Krais & Hoffmann (reprint B. Schwabe, Basel 1948).

Bataille, Georges. 1957. *L'erotisme*. Paris: Minuit.

Beauvoir, Simone de. 1949/1978. *Le deuxième sexe*, 2 vol. Paris: Gallimard.

Boas, Franz. 1940. *Race, Language and Culture*. New York: The Macmillan Company.

Breuil, Henri (Abbé). 1955. *Quatre cents siècles d'art pariétal*. Paris: Mame.

Cohen, Claudine. (2003) 2013. *La femme des origines. Images de la femmes dans la préhistoire occidentale*. Paris:Belin-Herscher.

Cohen, Claudine. 2009. "Darwin on Woman." *Comptes rendus Palevol, Biologies* 333: 157–165.

Cohen, Claudine. 2016. *Femmes de la préhistoire*. Paris: Belin.

Conkey, Margaret W., and Joan M. Gero, eds. 1991. *Engendering Archaeology: Women and Prehistory*. Cambridge, MA: B. Blackwell.

Dahlberg, Frances, ed. 1981. *Woman the Gatherer*. New Haven: Yale University Press.

Darwin, Charles. 1859 *On the Origin of Species*, 1st ed., with an introduction by Ernst Mayr. Cambridge, MA: Harvard University Press, facsimile reprint, 1964.

Darwin, Charles. (1871) 1877. *The Descent of Man and Selection in Relation to Sex*, 3rd ed. London: John Murray.

Dawkins, R. 1976. *The Selfish Gene*. Oxford: Oxford University Press.

Delporte, Henri. 1979. *L'image de la femme dans l'art préhistorique*, 2nd ed., 1993. Paris: Picard.

Dixson, A.F. 1998. *Primate Sexuality: Comparative Studies of the Prosimians, Monkeys, Apes, and Humans*. Oxford: Oxford University Press.

Engels, Friedrich. 1884. *Der Ursprung der Familie, des Privateigenthums und des Staats* Hottingen-Zurich. Engl. Tr. *The Origin of the Family, Private Property, and the State*, Ernest Untermann, trans. Chicago: Charles H. Kerr & Co., 1902.

French, Jennifer. 2015. "The Demography of the Upper Palaeolithic Hunter-Gatherers of South West France. A Multi Proxy Approach Using Archaeological Data." *Journal of Anthropological Archaeology* 39: 193–209.

Freud, Sigmund. 1918. *Totem and Taboo*. New York: A. A. Brill.

Galton, Francis. 1909 *Essays in Eugenics*. London: Eugenics Education Society.

Girard, René. 1972. *La violence et le sacré*. Paris: ed. Bernard Grasset. Engl. Transl. *Violence and the sacred*. Baltimore: The Johns Hopkins University Press, 1977.

Girard, René. 1978. *Quelques choses cachées depuis la fondation du monde*, Paris: ed. Bernard Grasset. Engl. Transl. *Things Hidden Since the Foundation of the World*. Stanford, CA: Stanford University Press, 1987.

Godelier, Maurice. 2004. *Les métamorphoses de la parenté*. Paris: Fayard.

Guthrie, Dale. 2005. *The Nature of Prehistoric Art*. Chicago: University of Chicago Press.

Haeckel, Ernst. 1868. *Natürliche schöpfungsgeschichte* Berlin: Reimer. Engl. Transl. *The History of Creation*, 2 vol. London: HS King & Son, 1876.

Haeckel, Ernst. (1874). *Anthropogenie oder, Entwickelungsgeschichte des menschen*. Leipzig: Engelmann. Engl. Tr. *The Evolution of Man*. New York: Appleton & Co, 1902.

Haraway, Donna. 1989. *Primate Visions. Gender, Race and Nature in the World of Modern Science*. London: Routledge.

Hawkes, Kristen, James F. O'Connell, Nicholas G. Blurton Jones, Helen Alvarez, and Eric L. Charnov. 1998. "Grandmothering, Menopause, and the Evolution of Human Life Histories." *Proceedings of the National Academy of Sciences* 95(3): 1336–1339.

Herrscher, Estelle. 2013. "Isotopic Evidence of Breastfeeding and Weaning Modalities from Archaeological Bones." *Cahiers de nutrition et de diététique* 48(2): 75–85.

Holloway, Ralph. 1986. "Human Paleontological Evidence Relevant to Language Behavior," *Human Neurology* 2:105–114.

Hrdy, Sarah. 1999. *The Woman That Never Evolved*. Cambridge, MA: Harvard University Press.

Knight, Chris. 1991. *Blood Relations, Menstruations and the Origins of Culture*. New Haven and London: Yale University Press.

Laming Emperaire, Annette. 1962 *La signification de l'art rupestre préhistorique, methodes et applications*. Paris: Picard.

Lee, Richard. 1979. *The !Kung San, Men, Women and Work in Foraging Society*. Cambridge: Cambridge University Press.

Lee, Richard, and Irven De Vore, eds. 1968. *Man the Hunter*. Chicago: Aldine.

Leroi-Gourhan, André. 1964. *Les religions de la préhistoire*. Paris: P.U.F.

Leroi-Gourhan, André. 1965. *La préhistoire de l'art occidental*. Paris: Mazenod.

Levi-Strauss, Claude. 1949. *Les structures élémentaires de la parenté* [*The Elementary Structures of Kinship*, translated by James Harle Bell, John Richard von Strummer, and Rodney Needham. Boston: Beacon Press, 1969]. Paris: PUF.

Lévi-Strauss, Claude. 1956. "The Family." In *Man, Culture and Society*, edited by Harry L. Shapiro, 261–285. New York: Oxford University Press.

Lovejoy, Owen. 1981. "The Origin of Man." *Science* 211:341–350.

Lubbock, John. 1865. *Pre-historic Times*. London: Williams and Norgate.

Lubbock, John. 1870. *The Origin of Civilization and the Primitive Condition of Man*. New York: Appleton and Co.

Luquet, Georges-Henri. 1926. *L'art et la religion des hommes fossiles,*. Paris: Masson.

Lyell, Charles. 1863. *The Geological Evidences of the Antiquity of Man with Remarks on the Origin of Species by Variation*. London: John Murray.

McLennan, John Ferguson. 1865. *Primitive Marriage, an Inquiry into the Origins of the Form of Capture in Marriage Ceremony*. Edinburgh: A. and C. Black.

Malinowski, Bronislaw. 1927. *The Father in Primitive Psychology*. London: Kegan Paul, Trench, Trubner and Co.

Mansur, M. E., R. Pique, and Asumpcio Vila. 2007. "Etude du rituel chez les chasseurs-cueilleurs, apport de l'ethnoarchéologie des sociétés de la Terre de Feu." In *Chasseurs-cueilleurs, comment vivaient nos ancêtres du paléolithique supérieur*, sous la dir. de S. A. de Beaune, CNRS editions.

Mellars, Paul A., K. Boyle, O. Bar-Yosef, and C. Stringer, eds. 2007. *Rethinking the Human Revolution: New Behavioural and Biological Perspectives on the Origin and Dispersal of Modern Humans*. Cambridge: McDonald Institute for Archaeological Research.

Mellars, Paul A., and C. Stringer, eds. 1989. *The Human Revolution. Behavioural and Biological Perspectives in the Origins of Modern Humans*. Edinburgh: Edinburgh University Press.

Miller, G. F. 2000. *The Mating Mind: How Sexual Choice Shaped the Evolution of Human Nature*. London: Heinemann.

Morgan, Lewis H. 1877. *Ancient Society*. London: Macmillan and Co.

Mortillet, Gabriel de. 1869. "Essai d'une classification des cavernes et des stations sous abri, fondée sur les produits de l'industrie humaine." *Comptes rendus de l'Académie des Sciences*, t. LXVIII.

Raphael, Max. 1945 *Prehistoric Cave Paintings*. Princeton: Pantheon Books.

Sahlins, Marshall.1977. *The Use and Abuse of Biology: An Anthropological Critique of Sociobiology*. Ann Arbor: University of Michigan Press.

Shennan S. 2001. "Demography and Cultural Innovation: A Model and Its Implication for the Emergence of Modern Human Culture." *Cambridge Archaeoogical Journal* 11:5–16.

Slocum, Sally. 1974. "Woman the Gatherer." In *Toward an Anthropology of Women*, edited by Rayna Reiter, 36–50. New York: Monthly Review Press.

Spencer, Herbert. 1874–1896. *Principles of Sociology*, 3 volumes. London: William and Norgate.

Stocking, George W., Jr. 1968. *Race, Culture and Evolution, Essays in the History of Anthropology*. New York: Free Press.

Stringer, Chris. 2013. *Lone Survivors, How We Came to Be the Only Humans on Earth*. London: Griffin.

Tanner, Nancy. 1981. *On Becoming Human: a Model of the Transition from Ape to Human and the Reconstruction of Early Human Social Life*. New York: Cambridge University Press.

Testart, Alain. 1986. *Essai sur la division du travail chez les anciens chasseurs-cueilleurs*. Paris: Éditions de l'EHESS.

Testart, Alain. 2014 *L'Amazone et la cuisinière*. Paris: Gallimard.

Tylor, Edward. 1865. *Researches into the Early History of Mankind*. London: John Murray.

Tylor, E. 1873–1877. *Primitive Culture. Researches Into the Development of Mythology, Philosophy, Religion, Art, and Custom*, 2 vol. London: John Murray.

Vila-Mitjà, Assumpcio, and G. Ruiz Del Olmo. 2001. "Informacion etnologia y analisis de la reproduccion social: el caso Yamana." *Revista Española de Antropologia Americana* 31:275–291.

Vila-Mitjà, Assumpció, Jordi Estévez, Daniel Villatoro, and Jordi Sabater-Mir. 2010. "Archaeological Materiality of Social Inequality Among Hunter-Gatherer Societies." In *Archaeological Invisibility and Forgotten Knowledge, Conference Proceedings, University of Łódź, Poland, 5th–7th September 2007*, edited by Karen Hardy, British Archaeological Reports, International Series 2183: 202–210.

Waal, Frans de, and Lanting Frans. 1997. *Bonobo: the Forgotten Ape*. Berkeley: University of California Press.

Washburn, Sherwood L., and C. Lancaster. 1968. "The Evolution of Hunting." In *Man the Hunter*, edited by Richard Lee and Irven De Vore, 293–303. Chicago: Aldine.

Westermarck, Edward. A. 1891. *The History of Human Marriage*. London: Macmillan.

White, Randall, and Michael Bisson. 1998. "Imagerie féminine du Paléolithique, L'apport des nouvelles statuettes de Grimaldi." *Gallia Préhistoire* 40(1): 95–132.

Wilson, E. O. 1975. *Sociobiology: The New Synthesis*. Cambridge, MA: Harvard University Press.

Zihlman, Adrienne L. 1995. "Misreading Darwin on Reproduction: Reductionism in Evolutionary Theory." In *Conceiving the New World Order: The Global Politics of Reproduction*, edited by Faye D. Ginsburg and Rayna Rapp, 425–443. Berkeley: University of California Press.

Zihlman, Adrienne. 1997. "The Paleolithic Glass Ceiling." In *Women in Human Evolution*, edited by Lori D. Hager, 91–94. London: Routledge.

Zihlman, Adrienne L. 1981. "Women as Shapers of the Human Adaptation." In *Woman the Gatherer*, edited by Frances Dahlberg, 75–120. New Haven: Yale University Press.

Zwang, Gérard. 2002. *Aux origines de la sexualité humaine*. Paris: Presses Universitaires de France.

Contributors

Markus Bastir, Museo Nacional de Ciencias Naturales (CSIC)

Fred L. Bookstein, University of Vienna

Claudine Cohen, École des Hautes Études en Sciences Sociales

Richard G. Delisle, University of Lethbridge

Robin Dennell, University of Exeter

Rob DeSalle, Sackler Institute for Comparative Genomics, American Museum of Natural History

Emma M. Finestone, CUNY Graduate Center and NYCEP

Huw S. Groucutt, Oxford University

Gabriele A. Macho, Oxford University

Siobhan Mc Manus, CEIICH-UNAM

Apurva Narechania, Institute for Comparative Genomics, American Museum of Natural History

Michael D. Petraglia, Max Planck Institute for the Science of Human History

Thomas W. Plummer, Queens College, CUNY Graduate Center and NYCEP

Jelle W. F. Reumer, Utrecht University

Jeff Rosenfeld, Rutgers University

Jeffrey H. Schwartz, University of Pittsburgh

Dietrich Stout, Emory University

Ian Tattersall, American Museum of Natural History, Division of Anthropology

Alan R. Templeton, Washington University in St. Louis, and the University of Haifa

Michael Tessler, Richard Gilder Graduate School, American Museum of Natural History

John de Vos, Naturalis Biodiversity Center

Peter J. Waddell, Massey University and the Ronin Institute.

Martine Zilversmit, Institute for Comparative Genomics, American Museum of Natural History

Index

Printed in the United States
by Baker & Taylor Publisher Services